Coastal Morphodynamics

Special Issue Editor
Gerben Ruessink

MDPI • Basel • Beijing • Wuhan • Barcelona • Belgrade

MDPI

Special Issue Editor
Gerben Ruessink
Utrecht University
The Netherlands

Editorial Office
MDPI AG
St. Alban-Anlage 66
Basel, Switzerland

This edition is a reprint of the Special Issue published online in the open access journal *Journal of marine science and engineering* (ISSN 2077-1312) in 2015–2016 (available at: http://www.mdpi.com/journal/jmse/special_issues/coastal_morphodynamics).

For citation purposes, cite each article independently as indicated on the article page online and as indicated below:

Lastname, F.M.; Lastname, F.M. Article title. *Journal Name*. **Year**. *Article number*, page range.

Image courtesy of Name

ISBN 978-3-03842-676-9 (Pbk)
ISBN 978-3-03842-675-2 (PDF)

Table of Contents

About the Special Issue Editor

Gerben Ruessink, Professor of Wave-dominated Coastal Morphodynamics at Utrecht University, received the Ph.D. degree in Physical Geography from Utrecht University, Utrecht, The Netherlands, in 1998. Together with his research group, and through national and international cooperation, he studies the natural and human-impacted morphological evolution of sandy beaches and dunes in response to wave-induced and aeolian processes. He uses a process-based approach, in which he often combines in-situ field data, remote sensing observations, laboratory experiments, and numerical models. His current research interests include aeolian process dynamics and seasonal to yearly exchange of sand between the beach and the dunes. He has (co-)authored over a hundred peer-reviewed journal papers and has given keynote speeches at international coastal conferences. In 2014 he was awarded a personal senior-career grant by the Netherlands Organization for Scientific Research for his innovative lines of coastal research.

Preface to "Coastal Morphodynamics"

Coasts are often beautiful landscapes with high biodiversity and provide a large and rapidly grow-ing proportion of the world's population with living and working environments, recreation, food, and drinking water. Coasts are also one of the most dynamic natural features on Earth and are under increasing pressure by human activities and climate change. This book is the printed edition of the Special Issue on Coastal Morphodynamics, launched in 2015 by the Journal of Marine Science and En-gineering and edited by Prof. Dr. Gerben Ruessink from Utrecht University. The eleven papers reflect present-day understanding of the natural and human-impacted behaviour of sandy beaches, barrier island systems, salt marshes and rock coasts based on in-situ field observations, remote-sensing data, laboratory experiments and numerical modelling. The solid understanding of coastal morphody-namics, as presented in the book, is critical for the sustainable management of our coasts.

Gerben Ruessink
Special Issue Editor

Journal of
Marine Science and Engineering

MDPI

Article

Mesoscale Morphological Change, Beach Rotation and Storm Climate Influences along a Macrotidal Embayed Beach

Tony Thomas [1,*], Nelson Rangel-Buitrago [2], Michael R. Phillips [1], Giorgio Anfuso [3] and Allan T. Williams [1,4]

[1] Coastal and Marine Research Group, University of Wales Trinity Saint David (Swansea), Mount Pleasant, Swansea, Wales SA1 6ED, UK; mike.phillips@uwtsd.ac.uk (M.R.P.); allan.williams@uwtsd.ac.uk (A.T.W.)

[2] Facultad de Ciencias Básicas, Programa de Física, Grupo de Geología, Geofísica y Procesos Litorales, Km 7 Antigua vía Puerto Colombia, Barranquilla, Atlántico 080020, Colombia; nelsonrangel@mail.uniatlantico.edu.co

[3] Ciencias de la Tierra, Universidad de Cadiz, Puerto Real 11510, Spain; giorgio.anfuso@uca.es

[4] CICA NOVA, Nova Universidade de Lisboa, Lisboa 1069-050, Portugal; Allan.williams@virgin.net

* Author to whom correspondence should be addressed; tony.thomas@uwtsd.ac.uk; Tel.: +44-1792481000.

Academic Editor: Gerben Ruessink
Received: 29 June 2015; Accepted: 25 August 2015; Published: 2 September 2015

Abstract: Cross-shore profiles and environmental forcing were used to analyse morphological change of a headland bay beach: Tenby, West Wales (51.66 N; −4.71 W) over a mesoscale timeframe (1996–2013). Beach profile variations were attuned with longer term shoreline change identified by previous research showing southern erosion and northern accretion within the subaerial zone and were statistically significant in both sectors although centrally there was little or no significance. Conversely a statistically significant volume loss was shown at all profile locations within the intertidal zone. There were negative phase relationships between volume changes at the beach extremities, indicative of beach rotation and results were statistically significant ($p < 0.01$) within both subaerial ($R^2 = 0.59$) and intertidal ($R^2 = 0.70$) zones. This was confirmed qualitatively by time-series analysis and further cross correlation analysis showed trend reversal time-lagged associations between sediment exchanges at either end of the beach. Wave height and storm events displayed summer/winter trends which explained longer term one directional rotation at this location. In line with previous regional research, environmental forcing suggests that imposed changes are influenced by variations in southwesterly wind regimes. Winter storms are generated by Atlantic southwesterly winds and cause a south toward north sediment exchange, while southeasterly conditions that cause a trend reversal are generally limited to the summer period when waves are less energetic. Natural and man-made embayed beaches are a common coastal feature and many experience shoreline changes, jeopardising protective and recreational beach functions. In order to facilitate effective and sustainable coastal zone management strategies, an understanding of the morphological variability of these systems is needed. Therefore, this macrotidal research dealing with rotational processes across the entire intertidal has significance for other macrotidal coastlines, especially with predicted climate change and sea level rise scenarios, to inform local, regional and national shoreline risk management strategies.

Keywords: mesoscale morphological change; beach rotation; storm climate

1. Introduction

Beaches situated in the lee of rocky outcrops or headlands, generally take some form of curvature known as curved, embayed, hooked, pocket and headland-bay beaches [1] and 51% of the world's

J. Mar. Sci. Eng. **2015**, *3*, 1006–1026

coastlines are representative of this morphology [2]. Along these coasts, nearshore wave energy is often high as waves are related to bathymetry and refraction/diffraction patterns [3]. Wave energy is focused on the headlands and dispersed into the bays, so headlands erode while the intervening bays fill up [4]. Shorelines on sand beaches are known to vary over a range of timescales [5], in response to both erosion and rotation events [6]. Severe shoreline movement can trigger the need for coastal mitigation measures especially if private or public dwellings are put at risk [7]. Therefore, understanding beach morphological variability is essential to support coastal risk assessment and help in the decision making process, especially in what concerns the implementation of mitigation measures in response to erosive events reported worldwide [8]. Shoreline rotation phenomenon can be defined as a landward or seaward movement at one end of a beach accompanied by the reverse pattern at the other end [7] and is known to be caused by variations in wave climate such as wave approach direction and energy flux [7,9–13].

Many researchers, have documented seasonal or short-term rotations [9,11,13], others have studied rotation at decadal scales [10,14–16]. Thomas *et al.* [12] provided a historic (Centurial) record of beach rotation, influenced by long term shifts in wind directional patterns that caused shoreline displacement resulting in up-drift erosion, down-drift accretion and subaqueous loss. Morphological responses of embayed beaches to storm and gale forcing have also been studied in the Northern [17–19] and Southern Hemispheres by amongst others [10,14]. The underlying causes of wave directional change have also been linked to subtidal mud bank and sandbank migration [20–22]. Unlike the macrotidal beach work carried out in this research, most rotation studies utilize variations in the location or volume of the subaerial zone to identify shoreline response; this was because almost all were studied in locations with microtidal or mesotidal ranges see for example [10,20,23–27]. The limited macrotidal research by Stone and Orford [28], Dehouk *et al.* [23], Maspataud *et al.* [29] and Thomas *et al.* [12,13] work within the present area of study also concentrated on subaerial regions. They all highlighted beach rotation despite limited wave exposure.

This paper builds upon Thomas *et al.* [16] work by analyzing both subaerial and intertidal zones of a macrotidal beach using mesoscale profile responses, manifested by differential longshore sediment translation expressed through rotation and realignment, when compared and contrasted with environmental forcing to analyse cause and effect. Evaluation of results identified changes in coastal processes and led to development of temporal and spatial regression models representing functions of intertidal rotation. While similar responses have been obtained worldwide within the subaerial zone these intertidal relationships have important consequences for embayed beach management strategies.

J. Mar. Sci. Eng. **2015**, *3*, 1006–1026

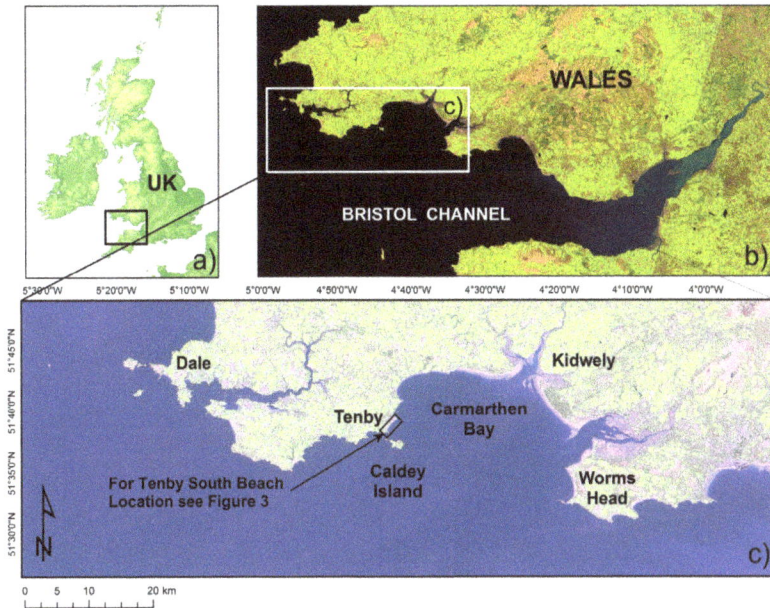

Figure 1. Locality of the study area: (**a**) United Kingdom; (**b**) Bristol Channel; (**c**) Carmarthen Bay.

2. Physical Background

The outer Bristol Channel on the west coast of Great Britain constitutes a large body of partially enclosed tidal water [30], with a tidal range of up to 12 m [31] (Figure 1a,b). Carmarthen Bay located on the Channel's northwestern margin is a relatively large embayment (Figure 1c), formed as a consequence of differential erosion and is mainly characterised by rocky cliffs and small embayments that contain pocket beaches [16,32–35]. The study area (Figure 2a) is a sub compartment of Carmarthen Bay delineated at its northern/southern ends by two Carboniferous limestone headlands—Tenby and Giltar respectively [36,37]. These two places epitomise a well-developed "honeypot" geared for tourism (Tenby) *versus* an important uninhabited conservation area (Giltar). The system comprises dunes (920 × 10^3 m^2), a shingle backshore and a wide sand intertidal zone. Sediment loss in the latter is *circa* 7000 m$^3 \cdot$year^{-1} [38,39] and the dune system follows the classic sequence of erosion in storms/high spring tides, although a dense *Ammophila arenaria* vegetation cover help retard erosion; the system being replenished when a more constructive wave regime occurs. Mean semi-diurnal tidal range is 7.5 m and predominant waves arrive from the south to west directions, with average height/periods respectively of 1.2 m and 5.2 s, which in high energy conditions can reach 5.5 m and 8.2 s [12]. Wave diffraction occurs due to Caldey and St Margaret's islands (Figure 2a) influencing a strong south to north longshore drift.

(a)

(b)

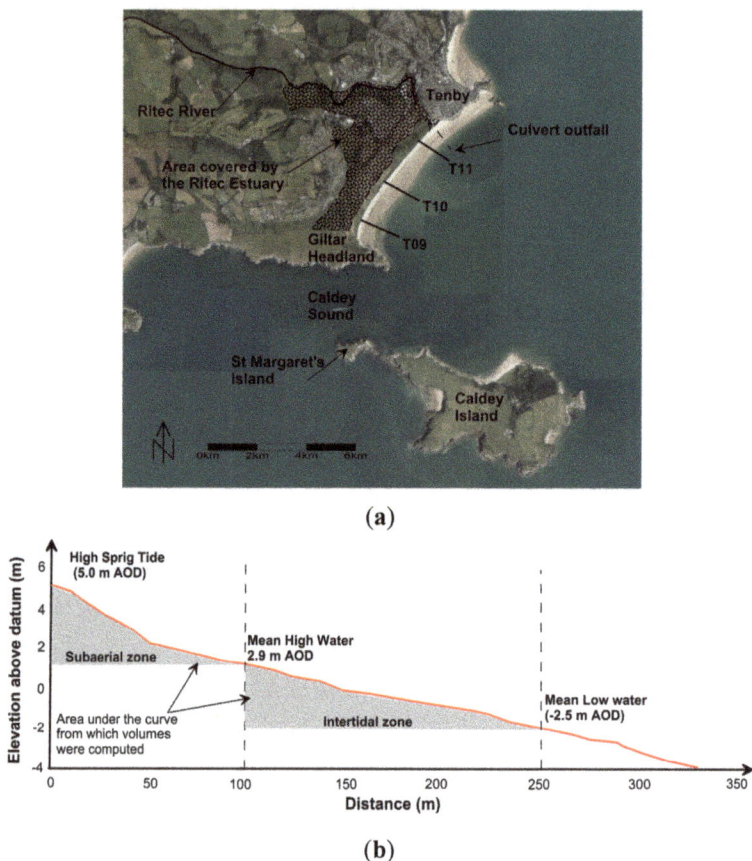

Figure 2. (a) Study area location plan including the topographical location of the representative cross shore profiles (T09–T11) from which beach level and volume change were calculated; and (b) a definition sketch showing the morphological zones from which comparative beach volumes were computed.

3. Methods

3.1. Beach Profile Monitoring (1996–2013)

Medium term changes, calculated using three profiles spaced at 580 m (Figure 2a), were representative of South (T09), Central (T10) and North (T11) beach sectors. Surveys were carried out during spring (April/May) and when available autumn (October/November), extending from the dune system control point, to low water (approximately 250 m). Profile locations enabled analysis of beach rotation by detailing the relationship between beach extremities.

The profiles were truncated to the high spring tidal level; sectional volumes, *i.e.*, the morphological variables, were then determined directly from the Regional Morphology Analysis Package (RMAP), where volume is calculated by extrapolating the area under the curve for one unit length ($m^3 \cdot m^{-1}$) of shoreline [40] (Figure 2b). Two areas were identified for detailed analysis, the sub-aerial (high spring tide mark to the mean high water mark) and the intertidal (high spring tide mark to the mean low water mark) shore zones. Profile data were collected using a total station with an accuracy of ±5 mm + 3 ppm vertically. Profile origins provided control points, referenced directly to the British OS Grid

J. Mar. Sci. Eng. **2015**, *3*, 1006–1026

Reference system. Beach profiles were generally surveyed during spring (April) and autumn (October), winter surveys from 1997 to 1999 were not available, and therefore, 22 surveys representing 17 years of data (1996–2013), were presented for analysis. Beach volume change within each morphological zone was used to characterize beach rotation, a methodology utilized by Klein *et al.* [9] who used sub-aerial beach volume change to similarly characterize rotation processes and Thomas *et al.* [16] who used both subaerial and intertidal volumes to assess rotation albeit with a much smaller dataset (1999–2007).

Clearly, observed changes in beach morphology cannot solely be related to wave direction and longshore drift [41]. Cross shore processes also induce profile readjustments that are non-rotational responses [42]. Therefore, rotational (related to longshore drift) and non-rotational (related to cut and fill or cross-shore transfer) have to be decoupled or smoothed out of the data for study of rotation phenomenon. To achieve this and assess medium timescale rotation, it was necessary to remove high frequency cut and fill (cross-shore) noise from the volume dataset. A method similar to that developed by Short and Trembanis [14] was implemented. The volume record was transformed into the standard normal form [43], using $z = x - x^-/\sigma$, where, z = normalised value, x = data volume record for each profile, x^- = average value for that profile, σ = standard deviation. Temporal mean survey volumes were averaged along the beach (representing the cut and fill behaviour). Spatial average values (b) were subtracted from the local normalised volume (a) to reveal a time-series where high frequency behaviour has been removed. Residual volumes were converted into dimensional units using $x = (z \times \sigma) + x^-$.

3.2. Wind, Wave and Storm Characterization Data

In this research direct comparisons were made between subaerial and intertidal volume change and environmental forcing agents (wind and wave climate) captured by the Turbot Bank wave rider buoy (51.603 N 5.100 W). The buoy owned and maintained by the UK Met Office records wave height, period and direction at 1 h intervals. The 17 year dataset supplied by the Met Office contained 121,452 independent values. In this work, a storm is defined as a climatic event during which the significant wave height (Hs) exceeds a threshold over a minimum during a specific time. Dolan and Davis [44] Storm Power Index was used to classify coastal storms. This index was calculated according to: Hs^2 td where, Hs = significant wave height and td = storm duration in hours. A storm wave height represented rare events in Tenby area with only 8% of total amount in the 17 years using the methodology proposed by Dorsch *et al.* [42]. This value reflects the wave height at which erosion starts to affect nearby areas, according to previous regional research findings [13,45]. The minimum storm duration was set at 12 h, in this way the storm affected the coast at least during a complete tidal cycle and the lapse time between successive storms was set at one day in order to create de-clustered, independent sets of storms [42,46,47]. Once storms were recognised, they were categorised by means of the natural breaks function analysis [48], into five classes from Class I (weak) to Class V (extreme) events.

4. Results

4.1. Beach Level Change (1996–2013)

Figure 3 shows individual cross-shore profile envelopes between 1996 and 2013; all three profiles are concave and indicative of two beach states: A dissipative/intermediate mid to low tidal zone and intermediate/reflective high tidal zone [2]. The greatest variance beach level occurs within the high tidal zone where the standard deviation (σ) is at its maximum value for all three profiles (σ = 0.826 m, 0.605 m and 1.071 m three respectively). The standard deviation is at its minimum value within the mid tidal zone of all profiles (σ = 0.139 m, 0.080 m and 0.163 m respectively). When first and last cross shore profiles are compared, T09 (south) highlights falling beach levels across the entire profile during the 17 year period of assessment (Figure 3a), T10 (central; Figure 3b) highlights erosion in both subaerial and lower intertidal zones and stability in the upper intertidal zone. Whereas, T11 (North;

Figure 3c) showed accretion in the sub aerial and upper intertidal zone, contrasted against erosion in the lower intertidal zone, with the point of oscillation near the MSL contour.

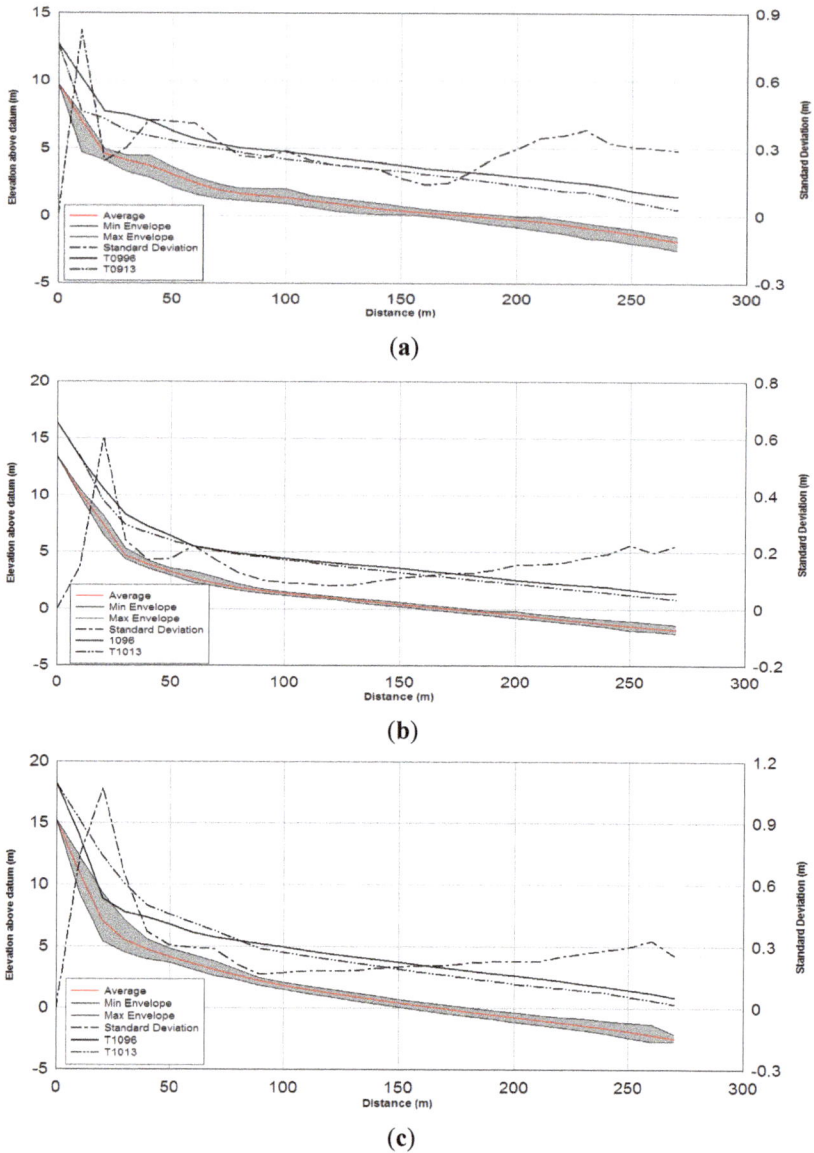

(a)

(b)

(c)

Figure 3. Cross shore profile envelopes, and standard deviations for the period 1996–2013, along with the first (1996) and last (2013) cross shore profiles offset by 3 m for clarity: (a) Transect 09; (b) Transect 10; and (c) Transect 11.

J. Mar. Sci. Eng. **2015**, *3*, 1006–1026

In order to quantitatively compare differences in beach level between the first (1996) and last (2013) surveys paired *t* tests were performed and the results presented in Table 1. The calculated *t* statistic (t_{calc}) when compared to tabulated *t*-values (t_{tab}) according to the degrees of freedom (df). Results showed $t_{calc} > t_{tab}$ indicating that there was a significant difference in beach level at profile locations T09 (south) and T10 (central) at 99% confidence and negative signs for t_{calc} indicated that beach levels had fallen between 1996 and 2013. Conversely, $t_{calc} < t_{tab}$ at profile location T11 suggested that there was no significant difference in beach level and the small positive t_{calc} value indicative of a slight increase in levels.

Table 1. Results of paired *t* tests—Surveys 1 and 22.

Profile	Mean	Std. Deviation	Std. Error Mean	95% Confidence		t_{calc}	df	Sig	t_{tab}	
				Lower	Upper				0.05	0.01
T09	0.677	0.437	0.083	0.50758	0.84640	−8.199	27	0.000	2.052	2.771
T10	0.338	0.258	0.049	0.23779	0.43771	−6.933	27	0.000	2.052	2.771
T11	0.000	1.000	0.189	−0.38766	0.38789	0.001	27	1.000	2.052	2.771

4.2. Beach Volumes (1996–2013)

Table 2: Produced directly from the RMAP programme show the volumes ($m^3 \cdot m^{-1}$) and inter-survey volumes for both subaerial and intertidal zones and data used to produce Figure 4. Subaerial volumes showed similar trends to the historic data *i.e.*, erosion in the south (T09) and accretion in the north (T10), regression models showed that a significant relationship existed between volume change and time, with R^2 values that explained 84% and 74% of data variation (y = −0.006x + 298.93 and y = 0.0057x − 104.84 respectively; Figure 4a). The historic central sector variability is also confirmed by a regression model that explained almost none of the data variation, suggesting that there was no relationship between central volume variation and time (y = −0.003x + 170.98). However, results are influenced by the location of the profile (*i.e.*, within the region of rotation). All profiles showed similar erosion trends and a significant relationship between volume change and time within the intertidal zone and R^2 values that explained 71% (T09), 79% (T10) and 75% (T11) data variation (y = −0.0316x + 2091.5, y = −0.0189x + 1602.5 and y = −0.0189x + 1602.5 respectively; Figure 4b).

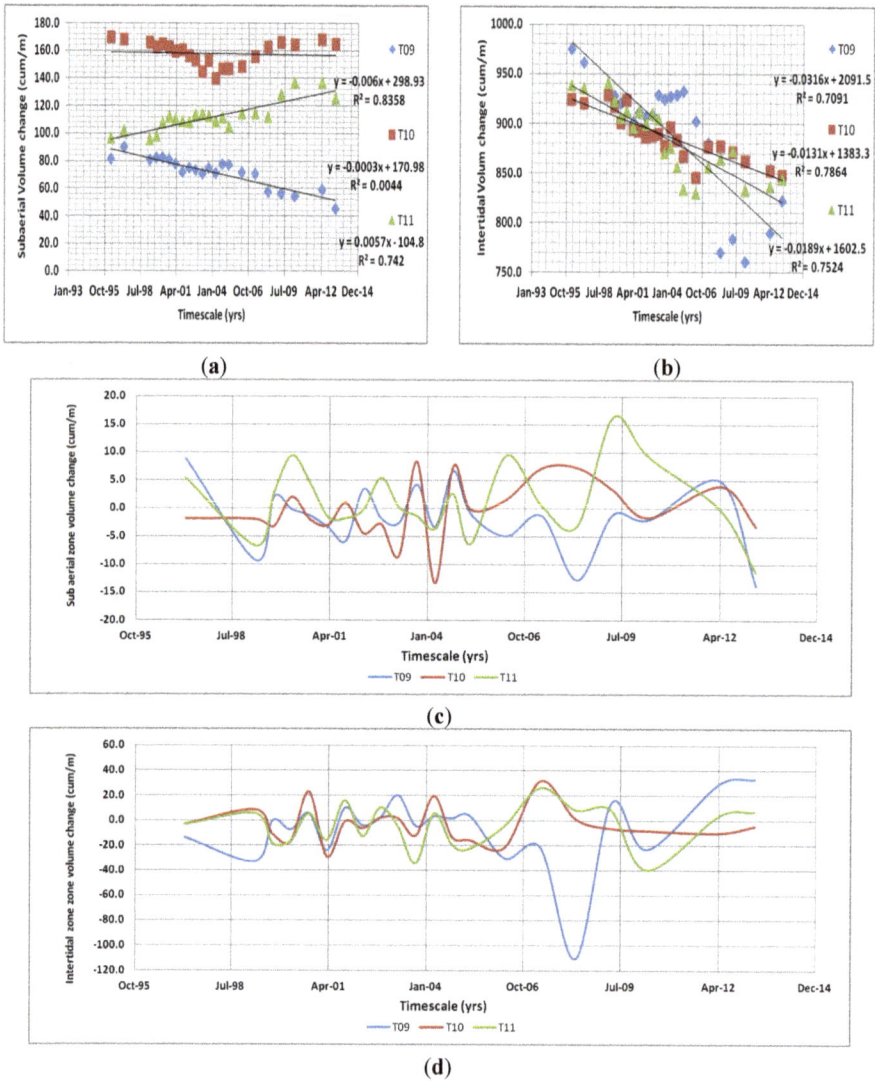

Figure 4. Analyses of beach volume trends between 1996 and 2013: (**a**) temporal subaerial volume change; and (**b**) temporal intertidal volume change. Graphical representations depicting temporal inter-survey volume change between 1996 and 2013: (**c**) subaerial zone; and (**d**) intertidal zone.

Table 2. Subaerial and intertidal volume change between 1996 and 2013.

Timescale		Subaerial Zone Volumes (m³·m⁻¹)			Subaerial Zone Inter-Survey Volumes (m³·m⁻¹)			Intertidal Zone Volumes (m³·m⁻¹)			Intertidal Zone Inter-Survey Volumes (m³·m⁻¹)		
on	off	T09	T10	T11	T09	T10	T11	T09	T10	T11	T09	T10	T11
	Apr-96	81.0	169.5	96.5				974.6	923.5	938.4			
Apr-96	Apr-97	89.7	167.5	101.8	8.7	−1.9	5.3	960.8	920.1	935.0	−13.8	−3.4	−3.4
Apr-97	Apr-99	80.3	165.5	95.1	−9.4	−2.0	−6.7	927.6	928.3	940.6	−33.2	8.3	5.6
Apr-99	Oct-99	82.2	162.2	97.7	1.9	−3.3	2.6	927.1	917.1	922.2	−0.5	−11.3	−18.4
Oct-99	Apr-00	81.9	164.2	107.0	−0.3	2.0	9.4	919.6	900.1	905.7	−7.5	−16.9	−16.5
Apr-00	Oct-00	80.6	162.1	111.5	−1.4	−2.0	4.5	925.1	922.7	911.3	5.5	22.6	5.6
Oct-00	Apr-01	77.1	159.1	110.0	−3.5	−3.1	−1.6	901.0	893.3	895.4	−24.1	−29.4	−15.9
Apr-01	Oct-01	71.3	160.0	108.2	−5.8	0.9	−1.8	910.7	892.2	911.5	9.7	−1.1	16.1
Oct-01	Apr-02	74.6	155.5	107.9	3.3	−4.5	−0.3	906.6	885.7	898.9	−4.1	−6.5	−12.7
Apr-02	Oct-02	72.7	152.6	113.3	−1.9	−2.9	5.4	908.7	887.5	909.0	2.1	1.8	10.2
Oct-02	Apr-03	70.1	144.1	113.4	−2.6	−8.5	0.1	928.6	889.1	904.4	19.9	1.6	−4.7
Apr-03	Oct-03	74.2	152.4	112.1	4.1	8.3	−1.4	924.0	876.8	869.8	−4.6	−12.4	−34.6
Oct-03	Apr-04	70.8	139.0	108.4	−3.4	−13.4	−3.7	927.0	896.1	875.4	3.0	19.3	5.6
Apr-04	Oct-04	77.4	146.5	110.9	6.6	7.4	2.5	928.4	882.8	856.0	1.4	−3.3	−19.4
Oct-04	Apr-05	76.3	146.2	104.6	−1.0	−0.3	−6.3	932.1	866.9	833.2	3.7	−15.9	−22.7
Apr-05	Apr-06	71.3	147.8	114.0	−5.0	1.5	9.4	901.8	845.1	829.1	−30.4	−21.7	−4.1
Apr-06	Apr-07	70.0	154.9	114.4	−1.4	7.1	0.4	880.0	876.4	855.4	−21.8	31.3	26.3
Apr-07	Apr-08	57.2	162.1	111.3	−12.8	7.2	−3.1	769.5	877.1	863.8	−110.5	0.7	8.4
Apr-08	Apr-09	56.0	165.5	127.5	−1.2	3.3	16.2	783.1	870.5	872.3	13.6	−6.6	8.5
Apr-09	Apr-10	53.9	163.8	136.7	−2.0	−1.7	9.2	760.0	862.0	832.3	−23.1	−8.5	−40.0
Apr-10	Apr-12	58.7	167.7	136.5	4.8	3.9	−0.2	789.0	851.8	836.2	29.0	−10.2	3.9
Apr-12	Apr-13	44.9	164.4	125.1	−13.8	−3.3	−11.4	821.7	847.0	843.3	32.7	−4.9	7.1

Negative values = erosion and positive values = accretion.

The inter-survey volume variation within the subaerial zone is represented graphically in Figure 4c, the beach volumes fluctuated between erosion and accretion on an almost annual basis at all profile locations. The southern and northern volume changes tended to be out of phase suggesting that when the southern sector erodes the northern sector follows similar erosion trends up to one year later and *vice versa*, with southern volumes fluctuating mostly below zero (erosion) and the northern volumes well above (accretion). Centrally, volumes fluctuated below zero between 1996 and 2005, and well above zero up until 2010, thereafter, all profile volumes dip well below zero. The inter-survey volume variation within the intertidal zone is also represented graphically in Figure 4d. Similar to the subaerial zone, volumes also fluctuated just above and below zero at all profile locations but tended to be in phase with one another up until 2005. Thereafter the southern sector eroded and central/northern sectors accreted, before all sector accreted towards the end of the assessed period. The accretive episode coincided with the erosion shown in the subaerial zone during the same period and concurs with Thomas *et al.*'s [49] work at Pendine, west Wales, where evidence showed that during storms and gales that coincide with the high spring tidal range the subaerial zone erodes, deposits sediment within the intertidal zone, from where longshore drift from south towards north erodes the intertidal zone until a similar event occurs reversing the trend.

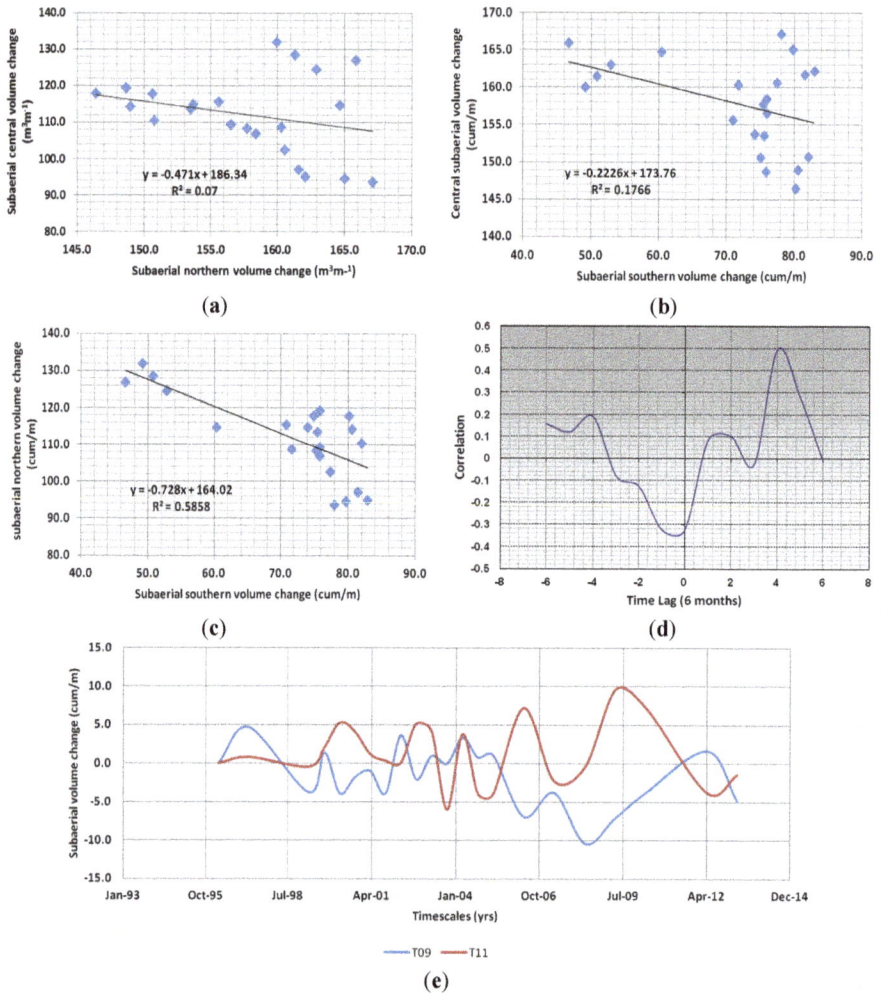

Figure 5. Analyses of transformed beach subaerial volume trends 1996–2013: (**a**) between south and north beach extremities; (**b**) between southern and central zones; (**c**) between northern and central zones; (**d**) cross-correlation results between south and north beach extremities; and (**e**) a graphical representation depicting temporal inter-survey volume change.

4.3. Beach Rotation (1996–2013)

The cross shore signal was removed from subaerial and intertidal volumes (Table 2) using the routine described by [14] and Figures 5 and 6 produced. Within the subaerial zone negligible negative relationships existed between the central and northern/southern sectors with R^2 values that explained almost none of the data variation ($y = -0.171x + 186.34$ and $y = -0.226x + 173.76$ respectively, Figure 5a,b). However, it is the significant relationship ($R^2 = 59\%$) that existed between profiles T09 (extreme south) and T11 (extreme north) that is of most interest, as this indicates a negative phase relationship between accretion/erosion patterns between southern and northern ends of the beach (*i.e.*, beach rotation). This was given by the regression equation $y = -0.728x + 164.02$ (Figure 5c). To investigate stronger potential correlations between south and north beach extremities, time lagged cross-correlations of volume changes between profiles T09 and T11 were calculated and represented in Figure 5d. Results show no improvement in correlation but a reversal in trend at a two time lags (in the positive direction), which indicates that southerly volume variations lag behind northern variations by one year. When a reversal in trend occurs (in the negative direction), northern variations lag behind southern change by four time lags (*i.e.*, two years). The volume changes at the beach extremities are represented graphically in Figure 5e and highlight three rotational periods in 2001, 2004 and 2012 respectively. These results show that even though there is limited exposure to waves within the subaerial zone of this macrotidal beach, rotational response can still be detected.

Within the intertidal zone a negligible positive relationship existed between the central and northern sectors and once again the R^2 value explained almost none of the data variation ($y = 0.5227x + 420$; $R^2 = 10\%$; Figure 6a) indicating that when variations took place in the northern sector similar changes also occurred centrally. In contrast, a high negative relationship existed between southern and central sectors indicating that when changes occur in the southern sector the opposite would be true in the central sector ($y = 0.1793x + 1046.8$, R2 = 62%; Figure 6b). However, it is the significant relationship ($R^2 = 70\%$) that existed between profiles T09 (extreme south) and T11 (extreme north) that is of most interest, as this indicates a negative phase relationship between accretion/erosion patterns between southern and northern ends of the beach (*i.e.*, intertidal beach rotation). This was given by the regression equation $y = -0.316x + 1165.2$ (Figure 6c). To investigate stronger potential correlations between south and north beach extremities, time lagged cross-correlations of volume changes between profiles T09 and T11 were calculated and represented graphically (Figure 6d). Results show no improvement in correlation but a reversal in trend at a four time lags (in the positive direction), indicating that southerly volume variations lag behind northern variations by two years. When a reversal in trend occurs (in the negative direction), northern variations lag behind southern change by three time lags (*i.e.*, 18 months). The volume changes at the beach extremities are represented graphically in Figure 6e and highlight an almost cyclical rotational behavioural pattern throughout the assessed period. These results show that rotation phenomena are not exclusive to subaerial sectors of macrotidal beaches.

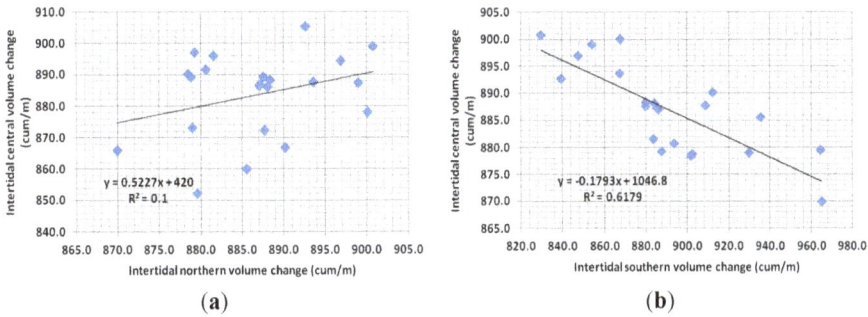

(a)

(b)

Figure 6. *Cont.*

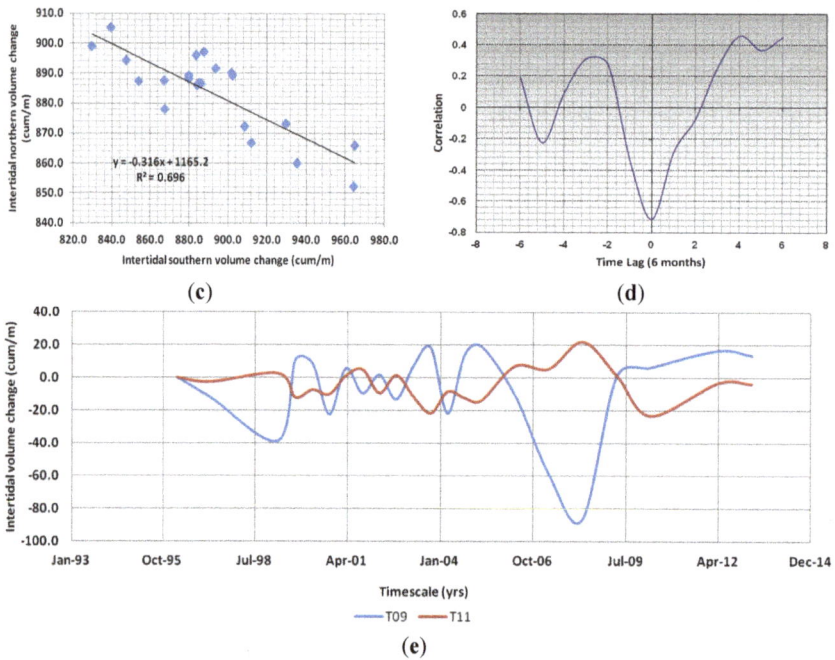

Figure 6. Analyses of transformed beach intertidal volume trends 1996–2013: (**a**) between south and north beach extremities; (**b**) between southern and central zones; (**c**) between northern and central zones; (**d**) cross-correlation results between south and north beach extremities; and (**e**) a graphical representation depicting temporal inter-survey volume change.

4.4. Wave Climate and Storms (1996–2013)

Data showed clear cyclic patterns when monthly average significant wave height (H_s) was assessed. Waves were usually low ($H_s < 1.4$ m) in May–August period (late spring to summer), reaching minimum values in July ($H_s = 1.24$). During the winter season, waves rapidly increased in height, reaching peak values ($H_s = 2.4$ m) in December–January.

Regression analysis showed that both monthly and annually averaged wave heights decreased throughout the period of assessment (-0.001 m·year^{-1} and -0.02 m·year^{-1} respectively). However, low recorded values of Pearson coefficient revealed that these trends are not statistically significant ($p > 0.05$). Similar results were obtained using the Mann-Kendall trend test and the Wilcoxon rank-sum test are in common use in similar studies [50–52]. This data evidenced quasi-periodic 4 year behavioural patterns in the recurrence of high wave height values. A spectral analysis of time series of extreme waves, based on the Fourier transformation [4] indicated a cyclic trend of 3 years.

In total 267 storm events were recorded during the period of assessment. Classes I (weak) and II (moderate) accounted for, respectively, 47% and 26% of records. These values were similar to [44,47,53–55] studies carried out in USA and Spain. Class III (significant), constituted 18% of the record and Classes IV (severe) and V (extreme) accounted for 4% and 6% respectively (Table 3).

Associated average wave height and storm duration values presented important variations (Table 2) and average wave period ranged from 8 (Class I) to 9.3 s (Class V). Storm power values were larger than Dolan and Davis (1992) [44] because of the major threshold of storm wave height selected in this study and longer storm durations. Variability patterns of storm duration and Storm Power Index were very similar to that found for the number of storms. This is because the stormy season

12

J. Mar. Sci. Eng. **2015**, 3, 1006–1026

(winter period) presented a greater number of storms and their overall duration resulted in elevated storm power.

Table 3. Storm classification statistics.

Class	Range	Frequency		Wave Height (Avg)	Period (Avg)	Duration (Avg)	Storm Power (Avg)
		N	%				
I	<600	125	47	4.7	8.0	16.7	373.2
II	601–1236	69	26	5.5	8.6	30.2	875.3
III	1237–2022	48	18	6.5	9.1	42.2	1679.3
IV	2023–3529	10	4	6.8	9.1	63.1	2753.7
V	>3529	15	6	7.0	9.3	96.5	4719.7

4.5. Environmental Forcing and Morphological Change (1998–2013)

Storm conditions and subaerial volume changes were compared and presented graphically in Figure 7a. During months where there is increased storm activity either at the start or end of the period both southern and northern shores (1999 and 2008 respectively) erode and when there are reductions in storm activity; the southern shore erodes and northern accretes (2000–2002, 2006–2007 and 2009–2010 respectively). With the exception of winter 2010 to summer 2012, it appears that southern shores are only stable or accretive during periods when there is no storm activity. The most significant gains in the northern sector occur when the wind is south-southwest (*i.e.*, weakly above zero) for example, winter 2008 to summer 2010. The only anomaly is the significant erosion took place in both sectors towards the end of the assessment period, although the wind direction may have been an influence that was mostly emanating from the south east. The intertidal zone behaved differently under storm conditions (Figure 7a), probably influenced by sediment inputs across shore. Increasing storm occurrence mostly led to southern erosion and northern accretion, easterly orientated winds resulted in accretion in both sectors and periods with little or no storm occurrences resulted in a northern loss, probably as a result of onshore sediment movement. No real trends appeared in the data and the system appears to react independently of storm events and while they undoubtedly have a major influence in this sediment-limited environment they may act to trigger major configuration changes and trend reversals and then subsequent storms even from a similar direction will trigger a reversal in trend that does not appear to be induced by external forcing this phenomenon was also highlighted by Cooper [18] in similar studies on the Irish coastline.

J. Mar. Sci. Eng. **2015**, 3, 1006–1026

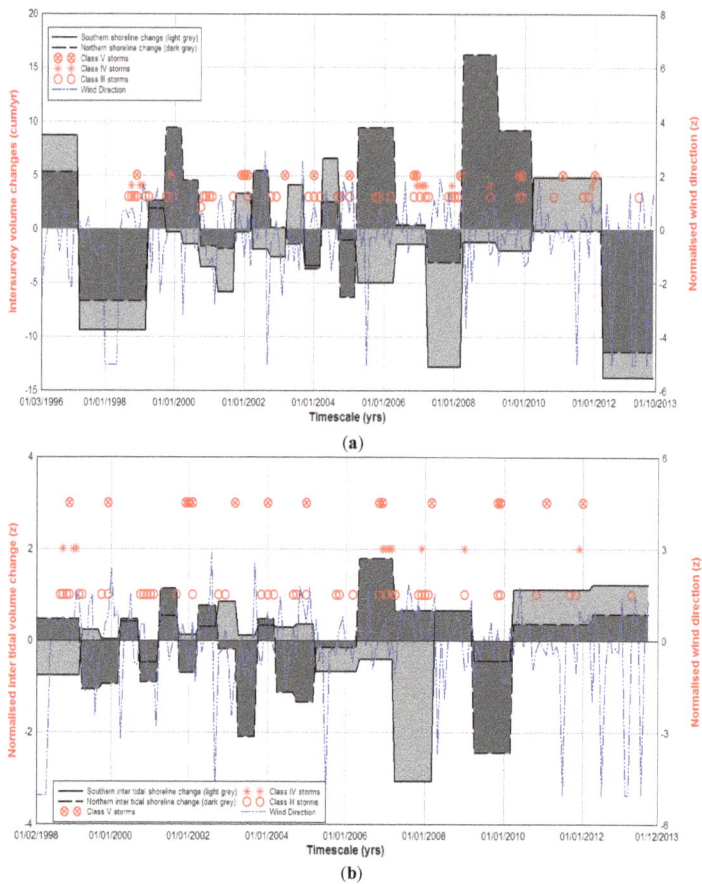

Figure 7. Graphical representations comparing normalised wind direction. Storm occurrence (class III, IV and V; 1996–2013) with: (**a**) subaerial volumes change; and (**b**) intertidal volume change.

5. Discussion

The study has examined available information on storm characterization, wave models and beach level and volume change over a 17-year period to establish if beach rotation and morphology changes, identified by Thomas *et al.* [16], continued at South Beach, Tenby, West Wales. Statistical tests suggest that there was significant beach level losses in both southern (T09) and central (T10) sectors ($p < 0.01$) and increasing beach levels in the northern sector (T11) albeit statistically insignificant ($p > 0.05$).

What is also of interest is that there are significant changes taking place (in terms of the beach level standard deviation) at the terminus of each beach profile suggesting that active beach profile extends into the subtidal zone. This is important for this particular littoral as sediment may be lost around the down drift headland and may explain the reason for continued beach level losses.

When temporal beach volume variations were examined within the subaerial zone, results agreed with the centurial trends found by Thomas *et al.* [12], southern (T09) and central (T10) erosion and northerly (T11) accretion ($R^2 = 84\%$. 1% and 79% respectively). Statistically there was very little correlation within the central region suggesting that changes are cyclic and overall stability showing a loss of <0.3 m^3·year^{-1}. With statistical significance all three assessed sectors eroded within the intertidal zone during the 17 year period of assessment ($R^2 = 71\%$ (T09). 79% (T10) and 75% (T11)).

This temporal trend of lowering beach level can be associated with a sediment deficiency from an offshore or up-drift location and/or changes to near-shore bathymetry. The macrotidal nature of the locality may also be an influence with the subaerial zone only affected during high tidal conditions.

When the cross shore component was removed from the data, negative phase relationships existed between south/north sectors within both assessed zones. This indicates that when one sector erodes the other accretes and *vice versa i.e.*, beach rotation (subaerial R^2 = 59%. Intertidal R^2 = 70%); there was also negative correlations between south/north and central beach sectors (*i.e.*, non-rotational) within the subaerial zone but with almost no statistical significance and once again the limited exposure to waves in this zone has influence. These results are not surprising given that the beach pivot point or region of rotation may not occur at the profile position but this contradicts Thomas *et al.* [16] findings, showing the profile position was closer to the beaches pivot point. There was a positive correlation between south and central sectors that suggested that when changes took place in the south similar changes took place in the central region and in the south the opposite would be true due to a negative relationship. Even though there was no improvement when cross-correlations were calculated, there was a trend reversal at a one year time lag in the subaerial zone and two years in the intertidal zone, confirmed in both cases by time-series analyses.

However, it is the intertidal results that are of most interest, this showed that with statistical significance that a clear pattern of rotation existed. This was surprising as the intertidal zone is *circa* 250 m wide. The centurial work [12] showed an almost consistent trend of beach rotation, eroding in the south and accreting in the north. They also showed that when the dune system eroded in the south the sediment was deposited within the intertidal zone and while some feedback was probable, most of the sediment moved alongshore, contributing to the northern sediment budget, with the overburden lost around the down drift headland (Tenby).

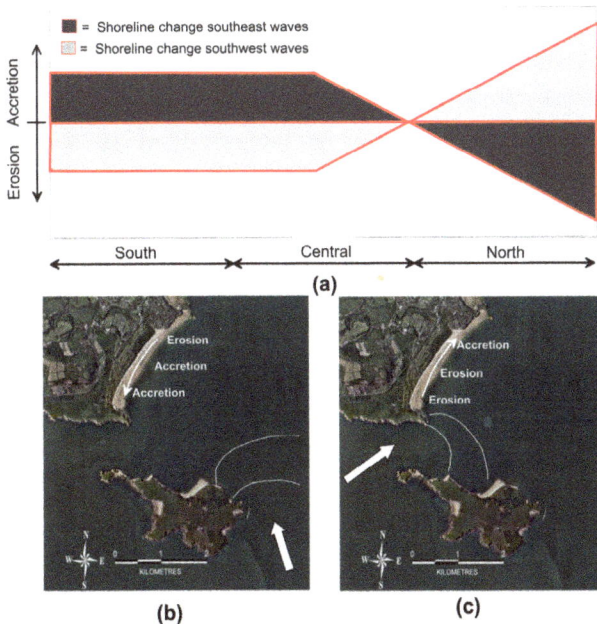

Figure 8. (**a**) A graphical illustration of the effects of southwesterly and southeasterly wave regimes have on Tenby South Sands; (**b**) a simplified conceptual model of wave propagation to nearshore based on waves from a south-westerly direction; and (**c**) a simplified conceptual model of wave propagation from a southeasterly direction.

Significant wave heights show clear cyclic patterns and these were attuned to the wave direction, where southwesterly winds dominate the winter climate with increased wave height. During summer there is a slight change toward east (from southwest) and lower wave heights. This is significant as feedback from north toward south has been shown to be reliant on easterly orientated waves that are sub-dominant in this region [13,16], which explains longer term beach rotation in one direction (southern erosion an northern accretion). Figure 8 was reproduced from Thomas [16] shows graphically (Figure 7a) the expected sediment movement along the bay when exposed to both dominant southwesterly waves and sub-dominant southeasterly waves and wave propagation is shown conceptually for both wave directions in Figures 8b and 8c respectively. In total, 267 storms occurred during the period of assessment and subsequent analysis highlighted both seasonal (summer/winter) and medium (3 yearly) cyclic behaviour. Twenty-eight percent of storms were classed between class III (severe) and V (extreme) and these are mainly generated by southwesterly wind regimes that cause south to north sediment movement. Southeasterly winds that produced counter drift generally occur during summer but with less intensity, explaining the longer term trends of south erosion and north accretion (*i.e.*, one directional rotation).

There was no quantitative correlation between storms and volume changes and qualitative assessment showed that the beach system is probably event driven. The shoreline reaction to a storm event or series of events may trigger either erosion or accretion that continues until another similar event triggers a reversal in trend. Again this would explain the longer term evolution of this embayment were the predominant environmental forcing is generated by Atlantic swell waves. Similar behaviour should be exhibited at other worldwide coastal locations and it is suggested that this work is repeated to establish specific responses; this would enable suitable coastal management policies to be developed in order to underpin intervention or no active intervention strategies and enable more effective use of limited resources.

6. Conclusions

Cross-shore profiles and environmental forcing were used to analyse morphological change of a headland bay beach: Tenby, West Wales (51.66 N; −4.71 W) over a mesoscale timeframe (1996–2013). Statistical tests showed that southern and central profile losses were significant and northern gains were insignificant when assessed across the entire profile. Beach volume variations were attuned with historic research within the subaerial zone given by statistically significant loss on southern shores and gain on northern shores, with the central region showing an insignificant loss. Volume loss was shown at all profile locations within the intertidal zone possibly influenced by sediment deficiencies either up drift or offshore. Beach rotation within both zones was established by statistically significant negative phase relationships at the beach extremities but not within the central region of rotation. Cross-correlations highlighted trend reversals suggesting that southern/northern sediment exchange lagged one another by up to two years. Qualitatively, time series analysis confirmed this rotational trend. There was little correlation between volume variation and storm occurrence, suggesting the system is event driven.

Wave height and storm events exhibited summer/winter cyclic trends that provided an explanation for longer term evolution at this location (*i.e.*, one directional rotation). In line with previous regional research, environmental forcing suggests that changes are influenced by variations in southwesterly wind regimes. Winter storms are more often than not generated by Atlantic southwesterly winds which cause both energetic waves and a south toward north sediment exchange. Southeasterly conditions that result in a trend reversal are generally limited to the summer period, where waves are fetch-limited and less energetic. Natural and man-made embayment beaches are common coastal features and many experience shoreline change jeopardising protective and recreational beach functions. In order to facilitate an effective and sustainable coastal zone management strategy, an understanding of the morphological variability of these systems is needed. Therefore, this macrotidal study's results have global implications, especially in response to predicted sea level

J. Mar. Sci. Eng. **2015**, *3*, 1006–1026

rise and climate change scenarios, and should be repeated elsewhere to inform the development of appropriate shoreline management strategies.

Acknowledgments: The authors would like to thank Welsh Assembly Government Aerial Photographs Unit, Welsh Government, Crown Offices, Cathays Park, Cardiff, Wales. CF10 3NQ for aerial photographs used in the research. The authors would like to thank the anonymous reviewers for their constructive comments and suggests that were very much appreciated and improved the content of this paper.

Author Contributions: Conceived and designed the experiments: TT NRB. Performed the experiments: TT NRB. Analyzed and checked data: TT NRB GA MRP AW. Wrote the paper: TT NRB GA MRP AW

Conflicts of Interest: The authors declare no conflict of interest.

References

1. Ojeda, E.; Guillen, J. Shoreline Dynamics and beach rotation of artificial embayed beaches. *Mar. Geol.* **2008**, *253*, 51–62. [CrossRef]
2. Short, A.D.; Masselink, G. Embayed and structurally controlled beaches. In *Handbook of Beach and Shoreface Morphodynamics*; Short, A.D., Ed.; John Wiley and Sons Ltd.: Chichester, UK, 1999; pp. 230–250.
3. Benedet, L.; Klein, A.H.F.; Hsu, J.R.C. Practical insights and applicability of empirical bay shape equations. In *Coastal Engineering Conference*; American Society of Civil Engineers: Reston, VA, USA, 2004; pp. 2181–2193.
4. Boashash, B. Theory of quadratic TFDs. In *TIme-Frequency Signal Analysis and Processing: A Comprehensive Reference*; Boashash, B., Ed.; Elsevier Ltd.: Oxford, UK, 2003; pp. 59–81.
5. Carter, R.W.G. *Coastal Environments: An Introduction to the Physical, Ecological and Cultural Systems of Coastlines*; Academic Press: London, UK, 1988; p. 617.
6. Harley, M.D.; Turner, I.L.; Short, A.D.; Ranasinghe, R. A reevaluation of coastal embayment rotation: The dominance of cross-shore *versus* alongshore sediment transport processes, Collaroy-Narrabeen Beach southeast Australia. *J. Geophys. Res.* **2011**, *116*. [CrossRef]
7. Bryan, K.R.; Foster, R.; MacDonald, I. Beach rotation at two adjacent headland-enclosed beaches. *J. Coast. Res.* **2013**, *118*, 2095–2100.
8. Silva, A.N.; Taborda, R.; Antunes, C.; Catalão, J.; Duarte, J. Understanding the coastal variability at Norte beach, Portugal. *J. Coast. Res.* **2013**, *118*, 2173–2178.
9. Klein, A.H.F.; Benedet Filho, L.; Schumacher, D.H. Short-term beach rotation processes in distinct headland bay systems. *J. Coast. Res.* **2002**, *18*, 442–458.
10. Ranasinghe, R.; McLoughlan, R.; Short, A.; Symonds, G. The southern oscillation index, wave climate and beach rotation. *Mar. Geol.* **2004**, *204*, 273–287. [CrossRef]
11. Ruiz de Alegria-Arzaburu, A.; Masselink, G. Storm response and beach rotation on a gravel beach, Slapton Sands, UK. *Mar. Geol.* **2010**, *278*, 77–99. [CrossRef]
12. Thomas, T.; Phillips, M.R.; Williams, A.T. Mesoscale evolution of a headland bay: Beach rotation Process. *Geomorphology* **2010**, *123*, 129–141. [CrossRef]
13. Thomas, T.; Phillips, M.R.; Williams, A.T.; Jenkins, R.E. Short-term beach rotation, wave climate and the North Atlantic Oscillation (NAO). *Prog. Phys. Geogr.* **2011**, *35*, 333–352. [CrossRef]
14. Short, A.D.; Trembanis, A.C. Decadal scale patterns of beach oscillation and rotation: Narabeen Beach, Australia—Time Series PCA and Wavelet Analysis. *J. Coast. Res.* **2004**, *20*, 523–532. [CrossRef]
15. Short, A.D.; Trembanis, A.C.; Turner, I.L. Beach Oscillation, Rotation and the southern Oscillation, Narabeen Beach, Australia. *Coast. Eng.* **2000**, 2439–2452. [CrossRef]
16. Thomas, T.; Phillips, M.R.; Williams, A.T.; Jenkins, R.E. Medium timescale beach rotation: Gale climate and offshore island influences. *Geomorphology* **2011**, *135*, 97–107. [CrossRef]
17. Cooper, J.A.G.; Jackson, D.W.T. Geomorphological and dynamic constraints on mesoscale coastal response to storms, Western Ireland. In *Coastal Sediments '07: Proceedings 6th International Symposium on Coastal Engineering and Science of Coastal Sediment Processes*; American Society of Civil Engineers: Reston, VA, USA, 2003; pp. 3015–3024.
18. Cooper, J.A.G. Temperate coasts. In *Applied Sedimentology*; Perry, C.M., Taylor, K., Eds.; Blackwell Publishing: Oxford, UK, 2007; pp. 263–301.
19. O'Connor, M.O.; Cooper, J.A.G.; Jackson, D.W.T. Morphological behaviour of headland-embayment and inlet-associated beaches, Northwest Ireland. *J. Coast. Res.* **2007**, *59*, 626–630.

20. Anthony, E.J.; Dolique, F. The influence of Amazon-derived mud banks on the morphology of sandy headland-bound beach in Cayenne, French Guiana: A short to long-term perspective. *Mar. Geol.* **2004**, *208*, 249–264. [CrossRef]
21. Anthony, E.J.; Gardel, A.; Dolique, F.; Guiral, D. Short-term changes in the planshape of a sandy beach in response to sheltering by a near shore mud bank, Cayenne, French Guiana. *Earth Surf. Processes Landf.* **2002**, *27*, 857–866. [CrossRef]
22. Thomas, T.; Phillips, M.R.; Williams, A.T.; Jenkins, R.E. A multi-century record of linked nearshore and coastal change. *Earth Surf. Processes Landf.* **2011**, *36*, 995–1006. [CrossRef]
23. Dehouck, A.; Dupuis, H.; Senechal, N. Pocket beach hydrodynamics: The example of four macrotidal beaches, Brittany, France. *Mar. Geol.* **2000**, *266*, 1–17. [CrossRef]
24. Dolique, F.; Anthony, E.J. Short-term profile changes of sandy pocket beaches affected by amazon-derived mud, Cayenne, French Guiana. *J. Coast. Res.* **2005**, *21*, 1195–1202. [CrossRef]
25. Pinto, C.A.; Tabord, T.; Andrade, C.; Teixeira, S. Seasonal and mesoscale variations at an embayed beach (Armacao De Pera, Portugal). *J. Coast. Res.* **2009**, *SI56*, 118–122.
26. Loureiro, C.; Ferreira, O.; Cooper, J.A.G. Contrasting morphological behaviour at embayed beaches in southern Portugal. *J. Coast. Res.* **2009**, *SI56*, 83–87.
27. Sedrati, M.; Anthony, E.J. A brief overview of plan-shape disequilibrium in embayed beaches: Tangier Bay (Morocco). *Mediterranee* **2007**, *108*, 125–130. Available online: http://mediteranee.revues.org/index190.html?file=1 (accessed on 21 September 2009). [CrossRef]
28. Stone, G.W.; Orford, J.D. Storms and their significance in coastal morpho-sedimentary dynamics. *Mar. Geol.* **2004**, *210*, 1–5. [CrossRef]
29. Maspataud, A.; Ruz, M.H.; Hequette, A. Spatial variability in post-storm beach recovery along a macrotidal barred beach, southern North Sea. *J. Coast. Res.* **2009**, *1*, 88–92.
30. Uncles, R.J. Physical properties and processes in the Bristol Channel and Severn Estuary. *Mar. Pollut. Bull.* **2010**, *61*, 5–20. [CrossRef] [PubMed]
31. Phillips, M.R.; Crisp, S. Sea level trends and NAO influences: The Bristol channel/Severn estuary. *Glob. Planet. Change* **2010**, *73*, 211–218. [CrossRef]
32. Lavernock Point to St Ann's Head SMP2. Available online: http://www.southwalescoast.org/contents.asp?id=55#SMP2MainDocument (accessed on 20 June 2011).
33. Toghill, P. *The Geology of Britain: An Introduction*; Crowhill Press Ltd.: Wiltshire, UK, 2000; p. 192.
34. Hunter, A.; Easterbrook, G. *The Geological History of the British Isles*; The Alden Group: Oxford, UK, 2004; p. 143.
35. Mackie, A.S.Y.; James, J.W.C.; Rees, E.I.S.; Darbyshire, T.; Philpot, S.L.; Mortimer, K.; Jenkins, G.O. The Outer Bristol Channel Marine Habitat Study. 2002. Available online: http://www.marlin.ac.uk/obc/pdfs/report/chapter%202.pdf (accessed on 12 December 2010).
36. Owen, T.R. *Geology Explained in South Wales*; David & Charles Publishers: Newton Abbott, UK, 1973; p. 177.
37. Hillier, R.D.; Williams, B.P.J. The alluvial OLD Red Sandstone: Fluvial Basins. In *The Geology of England and Wales*; Brenchley, P.J., Rawson, P.F., Eds.; The Geological Society: London, UK, 2006; pp. 155–172.
38. Duvivier, P. *South Beach, Tenby*; Posford Duvivier: Peterborough, UK, 1998; p. 19.
39. Gibbard, B. *Tenby South Beach Erosion: Review of Wind/Wave Data, Beach Monitoring and Modelling*; Royal Haskoning: Peterborough, UK, 2005; p. 39.
40. Morang, A.; Batten, B.K.; Connell, K.J.; Tanner, W.; Larson, M.; Kraus, N.C. *Regional Morphology Analysis Package (RMAP), Version 3: Users Guide and Tutorial*; Coastal and Hydraulics Engineering Technical Note ERDC/CHL CHETN-XIV-9. U.S. Army Corps of Engineers: Washington, DC, USA, 2009. Available online: http://chl.erdc.usace.army.mil/chetn/ (accessed on 6 June 2010).
41. Taveira-Pinto, C.A.; Taborda, R.; Andrade, C.; Teixeira, S.B. Seasonal and Mesoscale variations at an Embayed Beach (Armacao De Pera, Portugal). *J. Coast. Res.* **2009**, *25*, 118–122.
42. Dorsch, W.; Newland, T.; Tassone, D.; Tymons, S.; Walker, D. A statistical approach to modeling the temporal patterns of ocean storms. *J. Coast. Res.* **2008**, *24*, 1430–1438. [CrossRef]
43. Davis, J.C. *Statistics and Data Analysis in Geology*, 3rd ed.; Wiley and Sons: Chichester, UK, 2002; p. 638.
44. Dolan, R.; Davis, R.E. An intensity scale for Atlantic coast northeast storms. *J. Coast. Res.* **1992**, *8*, 352–364.
45. Thomas, T.; Lynch, S.K.; Phillips, M.R.; Williams, A.T. Long-term evolution of a sand spit, physical forcing and links to coastal flooding. *Appl. Geogr.* **2014**, *53*, 187–201. [CrossRef]

J. Mar. Sci. Eng. **2015**, *3*, 1006–1026

46. Morton, I.; Bowers, J.; Mould, G. Estimating return period wave heights and winds speeds using a seasonal point process model. *Coast. Eng.* **1997**, *26*, 251–270. [CrossRef]

47. Rangel, N.; Anfuso, G. Winter wave climate, storms and regional cycles: The SW Spanish Atlantic coast. *Int. J. Climatol.* **2013**, *33*, 2142–2156. [CrossRef]

48. Jenks, G.F.; Caspal, F.C. Error on choropletic maps: Definition, measurement, reduction. *Ann. Assoc. Am. Geogr.* **1971**, *61*, 217–244. [CrossRef]

49. Thomas, T.; Phillips, M.R.; Williams, A.T.; Jenkins, R.E. Rotation on two adjacent open coast macrotidal beaches. *Appl. Geogr.* **2012**, *35*, 363–376. [CrossRef]

50. Carter, D.J.T.; Draper, L. Has the north-east Atlantic become rougher? *Nature* **1988**, *332*, 494. [CrossRef]

51. Bacon, S.; Carter, D.J.T. A connection between mean wave height and atmospheric pressure gradient in the North Atlantic. International. *J. Climatol.* **1993**, *13*, 423–436. [CrossRef]

52. Allan, J.C.; Komar, P.D. Are ocean wave heights increasing in the eastern North Pacific? *EOS* **2000**, *47*, 561–567. [CrossRef]

53. Moritz, H.; Moritz, H. Evaluating extreme storm power and potential implications to coastal infrastructure damage, Oregon Coast USA. In Proceedings of the 9th International Workshop on Wave Hindcasting and Forecasting, Victoria, BC, Canada, 24–29 September 2006.

54. Mendoza, E.T.; Jimenez, J.A. Coastal storm classification on the Catalan littoral (NW Mediterranean). *Ing. Hidrául. Méx.* **2008**, *23*, 23–34.

55. Rangel-Buitrago, N.; Anfuso, G. Coastal storm characterization and morphological impacts on sandy coasts. *Earth Surf. Processes Landf.* **2011**, *36*, 1997–2010. [CrossRef]

Journal of
Marine Science and Engineering

MDPI

Brief Report

Wave and Hydrodynamic Modeling for Engineering Design of Jetties at Tangier Island in Chesapeake Bay, USA

Lihwa Lin *, Zeki Demirbilek, Donald Ward and David King

Coastal and Hydraulics Laboratory, U.S. Army Engineer Research and Development Center, 3909 Halls Ferry Road, Vicksburg, MS 39180, USA; Zeki.Demirbilek@usace.army.mil (Z.D.); Donald.Ward@usace.army.mil (D.W.); David.King@usace.army.mil (D.K.)

* Author to whom correspondence should be addressed; Lihwa.Lin@usace.army.mil; Tel.: +1-601-634-2704.

Academic Editor: Gerben Ruessink

Received: 8 September 2015; Accepted: 17 November 2015; Published: 1 December 2015

Abstract: The protection of a boat canal at the western entrance of Tangier Island, Virginia, located in the lower Chesapeake Bay, is investigated using different structural alternatives. The existing entrance channel is oriented 45 deg with respect to the local shoreline, and exposed directly to the lower Bay without any protection. The adjacent shoreline has experienced progressive erosion in recent decades by flooding due to severe storms and waves. To protect the western entrance of the channel and shoreline, five different jetty and spur combinations were proposed to reduce wave energy in the lee of jetties. Environmental forces affecting the proposed jettied inlet system are quantified using the Coastal Modeling System, consisting of a spectral wave model and a depth-averaged circulation model with sediment transport calculations. Numerical simulations were conducted for design wave conditions and a 50-year return period tropical storm at the project site. Model results show a low crested jetty of 170-m length connecting to the north shore at a 45-deg angle, and a short south spur of 25-m long, provide adequate wave-reduction benefits among the five proposed alternatives. The model simulation indicates this alternative has the minimum impact on sedimentation around the structured inlet and boat canal.

Keywords: coastal modeling; jetty design; Tangier Island; Chesapeake Bay

1. Introduction

Tangier Island (75°59.4′ W, 37°49.8′ N) is the southernmost of a string of islands. The shallower Tangier Sound separates the lower Chesapeake Bay on the west from the east bay (Figure 1). The island, approximately 8 km (5 miles) long by 3.2 km (2 miles) wide, is located in the Virginia portion of Chesapeake Bay, 36 km (20 miles) southwest of Crisfield and 112 km (70 miles) north of Norfolk. Tangier Island is comprised of a few low fine-grained sand ridges with intervening marshlands having numerous islets and tidal creeks. The highest elevations of the island are only a few meters above the mean tide level (MTL). The small populated areas are primarily three interconnected ridges on the south-central portion of the island.

Figure 1. Location of Tangier Island (small rectangular box) in Chesapeake Bay (large box).

Tangier Island boat canal is a narrow light-draft channel that runs east–west across the mid-section of the island (Figure 2). It is approximately 2.3 km (1.5 miles) long, 80 m (265 ft) wide, and 4 m (13 ft) deep for small-boat traffic. Numerous mooring docks and seafood processing sheds along both sides of the canal are the main infrastructure of local fishing and crabbing industries.

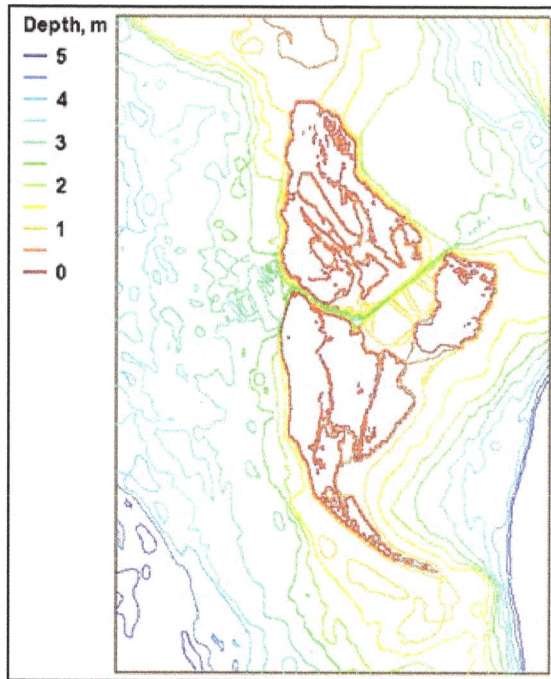

Figure 2. Depth contours in the boat canal crossing through the mid-section of Tangier Island.

The western side of Tangier Island is exposed to large incident waves (up to 2-m wave height) generated during storms in the Chesapeake Bay from the northwest through southwest quadrants. The western shoreline has long experienced progressive flooding and erosion during storms. Due to prevailing wind patterns, the littoral transport along the west shorelines of the island is directed toward the south. During storms, large waves with strong currents and high water frequently enter the western entrance of the boat canal, causing damage to shorelines and structures.

The U.S. Army Corps of Engineers (USACE) Research and Development Center (ERDC) has conducted a numerical modeling study of waves and hydrodynamics for jetty alternatives intended to protect shorelines and reduce wave energy in the western portion of the canal. The primary goal of the study was to develop a quantitative estimate of waves and wave reduction in the canal for alternatives with minimal effects on channel dredging requirements and boat traffic in the channel.

2. Local Environmental Conditions

Environmental forces that normally impact the western entrance of Tangier Island boat canal are wind-waves, currents, and water levels. These natural forcings consist of metocean events including summer storms, northeasters, and tropical events, which can impact the Chesapeake Bay and reach Tangier Island from different directions. Seasonal wind patterns vary over the bay. In the winter, the dominant winds are from the north and northwest; they are from the southwest in the summer, with local breezing shifting the wind direction on a daily basis. Larger waves generally occur during northeasters and tropical storms, when high winds blow across the bay.

The west shoreline of Tangier Island is exposed to open water in the lower Bay area, where strong wind can generate large waves. Figure 3 shows two sample wind roses during for 2011 and 2012 from NOAA station 8632837 (37°32.3′ N, 76°0.9′ E) at Rappahannock Light, VA, approximately 35 km (22 mile) south of Tangier Island in the lower Bay. Winds stronger than 10 m/s (~20 kt) mostly follow a longer fetch along the north–south direction in the Bay. During northeasters with sustained winds of 15 to 20 m/s (30 to 40 kt), local wave heights ranging from 1.5 to 2.5 m (5 to 8 ft) can be expected along the west side of Tangier Island.

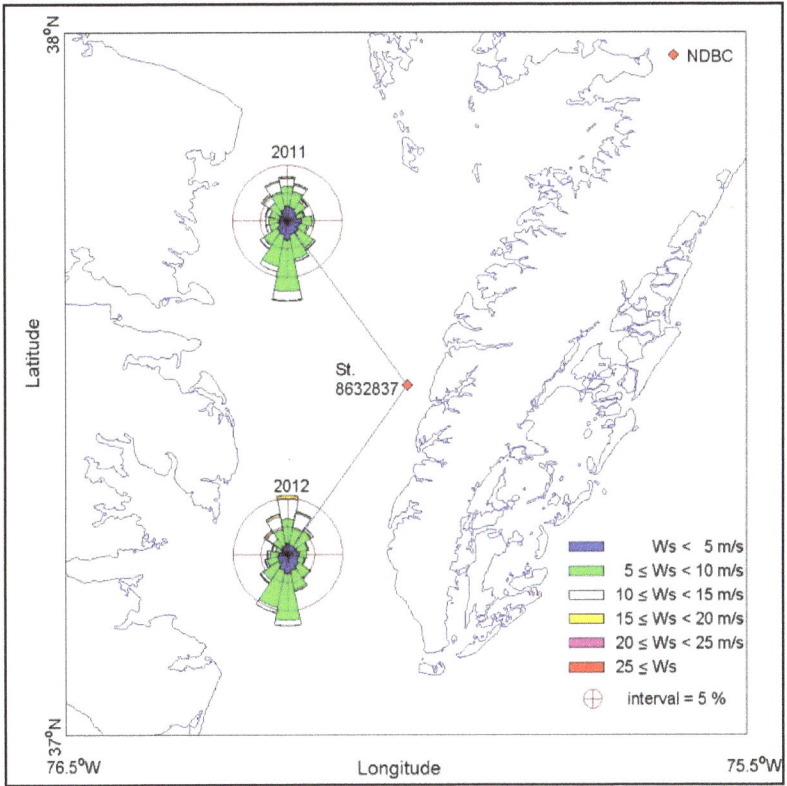

Figure 3. Wind roses for year 2011 and 2012 at NOAA Station 8632837.

Water level fluctuations in the Chesapeake Bay are dominated by oceanic tides interacting with the Bay. Tides at Tangier Island are semi-diurnal, with a 0.6-m mean tidal range. Abnormal water levels or storm surge can occur during tropical events. In the lower Chesapeake Bay, storm surges above the mean water level for 50-year and 100-year recurrence intervals are estimated at 1.5 m and 1.8 m (5 and 6 ft), respectively [1]. The relative sea level rise estimate due to absolute sea level change and land subsidence in the Chesapeake Bay ranges between 3 and 6 mm/year [1].

Historically, the Chesapeake Bay froze more often during the 19th and early 20th centuries, but rarely in the last two decades as a result of regional warming, at approximately +1 °C (+2 °F) per decade. The lower Bay may briefly become covered by thinner ice during a severe winter season. The Bay icing is not considered in the present study.

3. Structural Alternatives

The primary area of interest in this modeling study is the west channel section of the Tangier Island boat canal shown in Figure 4. This narrow canal is the only navigation route that cuts through the middle of Tangier Island and connects the east and west sides of the island. The average west channel base width is 18.3 m (60 ft), the top width is 30.5 m (100 ft), and the channel depth varies from 2.3 to 4 m (7.5 to 13 ft). The narrowest cross section (bank-to-bank) is 70 m (230 ft).

Figure 4. Western part of Tangier Island boat canal.

Five structural alternatives with a north jetty connecting to the north shoreline were evaluated, where the north jetty is either a straight or a dogleg structure. Due to cost constraints, the total length of the north jetty is limited to 200 m (650 ft). Alternatives 1 and 2 consider a north jetty of different length. The north jetty is positioned as close to the channel as possible at a safe (for navigation) distance of 50 m to 100 m (164 ft to 328 ft) from the channel edges. Alternatives 3, 4, and 5 include an additional short spur attaching to the south shoreline.

Figure 5 shows depth color contour maps for the existing western entrance channel (channel position and centerline in black lines) and structural features considered in Alternatives 1 to 5 (Alts 1 to 5). In all five alternatives, both jetty and spur structures have a crest elevation of 1 m (3.3 ft) above MTL (roughly the same as MSL) and crest width of 4 m (13 ft). The selection of jetty/spur crest elevation and crest width is based on the cost-to-benefit ratio and federal funding available for construction of structures in the project. The crest and width of proposed jetty/spur structures may be raised and expanded in the future if a higher crest elevation is required. Alt 1 has a straight north jetty 85 m (280 ft) long that is normal to the north shoreline. Alt 2 has a dogleg north jetty 170 m (560 ft) long with a bayward segment 85 m (280 ft) long that is parallel to the entrance channel. Alt 3 has the same dogleg north jetty as in Alt 2 and a short south spur 25 m (82 ft) long that points north (towards the channel). Alt 4 has the same dogleg north jetty as in Alt 2 and a south spur 25 m (82 ft) long directed towards the northwest (normal to south shoreline). Alt 5 has a straight north jetty identical to Alt 1 and a south spur identical to Alt 4. In addition to guiding the tidal flow along the navigation channel, the second purpose of the north jetty is to protect the shoreline east of the north jetty from wave action. The primary purpose of the south spur is to stabilize the shoreline south of the spur and reduce wave penetration into the western entrance channel. Table 1 presents a summary of existing channel and structural features for the five alternatives.

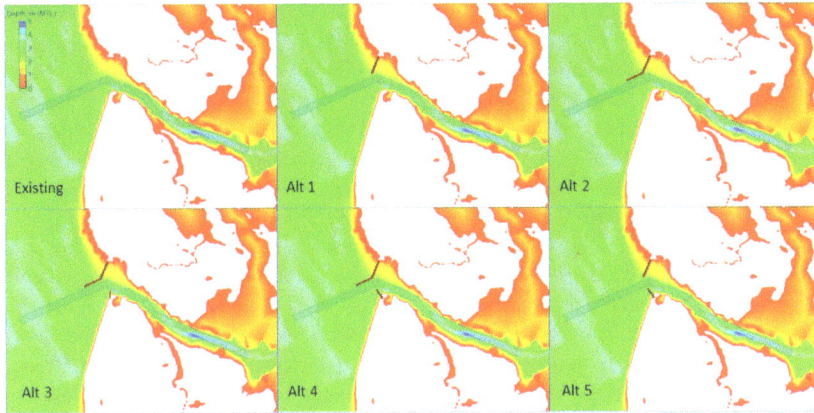

Figure 5. Depth color contours of the existing western entrance channel and Alts 1 to 5.

Table 1. Structural features of alternatives.

Alt	Straight North Jetty (85 m Long)	Dogleg North Jetty (170 m Long)	South Spur (25 m Long) Toward Channel	South Spur (25 m long) Normal to Shoreline
1	X			
2		X		
3		X	X	
4		X		X
5	X			X

4. Numerical Modeling Approach

The Coastal Modeling System (CMS) was used to calculate waves, currents, sediment transport, and morphology change [2]. It is an integrated modeling system that consists of a steady-state spectral wave model (CMS-Wave) and a two-dimensional, time-dependent circulation model (CMS-Flow), which includes sediment transport and bed change capabilities. The CMS uses the Surface-water Modeling System (SMS) interface for grid generation, model setup, and post-processing [3].

The CMS models can run on a grid with variable rectangular cells. To save computational time, large cells can be used in large-area applications away from the area of interest while fine-grid resolution is used in the area of interest. The CMS can also run on nested grids that include many large and small grids [4]. The most commonly applied grid nesting involves two model grids: a large grid (parent grid) and a small grid (child grid). The application of grid nesting can dramatically reduce the computational time as compared to a large grid with fine resolution for the entire model domain. A parent grid may be used to simulate regional processes such as wave generation and propagation in a large domain. A child grid can resolve more complex bathymetry and shoreline geometry in a smaller area for more accurate modeling of nearshore wave processes. Water levels and currents calculated from the parent grid are interpolated to the child grid boundaries. Wave spectra calculated from the parent grid are saved at selected locations along the offshore boundary of the child grid. Conventionally, it suffices to save a single-location spectrum from the large-domain parent grid for wave input and apply it to the entire sea boundary of a comparatively much smaller domain child grid. For a more inhomogeneous wave field, multiple locations of wave spectra may be saved from the parent grid and interpolated for more realistic wave forcing along the seaward boundary of the child grid. The main goal of grid nesting is to minimize computational time while not sacrificing modeling accuracy.

J. Mar. Sci. Eng. **2015**, 3, 1474–1503

The CMS has been applied in many coastal wave, circulation, and sediment transport studies including open coasts, inlets, bays, estuaries, and lakes [5]. Most recent bay applications include modeling waves, flow, and sediment transport of Braddock Bay in Lake Ontario [6], and Matagorda Bay and Galveston Bay in Texas [7,8].

The CMS uses physics-based theories to calculate wave generation, growth, and dissipation under variable wind/pressure fields in a bay-ocean system and, therefore, can simulate large-scale storm and hurricane events [9–14]. The use of the CMS in wave modeling is more realistic for bay applications and more accurate than classical fetch-based empirical curves as the wind wave generation and growth are strongly affected by the complex of bay geometry and varying bathymetry [15,16]. This is particularly true in the Chesapeake Bay because many river tributaries, navigation channels, shoals, islands, and peninsulas coexist in the Bay. The long narrow lake basin with the broader lower bay connecting to the Atlantic Ocean causes more complexity of wave action and flow circulation in the Chesapeake Bay. Because the western channel of Tangier Island is exposed to open water in the lower Bay, strong winds from the northwest and southwest quadrants can generate large waves in the area. During a tropical storm, Tangier Island may be threatened by high water as low atmospheric pressure and strong wind can trap water against the higher ground barriers and land to the east and north of the island.

Figure 6 shows the CMS modeling domain for the Chesapeake Bay region and corresponding depth contours. This bay-wide large grid domain, approximately 100 km by 300 km (60 miles by 180 miles), is referred to as the regional grid (parent grid), which has a constant grid cell size of 500 m by 500 m (1600 ft by 1600 ft). The depths in this grid vary from 0 to 45 m (0 to 150 ft). The CMS modeling includes a second domain, referred to as the local grid (child grid) for Tangier Island, which is shown in Figures 1 and 2. This grid has a finer resolution to represent details of Tangier Island shorelines and bathymetry. The domain of the child grid is approximately 5 km by 7 km (3 miles by 4.4 miles), with the local grid cell size varying from 3 m to 50 m (10 ft to 160 ft).

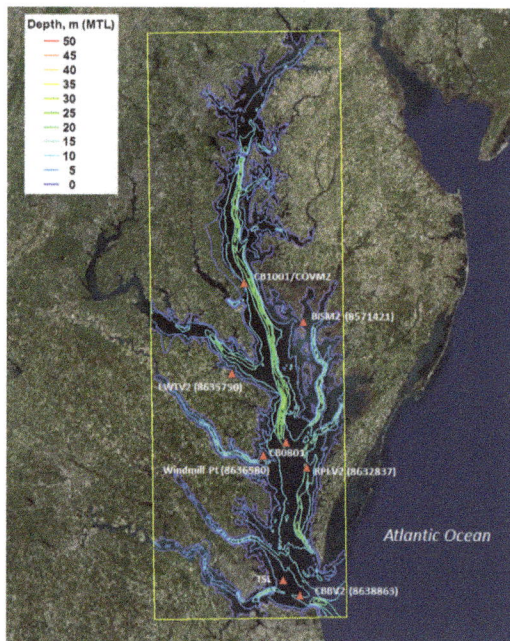

Figure 6. NOAA and VIMS coastal stations (red triangle) and CMS depth contours (m, MTL) for the CMS Chesapeake Bay regional grid domain (yellow box).

4.1. Model Calibration

The calibration of the CMS was conducted separately for CMS-Wave and CMS-Flow in the Chesapeake Bay regional grid. CMS-Wave was calibrated using wave spectral data collected at Thimble Shoal Light (TSL) gauge (see Figure 6), maintained by the Virginia Institute of Marine Science (VIMS) from 1988 to 1995 [16–18].

The calibration of CMS-Flow was conducted for August 2014. CMS-Flow was forced by hourly water level data, collected from NOAA Coastal Station 8638863 (Chesapeake Bay Bridge Tunnel, VA, USA), along the lower bay entrance boundary for flow exchange with the Atlantic Ocean. Atmospheric input to CMS-Flow was based on hourly wind data collected at NOAA Station 8632837 (Rappahannock Light, VA, USA). August 2014 was selected in CMS-Flow calibration because the wind speed for this month was on average the lowest during the year and so the effect of wind on tidal hydrodynamics is small for the calibration. Figure 7 shows thr wind data collected at NOAA 8632837 and 8638863 in 2014 [19]. A hydro time step of 15 s and spatially constant Manning coefficient of 0.02 were used in the model calibration. Figure 8 shows the model-data comparison water levels at NOAA Stations 8571421 (Bishops Head, MD, USA), 8635750 (Lewisetta, VA, USA), and 8636580 (Windmill Point, VA, USA). Model water levels and data are correlated well. Correlation coefficients between model water levels and data at Stations 8571421, 8635750, and 8636580 are equal to 0.98, 0.97, and 0.93, respectively.

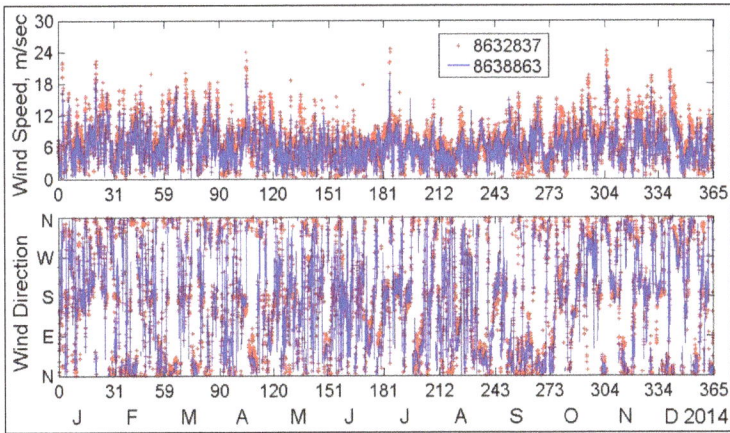

Figure 7. Wind measurements at Rappahannock Light, VA (8632837) and Chesapeake Bay Bridge Tunnel, VA (8638863) for 2014.

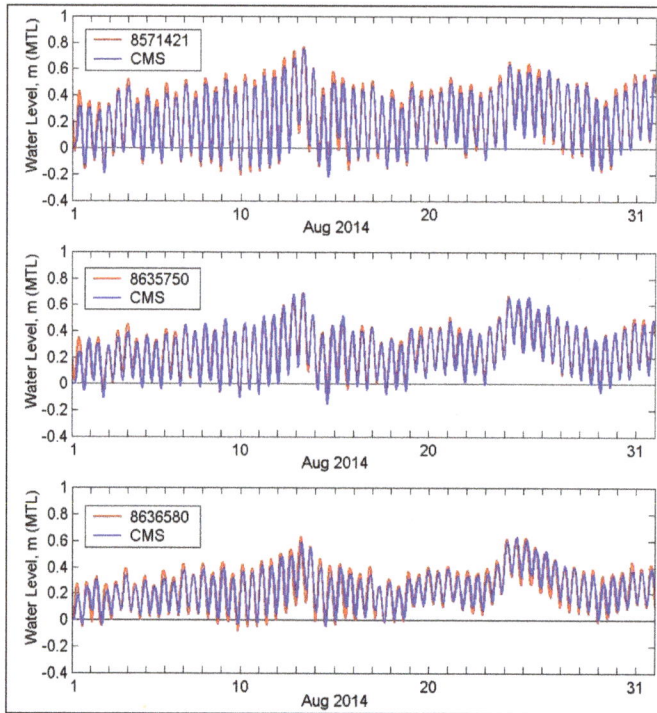

Figure 8. Model and measured water levels at Bishops Head, MD (8571421), Lewisetta, VA (8635750), and Windmill Point, VA (8636580) for August 2014.

Surface current measurements were available in the mid- and south Chesapeake Bay during 2014 from NOAA Stations CB0801 (Rappahannock Shoal Channel, VA, USA) and CB1001 (Cove Point LNG Pier, MD, USA). Figure 9 compares easting and northing current components calculated by CMS-Flow with data collected at CB0801 and CB1001. The positive current speed component along the East–West (E–W) line is directed to the east, and the positive component along the North–South (N–S) line is directed to the north. Because model currents are depth-averaged and the data are for surface currents, the magnitude of the model currents is generally smaller than in the data. Correlation coefficients between model current E–W components and data at CB0801 and CB1001 are equal to 0.27 and 0.88, respectively. Lesser correlation between E–W components and the data at CB0801 is likely due to more wind–wave interactions over weaker E–W components which are not simulated in the calibration. Correlation coefficients between model current N–S components and data at both CB0801 and CB1001 are equal to 0.89.

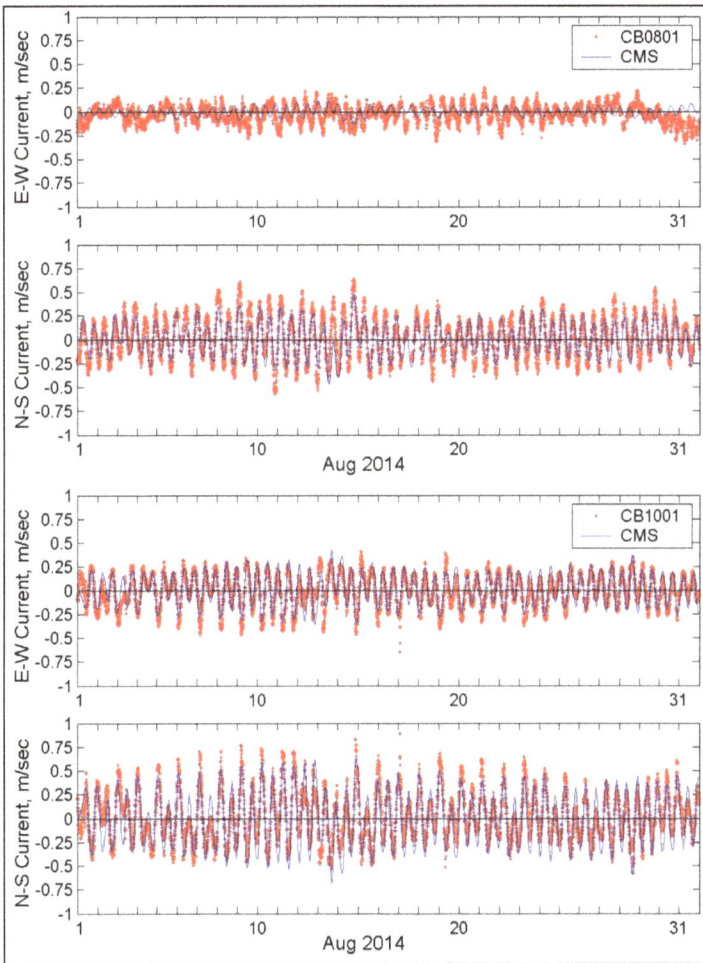

Figure 9. Comparison of calculated depth-averaged current components and measured surface current data at NOAA Stations CB0801 and CB1001 for August 2014.4.2.

4.2. Forcing Conditions

After model calibration, the modeling focused on structural design estimates for a 50-year return period for design storm conditions and a historical hurricane. For the 50-year design storm conditions, a constant wind speed of 20 m/s (40 kt) was selected based on a previous study by Basco and Shin (1993) from analysis of 1945–1983 storms at Patuxent Naval Air Station. The storms included both tropical events and northeasters. In the present study, wave generation was simulated for nine constant wind directions covering a westerly half-plane sector from north to south. These nine directions, each covering a 22.5-deg angle, present the design wind from the N, NNW, NW, WNW, W, WSW, SW, SSW, and S directions. Model simulations were conducted for two different water levels representing the observed mean tide level (WL = 0 m) and high water level (WL = 1.5 m or 5 ft) from nearby NOAA coastal stations located in the mid- and lower bay (Figure 6).

Figure 10 shows the example of the water level measurements for year 2012 recorded at three NOAA Stations: 8571421 (Bishops Head, MD), 8636580 (Windmill Point, VA), and 8638863 (Bay

J. Mar. Sci. Eng. **2015**, *3*, 1474–1503

Bridge Tunnel, VA). The maximum water level observed at Bay Bridge Tunnel Station was 1.5 m (5 ft), MTL, during Hurricane Sandy. Table 2 lists the 50-year design wind conditions with wind speed of 20 m/s and nine wind directions for two water levels (WL) of 0 and 1.5 m (5 ft). Table 2 also lists the approximate fetch length corresponding to each of the nine wind directions. In general, the wind over longer fetch generates greater wave height at the downwind end, although in reality this also depends on the water depth variation along the fetch and wave refraction over shallower areas before waves reach the oblique shoreline.

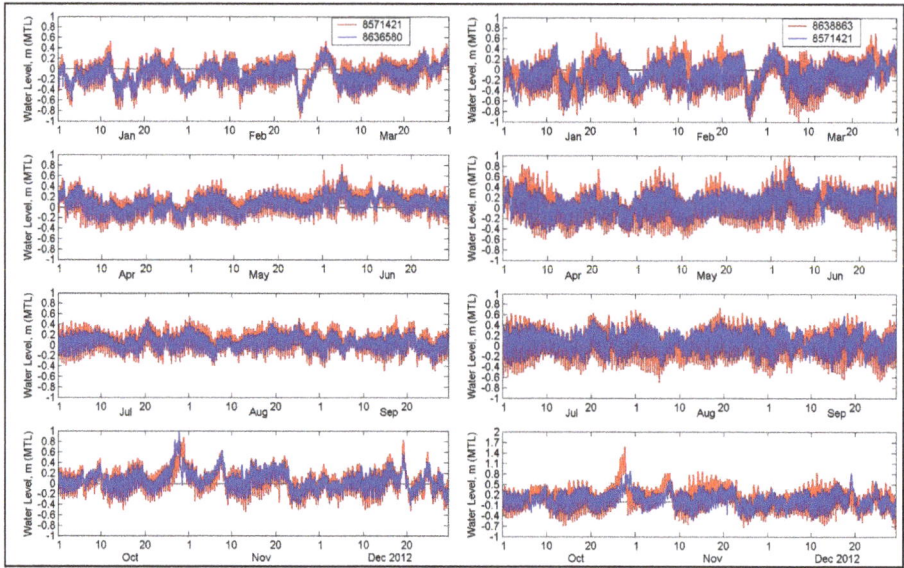

Figure 10. Water levels at Bishops Head (Station 8571421), Windmill Point (8636580), and Bay Bridge Tunnel (8638863) for 2012.

Table 2. Design wind and water level conditions for a 50-year return period northeaster storm.

Wind Dir (deg) *	N (0)	NNW (337.5)	NW (315)	WNW (292.5)	W (270)	WSW (247.5)	SW (225)	SSW (202.5)	S (180)
Fetch ** (km)	30	55	70	25	20	25	30	45	100
Wind Speed					20 m/s				
Mean Tide Level					WL = 0 m				
Mean High Water					WL = 1.5 m				

* Meteorological convention; ** Approximation.

Two water levels (WL) were used in wave simulations for the 50-year design wind conditions: were 0 and 1.5 m (5 ft) with respect to MTL. The WL = 0 m represented the mean water level corresponding to non-tropical storm (e.g., northeasters) design wind conditions, and WL = 1.5 m corresponded to the maximum storm surge for tropical storm (or hurricanes) design wind conditions. Since there is no water level measurement at Tangier Island, the selection of maximum storm surge during a hurricane is based on water level data collected in the last 50 years (after 1965) from three nearby NOAA coastal stations at Bishop Heads (Station ID 8571421), Lewisetta (8635750), and Windmill Point (8636580). These stations show that the maximum storm surge is approximately 1.5 m during Hurricanes Isabel (2003) and Sandy (2012) [13,20]. Therefore, WL = 1.5 m MTL was used as the maximum storm surge for the tropical storm in the 50-year design wind condition.

For a 50-year tropical event, Hurricane Isabel (September 2003) was selected and simulated in the entire Bay [20,21]. The strong east-to-west winds associated with Hurricane Isabel produced higher water levels along the west as well as the south side of the bay and a relatively lower water level along the east side of the bay. Figure 11 shows the examples of water level measurements from Bay Bridge Tunnel (8638863) and Windmill Point, VA (8636580) during Isabel in September 2003. The maximum water levels observed at Stations 8638863 and 8636580 were 1.87 m (6.1 ft) and 1.44 m (4.7 ft) MTL, respectively.

4.3. Design Wind and Water Level Simulations

Design wind and wave simulations were performed for nine wind directions and two water levels, 0 and 1.5 m (0 and 5 ft) MTL, listed in Table 2. The design wind conditions, representing the 50-year return period, were used for the existing western channel and Alts 1 to 5 (Figure 5). The 50-year wind condition was based on a previous study by Basco and Shin [22]. The simulations were first conducted with a regional grid, and results were then used as input in the local Tangier Island grid calculations.

A total of 108 simulations (nine wind conditions, two water levels, and six configurations) were performed to develop spatially varying estimates of the wind waves throughout the Chesapeake Bay. As an example, Figure 12 shows the bay-wide wave height fields calculated by the CMS for two wind directions from NW and SW (wind speed of 20 m/s or 40 kt) and two water levels of 0 and 1.5 m (5 ft) MTL.

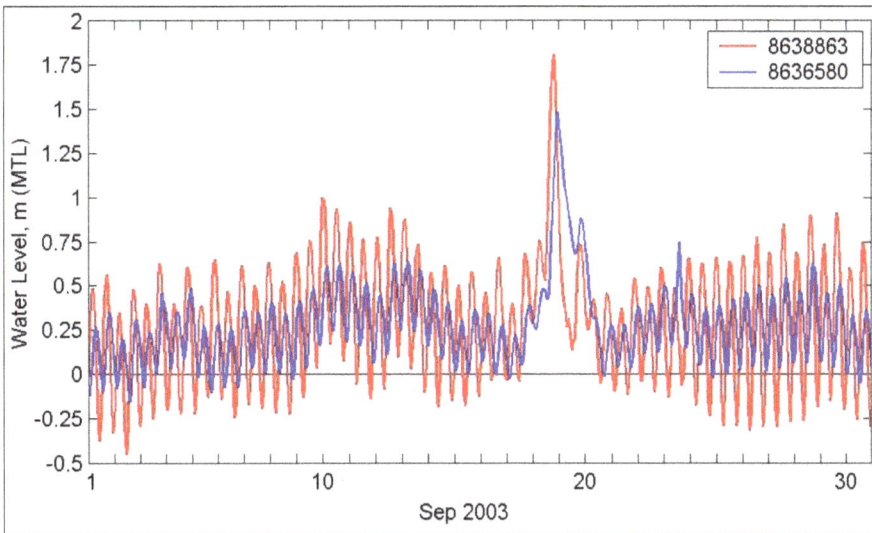

Figure 11. Water levels at Bay Bridge Tunnel (8638863) and Windmill Point (8636580), VA, for September 2003.

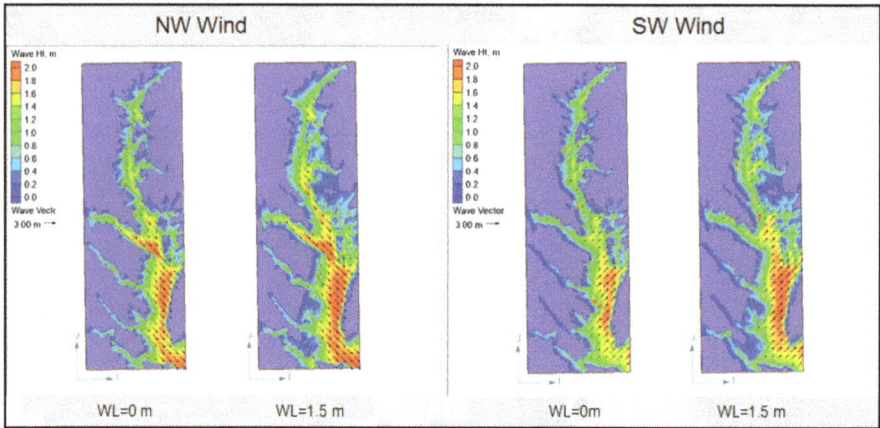

Figure 12. Calculated wave heights in Chesapeake Bay at two water levels (50-year design winds from NW and SW directions).

Wave model results from the regional grid (parent grid) for the entire Bay were used as input in the local grid (child grid) to develop the estimates of waves at the project site. Figures 13–16 show contour and vector plots of calculated wave height and direction for Existing, Alts 1, 2, and 3, respectively, for 50-year design wind speed of 20 m/s from NW and SW, and water levels of 0 m and 1.5 m MTL. The extent of wave penetration into the canal, and variation of wave heights along the channel centerline and north and south shorelines are shown in the color-coded plots of wave field in Figures 13–16. It should be noted that because CMS-Wave is a steady-state spectral model, the calculated wave field corresponds to a saturated sea state for each input wind condition. The saturated sea is also known as the developed sea that water surface waves cannot grow more under the specified wind input condition. This is possible in open water under constant wind conditions over a sufficiently long time period. In a limited fetch area (e.g., in a bay or lake) for the same wind conditions, wave generation can reach saturated state in a relatively shorter time. Because the Chesapeake Bay has a limited water body, the saturated sea calculated by the CMS-Wave should provide a good approximation of the design wind conditions. The calculated corresponding wave height is the maximum wave field for the design wind. The wave reduction analysis was performed for all simulations by comparing alternatives to the existing channel along three transact lines along the channel centerline and the north and south shorelines (Figure 17). Model wave heights were saved at 103 stations along these three transacts for further analysis (Figure 18).

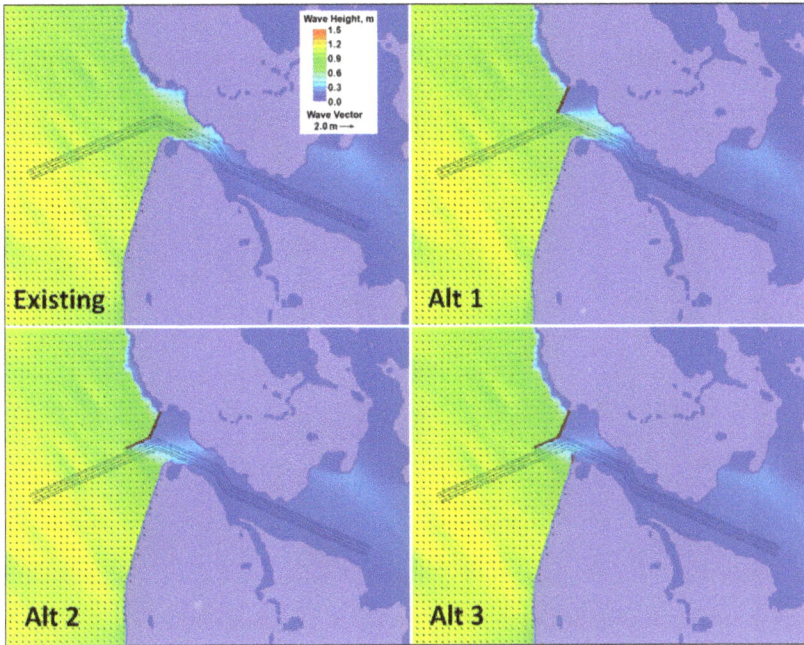

Figure 13. Example of model calculated wave heights in the west channel for the 50-year design winds from NW at 0 m water level.

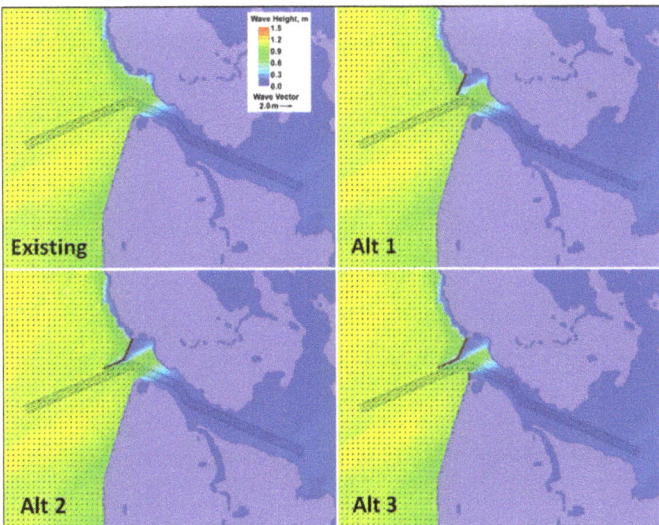

Figure 14. Example of model calculated wave heights in the west channel for the 50-year design winds from SW at 0 m water level.

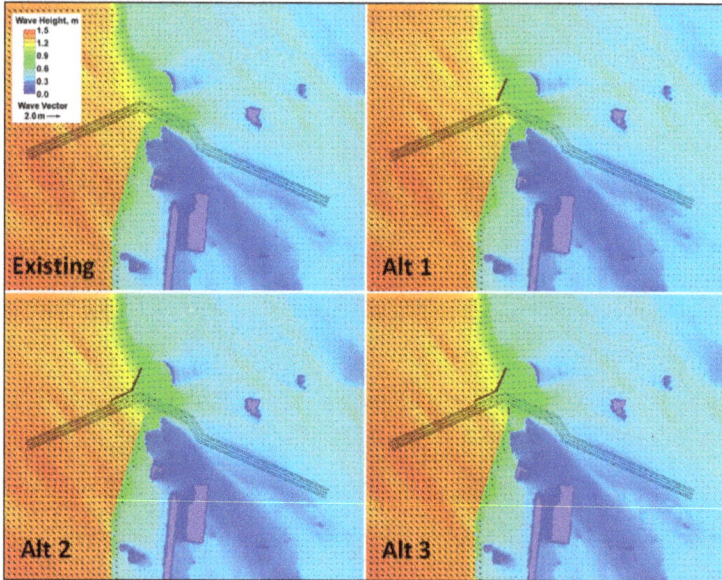

Figure 15. Example of model calculated wave heights in the west channel for the 50-year design winds from NW at 1.5 m water level.

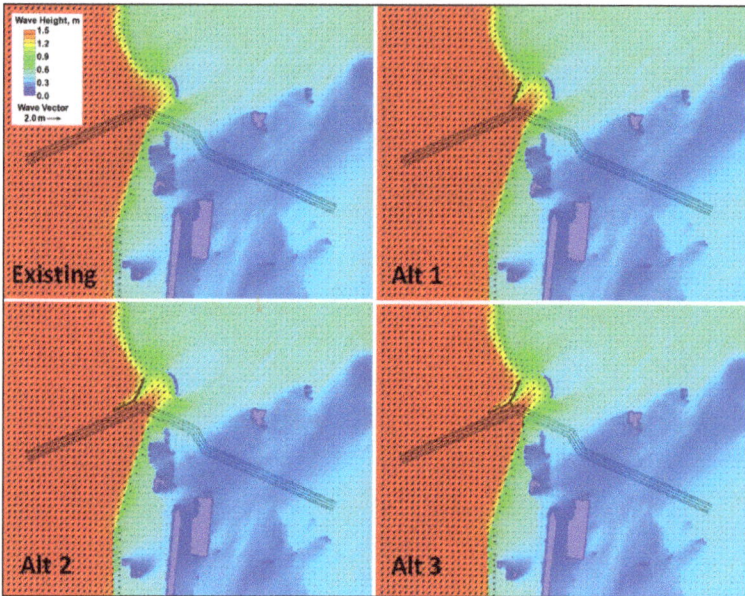

Figure 16. Example of model calculated wave heights in the west channel for the 50-year design winds from SW at 1.5 m water level.

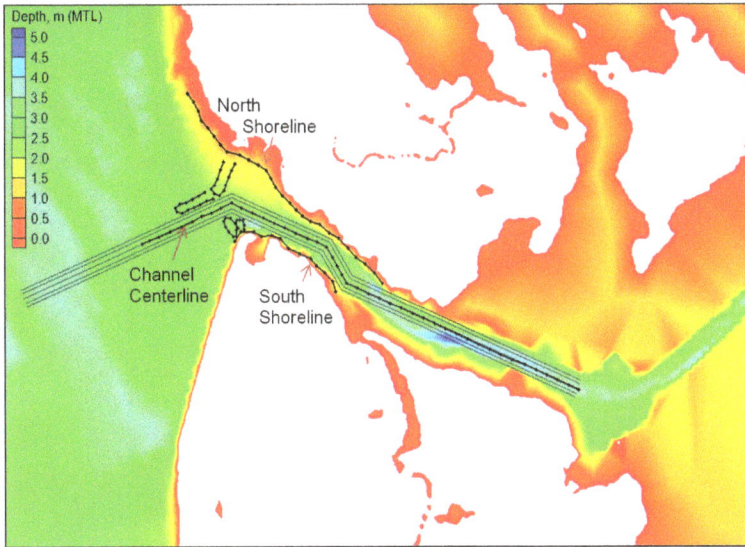

Figure 17. Three transects (lines) used to extract model wave heights.

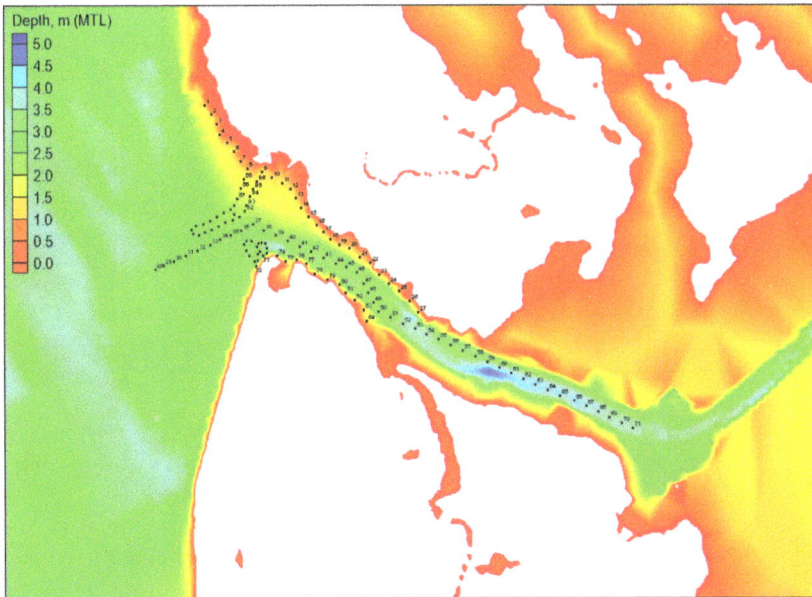

Figure 18. Point locations (stations) used to extract model wave heights.

5. Results and Discussion

5.1. Fifty-Year Design Wind Waves

The wave reduction analysis was performed for all simulations by comparing model results for the alternatives to the existing channel. Figures 19–24 show these comparisons of wave height variation along the west channel centerline for the NW, W, and SW directions. Figure 25 is an example of the calculated wave heights for Alt 4 along the west channel for all directions at 0 m water level. In these figures, incident waves are seen to decrease at Sta 32 (the western channel entrance). The long north jetty in Alts 2–4 has the location between Sta 32 and 38, whereas the short north jetty in Alts 1 and 5 has the location between Sta 36 and 38.

For all wind directions and both water levels investigated in this study, the analysis of wave height reduction from five alternatives is based on the wave height reduction factor calculated as the percentage of wave height reduction to the wave heights in the existing channel without the project condition:

$$\left| \frac{(\text{Wave Weight}, \text{Alterantive}) - (\text{Wave Weight}, \text{Existing Channel})}{(\text{Wave Weight}, \text{Existing Channel})} \right| \times 100\%$$

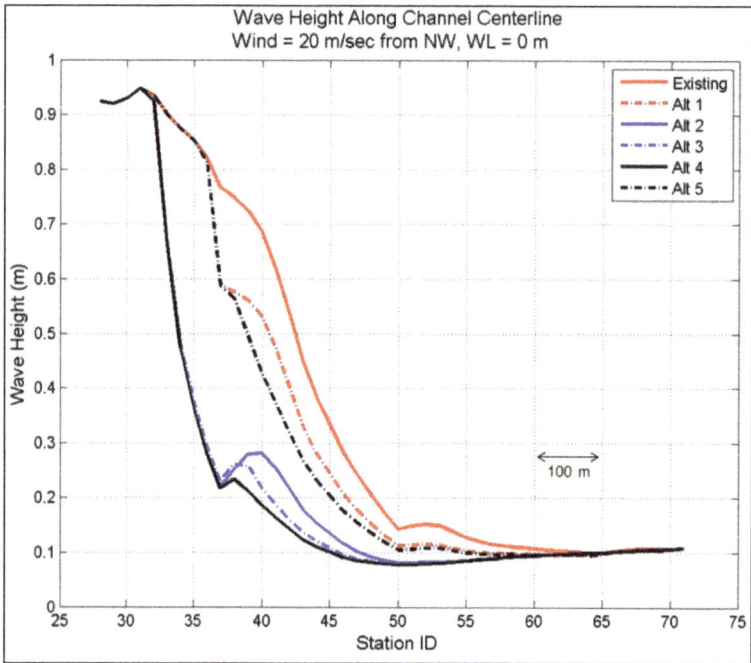

Figure 19. Model wave heights in the west channel for 50-year design wind from NW and WL = 0 m.

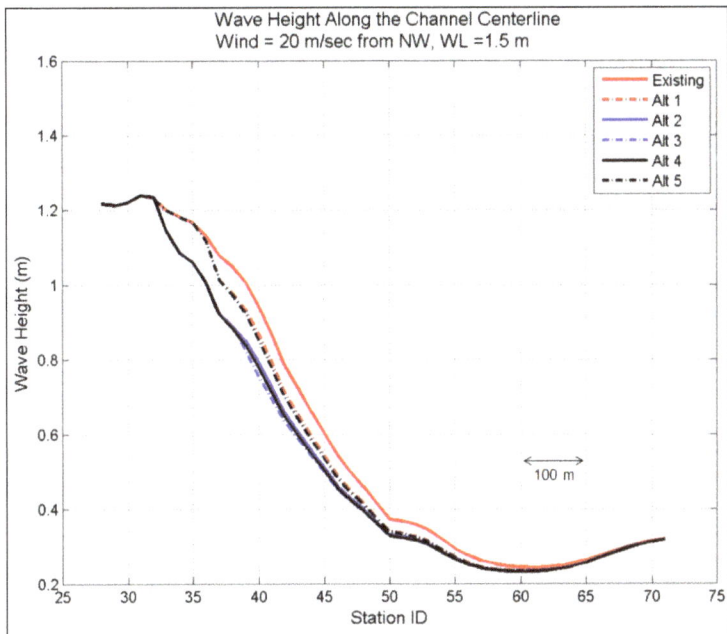

Figure 20. Model wave heights in the west channel for 50-year design wind from NW and WL = 1.5 m.

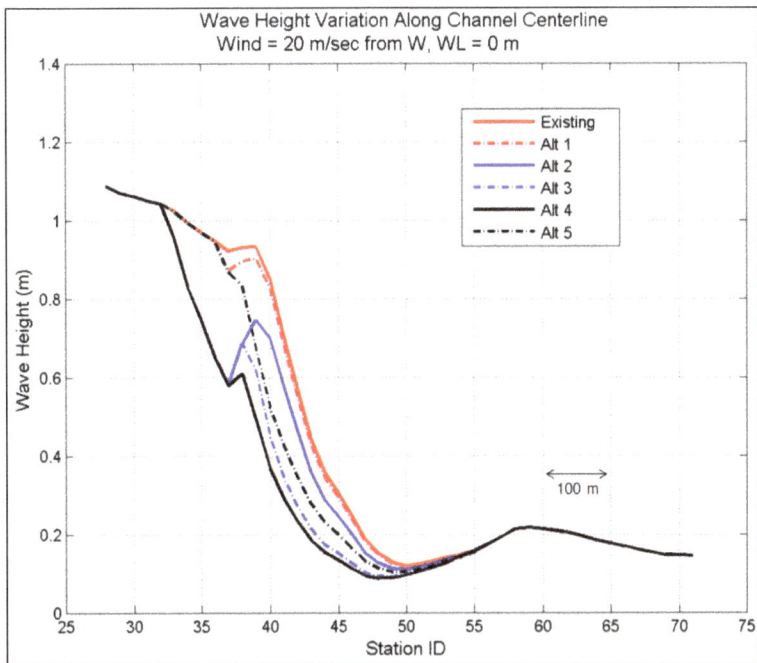

Figure 21. Model wave heights in the west channel for 50-year design wind from W and WL = 0 m.

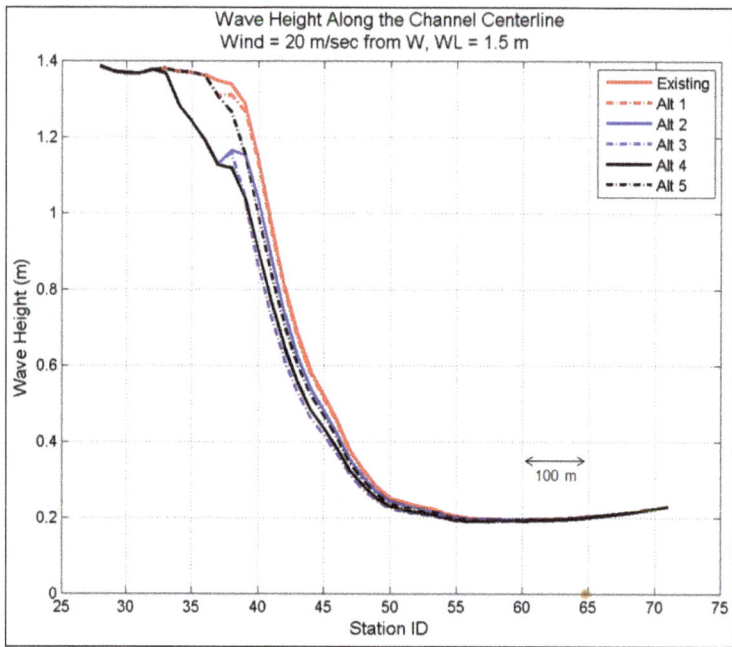

Figure 22. Model wave heights in the west channel for 50-year design wind from W and WL = 1.5 m.

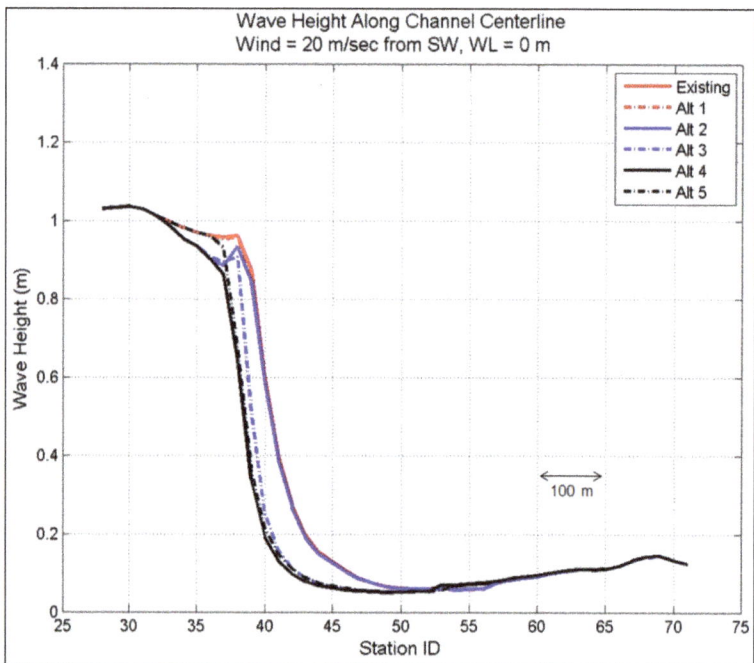

Figure 23. Model wave heights in the west channel for 50-year design wind from SW and WL = 0 m.

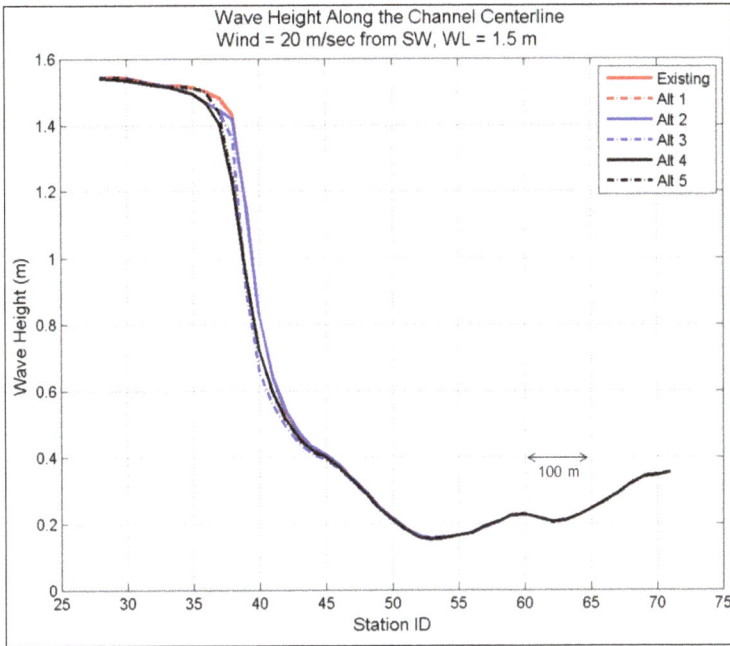

Figure 24. Model wave heights in the west channel for 50-year design wind from SW and WL = 1.5 m.

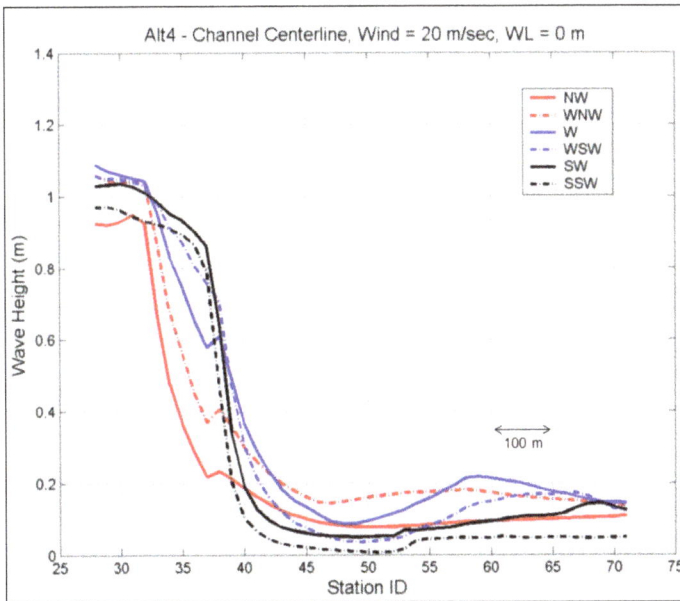

Figure 25. Model wave heights for Alt 4 along the west channel for 50-year design wind from six directions (NW, WNW, W, WSW, SW, and SSW) and WL = 0 m.

For example, Figures 26–28 show the wave height reduction factor along the channel centerline for Alternatives 1 to 5 for 50-year design winds from directions of NW,W,SW, respectively, and WL = 0 m.

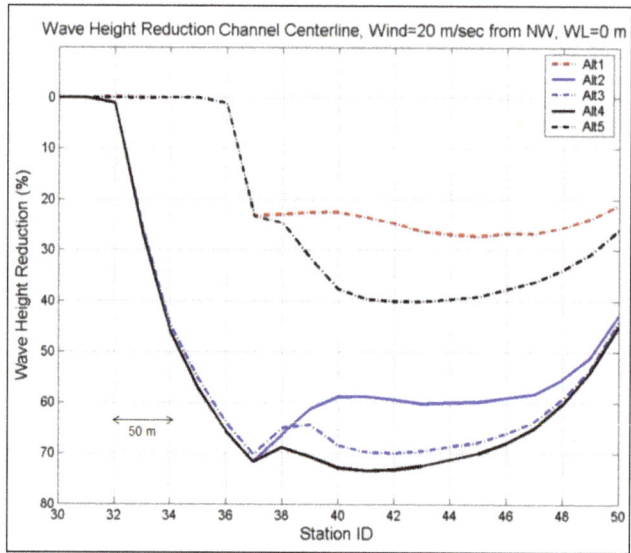

Figure 26. Calculated wave height reduction for Alts 1–5 along the west channel centerline for 50-year design wind from NW and WL = 0 m.

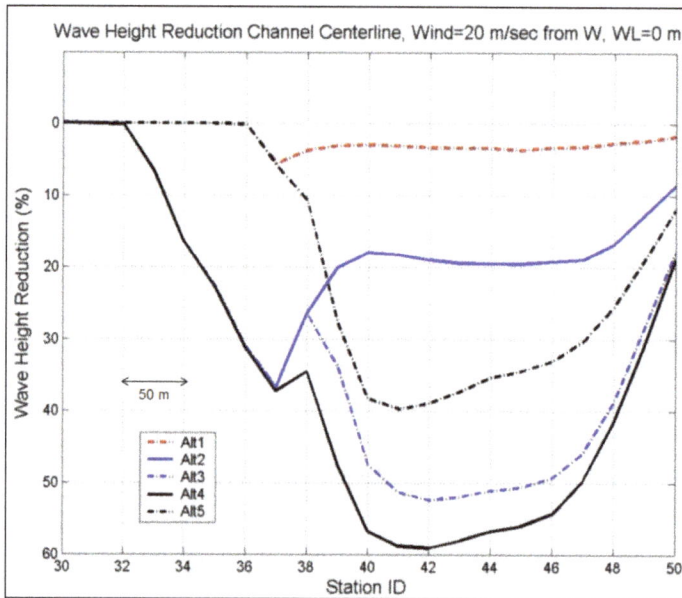

Figure 27. Calculated wave height reduction for Alts 1–5 along the west channel centerline for 50-year design wind from W and WL = 0 m.

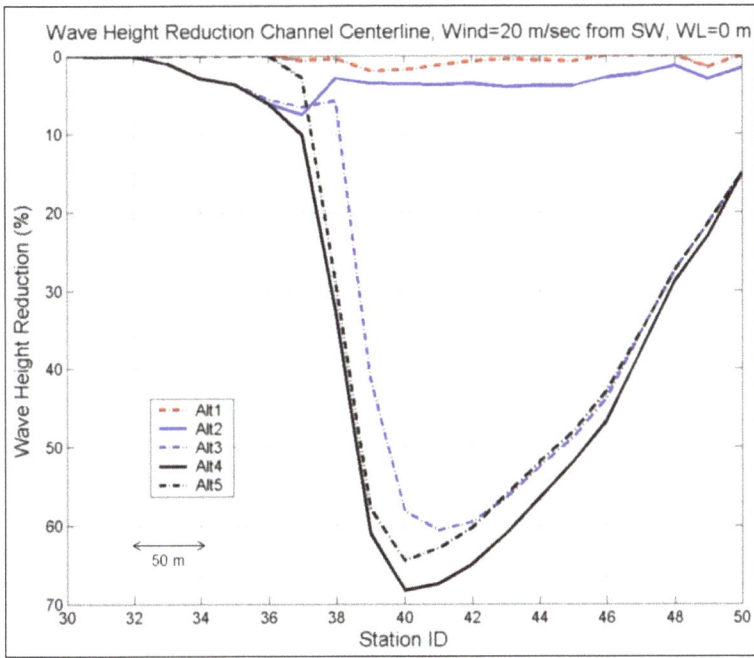

Figure 28. Calculated wave height reduction for Alts 1–5 along the west channel centerline for 50-year design wind from SW and WL = 0 m.

Among all the alternatives, Alts 3 and 4 produced the largest wave reduction for WL = 0 m and 1.5 m, respectively, along the west channel centerline and north and south shorelines. Larger waves are obtained for WL = 1.5 m as compared to WL = 0 m. It is noted that for WL = 1.5 m, the structures used in the alternatives will be submerged, losing much of their effectiveness to intercept and reduce wave energy propagating into the west and mid-sections of the channel. Under such extreme water level conditions, the wave height reduction cannot be used as a measure to rank the alternatives. Consequently, the ranking of alternatives is based on their performance (e.g., wave height reduction factor) calculated for WL = 0 m.

5.2. Hurricane Isabel

A similar analysis of waves was performed using Hurricane Isabel for the existing channel and five alternatives. The analysis was to estimate waves and water levels for a 50-year hurricane event. Because Hurricane Isabel approached the Chesapeake Bay from the southeast, the strong easterly winds associated with the hurricane produced elevated water levels along the west side of the bay and lowered the water level along the east side of bay. As a consequence, relatively lower waves had occurred at and around Tangier Island during Isabel. The wind and water level pattern associated with Hurricane Isabel was simulated for 17–23 September 2003. Both surface wind and pressure fields, used as input in the CMS, were generated from a PBL numerical model for tropical storms [21]. Figure 29 shows an example of the calculated maximum wave field in the Bay (regional grid), and at Tangier Island (local grid) for the existing channel configuration during Isabel. Model results indicated comparatively lower waves and water levels at Tangier Island than in the western portion of the bay. Figure 30 compares calculated high water levels (~1.5 m MTL) to data at NOAA Stations 8635750 (Lewisetta, VA) and 8636580 (Windmill Point, VA) during Isabel.

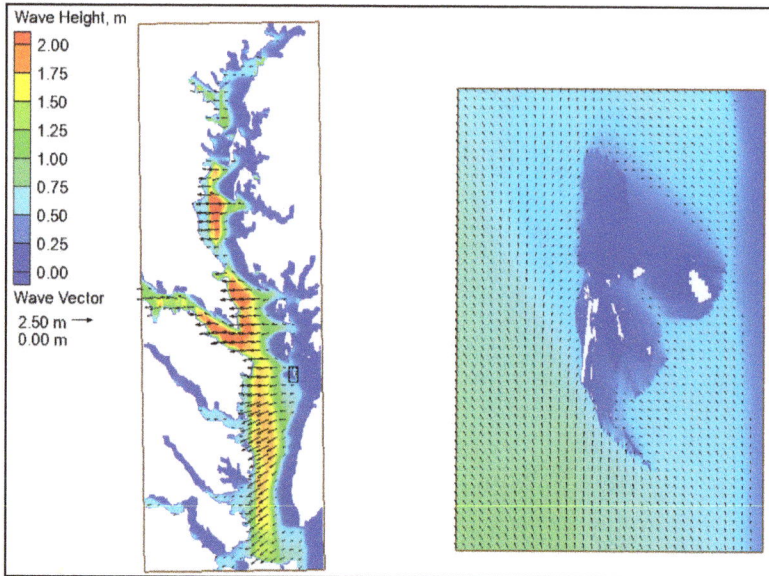

Figure 29. Calculated maximum wave height fields during Hurricane Isabel.

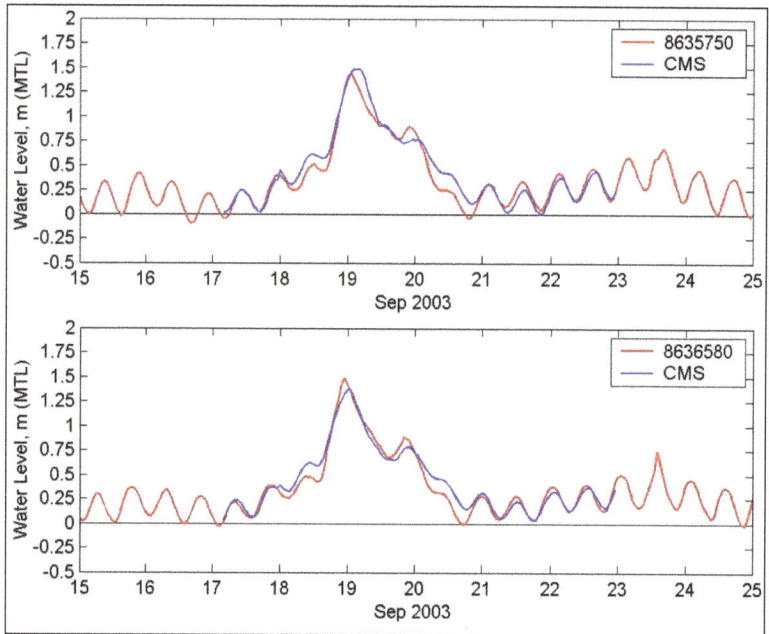

Figure 30. Calculated and measured water levels at NOAA Stations 8635750 (Lewisetta, VA) and 8636580 (Windmill Point, VA) under Hurricane Isabel.

5.3. Estimates for Structure Design

Table 3 presents the average of wave height reduction factors at WL = 0 m along the west channel centerline at Sta 30 to 50 (Figure 18) for five alternatives. Tables 4 and 5 present the average wave height reduction factors at WL = 0 m along the north shoreline (Sta 5 to 25) and along the south shoreline (Sta 74 to 80), respectively. The average of wave height reduction factor is provided for nine wind directions with the average of cases of all nine wind directions for each alternative. Results indicate the wave reduction for Alt 4 was greater than other alternatives for WL = 0 m.

Table 3. Average wave height reduction factors along channel centerline (Sta 30 to 50) at WL = 0 m.

Wind Dir	Alt 1	Alt 2	Alt 3	Alt 4	Alt 5
N	20.3	53.9	57.4	56.5	27.3
NNW	21.5	51.9	55.0	56.7	28.6
NW	16.4	48.6	52.0	54.0	22.9
WNW	5.6	31.2	38.5	40.7	14.4
W	2.2	16.7	31.4	35.1	18.5
WSW	1.1	7.7	30.9	36.1	27.1
SW	0.5	2.9	26.0	30.4	27.4
SSW	0.1	2.8	38.2	43.9	38.4
S	0.2	2.1	25.7	30.6	28.1
Average	7.5	24.2	39.5	42.7	25.9

Table 4. Average of wave height reduction factors along north shoreline (Sta 5 to 25) at WL = 0 m.

Wind Dir	Alt 1	Alt 2	Alt 3	Alt 4	Alt 5
N	38.4	57.5	61.6	61.2	44.9
NNW	39.6	55.9	59.0	60.5	46.2
NW	34.2	51.3	54.6	56.9	40.4
WNW	23.0	40.3	47.2	50.4	31.4
W	15.3	24.8	37.8	42.3	29.3
WSW	12.3	17.3	35.6	40.6	34.4
SW	11.2	15.5	37.2	40.6	36.9
SSW	10.8	15.7	40.5	46.9	40.8
S	11.4	20.0	45.6	50.9	42.1
Average	21.8	33.1	46.6	50.0	38.5

Table 5. Average of wave height reduction factors along south shoreline (Sta 74 to 80) at WL = 0 m.

Wind Dir	Alt 1	Alt 2	Alt 3	Alt 4	Alt 5
N	18.3	56.3	74.0	63.6	35.8
NNW	16.6	48.4	65.7	65.6	35.3
NW	8.7	34.9	52.2	53.2	24.9
WNW	2.4	17.4	36.7	40.3	17.3
W	1.6	11.0	37.7	42.7	25.4
WSW	0.7	4.4	41.6	45.9	36.5
SW	0.3	2.1	45.7	47.1	43.6
SSW	0	2.1	65.1	68.6	61.1
S	0	1.2	59.6	63.6	59.5
Average	5.4	20.0	53.1	54.5	37.7

With a higher water level scenario (WL = 1.5 m), the average wave reduction was less for all alternatives, about 25 percent less than the Existing Channel configuration (without project). Based on model results for 50-year wind forcing and Hurricane Isabel simulation, Alt 4 was more effective alternative overall in reducing wave energy propagation in the canal than the other alternatives at WL = 0 m. It is noted that at the higher water level WL = 1.5 m, the structures evaluated would be either partially or fully submerged, thereby diminishing their effectiveness to intercept and reduce wave energy penetrating into the west and mid-sections of the channel. At this higher water level, wave height reduction cannot be used as a proper measure to evaluate the alternatives.

The bottom lines in Tables 3–5 give the wave reduction factors for each alternative averaged over nine wind directions along the channel centerline, north shoreline, and south shoreline, respectively. By averaging results for the centerline and north and south shoreline, the overall representative wave reduction rating was calculated in Table 6. Among all alternatives, Alt 4 yielded the highest representative wave reduction.

Table 6. Representative wave reduction ratings for Alts 1 to 5.

Alternative	Alt 1	Alt 2	Alt 3	Alt 4	Alt 5
Average Wave Reduction (%)	11.6	25.8	46.4	49.1	34.0

5.4. Channel Sedimentation

The proposed alternatives may affect the overall sedimentation pattern in the vicinity of the structures and throughout the channel reaches of Tangier Island. To address this concern, the CMS was used to simulate the sediment transport for Hurricane Isabel. The purpose of the sediment simulation was to provide a quick view of potential shoaling and erosion areas for a 50-year tropical storm condition. It was by no means to model the long-term morphology evolution at Tangier Island. In the absence of sediment grain sized distribution data, a constant sediment median size of 0.15 mm was assumed in this simulation.

Figure 31 shows the calculated spatially varying sediment accretion and erosion field for the existing west channel configuration (without project) in the simulation of Hurricane Isabel. Model results show sediment deposition immediately outside the west entrance channel and bottom erosion inside the west entrance channel. Figures 32 and 33 show calculated erosion and deposition fields in the simulation of Isabel for Alts 2 and 4, respectively. Model results showed sediment scouring in the channel near the tip of the breakwater and getting trapped inside the west entrance channel. More bottom erosion occurred between the north breakwater and south spur structure in Alts 4 and 5. The eroded sediment was carried by the stronger current into the bay, and the sediment deposition was insignificant inside the west channel entrance.

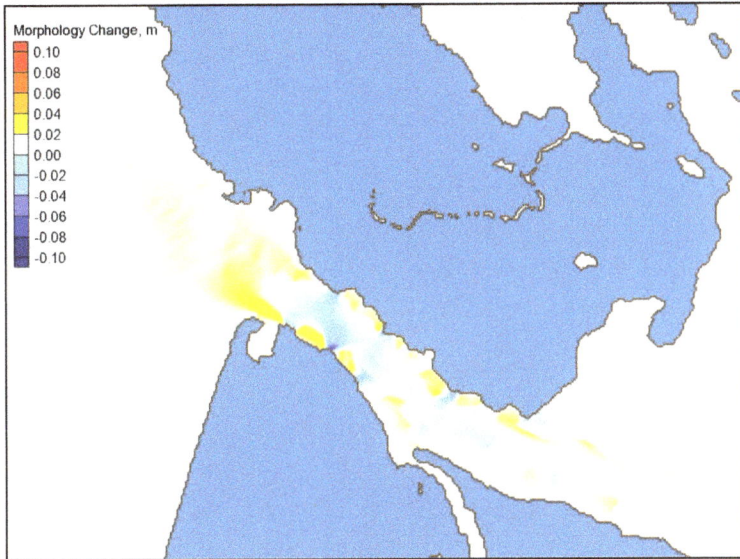

Figure 31. Model sediment accretion and erosion for the existing configuration under Hurricane Isabel.

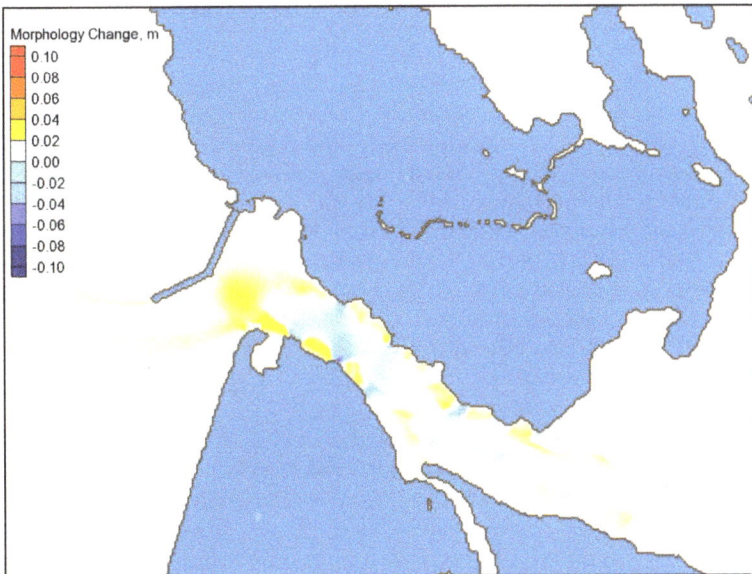

Figure 32. Calculated sediment accretion and erosion for Alt 2 under Hurricane Isabel.

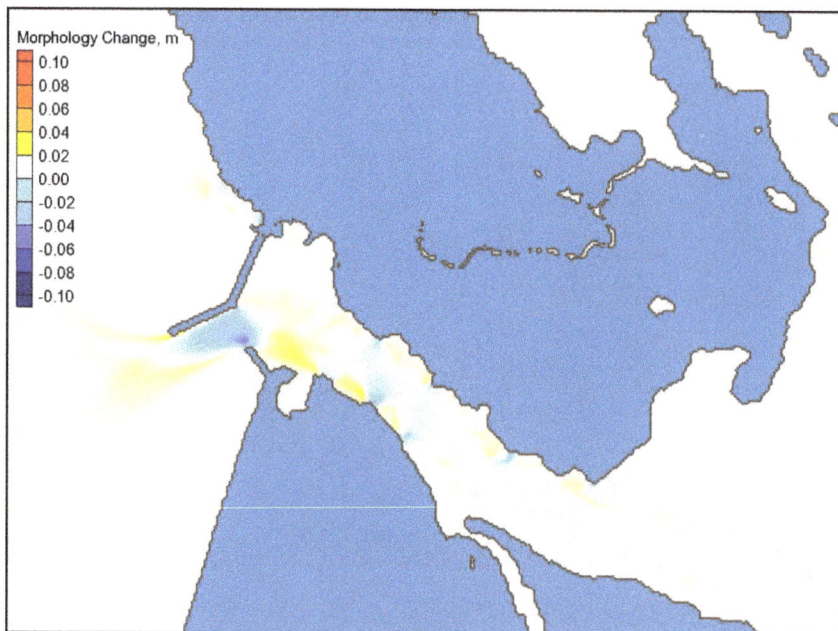

Figure 33. Calculated sediment accretion and erosion field for Alt 4 under Hurricane Isabel.

Overall, sediment transport results for the existing channel configuration and all five alternatives were similar, showing an insignificant morphology change (magnitude of either erosion or deposition less than 2 in. or 0.05 m). While the structures are intended to reduce wave energy in the channel, the currents increase and scour channel near the structures. Some additional settling of suspended sediments occurred away from the channel due to reduced wave-induced currents. Based on the model results for the 50-year return interval tropical storm (Hurricane Isabel), the depth-averaged current magnitude was less than 3 ft/s (1 m/s) in the channel and the maximum channel depth change was less than 2 in. (0.05 m). Therefore, the combined tidally-driven and wave-driven currents in the channel are usually below the threshold for the initiation of sediment motion. No significant effect of structure on channel sedimentation and channel infilling was apparent in the Hurricane Isabel simulation.

6. Summary and Conclusions

Numerical modeling of waves and currents was conducted to assess the impacts of these environmental forces on jetty alternatives to protect a shallow draft navigation channel entrance on Tangier Island, VA, located in the south Chesapeake Bay. The Coastal Modeling System (CMS), an integrated numerical tool that includes a spectral wave model, a two-dimensional depth-averaged hydrodynamic model with sediment transport calculations, is used to investigate the potential effects of waves and hydrodynamics on a relocation and replacement dock.

The existing channel geometry and five alternatives were investigated using the CMS, an integrated wave-hydro-sediment transport modeling system. The five alternatives evaluated consisted of a breakwater system that included a low crest jetty connecting to the north shoreline. Alts 3, 4, and 5 had an optional short structure (spur) joining to the south shoreline. Structural design estimates are based on findings of numerical wave and hydrodynamic modeling conducted for the 50-year design wind speeds, waves, and two water levels. The 50-year wind speed was considered as idealized condition and was based on a previous study by Basco and Shin [22]. Different structure alternatives were evaluated to determine an optimal design, as determined by the level of wave energy reduction

J. Mar. Sci. Eng. **2015**, *3*, 1474–1503

in the navigation channel. The hydrodynamic modeling study results (e.g., wave height, period, direction, and water level) along the western side of the proposed structure footprint were used in these preliminary wave control structural design calculations. These calculations include structural stability, run-up/overtopping, and transmission through and over the structure.

Overall, Alt 4 performed better than the other alternatives in its protection of the boat canal entrance. However, some of the other alternatives also provided considerable wave reduction benefits. The comparison of the alternatives shows that Alts 3, 4, and 5 with south spur jetty outperformed Alts 1 and 2 with no south spur for reducing the wave energy in the channel.

It should be noted that the geometry of the channel itself, even without any jetty structure, helps to dampen the propagating waves in the boat canal. For example, at Sta 50, located approximately 300 m (1000 ft) down the channel from the western entrance, the wave energy has dissipated to the extent that wave height is reduced to 10% to 20% of the height in the bay. Thus, while the CMS shows that Alt 4 provides the greatest wave reduction benefits amongst the five alternatives, multiple criteria may be used in the selection of the optimal alternative for the final design. In closing, this modeling study provides estimates of waves and currents necessary for a follow-up structural design study that will determine the final length, orientation, elevation, and width of the structure system and foundation requirements.

Acknowledgments: The authors would like to thank both of the USACE district, Norfolk, and ERDC Navigation RD & T program for supporting and funding the study. Special thanks go to Julie D. Rosati for her continual support and encouragement towards development and improvement of the CMS. Permission was granted by the Chief, USACE to publish this information.

Author Contributions: All authors have conceived, formulated, conducted, and contributed to the work described in this paper. L.L and Z.D. wrote the first version of the manuscript and D.W. and D.K. edited the paper. All authors read and approved the final manuscript.

Conflicts of Interest: The authors declare no conflict of interest.

References

1. Boon, J.D.; Brubaker, J.M.; Forrest, D.R. *Chesapeake Bay Land Subsidence and Sea Level Change—An Evaluation of Past and Present Trends and Future Outlook*; Virginia Institute of Marine Science Special Report, No. 425; VIMS, Gloucester Point: Norfolk, VA, USA, 2010.
2. Demirbilek, Z.J.; Rosati, J. *Verification and Validation of the Coastal Modeling System, Report 1: Summary Report*; ERDC/CHL Technical Report 11-10; U.S. Army Corps of Engineers Research and Development Center: Vicksburg, MS, USA, 2011.
3. Zundel, A.K. *Surface-Water Modeling System Reference Manual—Version 9.2*; Brigham Young University Environmental Modeling Research Laboratory: Provo, UT, USA, 2006.
4. Lin, L.; Rosati, J.; Demirbilek, Z. *CMS-Wave Model: Part 5. Full-Plane Wave Transformation and Grid Nesting*; ERDC/CHL CHETN-IV-81; U.S. Army Corps of Engineers Research and Development Center: Vicksburg, MS, USA, 2012.
5. Coastal Modeling System. Available online: http://cirpwiki.info/wiki/CMS (accessed on 8 September 2015).
6. Demirbilek, Z.; Lin, L.; Hayter, E.; O'Connell, C.; Mohr, M.; Chader, S.; Forgette, C. *Modeling of Waves, Hydrodynamics and Sediment Transport for Protection of Wetlands at Braddock Bay, NY*; ERDC TR-14-8; U.S. Army Corps of Engineers Research and Development Center: Vicksburg, MS, USA, 2015.
7. Lambert, S.S.; Willey, S.S.; Campbell, T.; Thomas, R.C.; Li, H.; Lin, L.; Welp, T.L. *Regional Sediment Management Studies of Matagorda Ship Channel and Matagorda Bay System, Texas*; ERDC/CHL TR-13-10; U.S. Army Corps of Engineers Research and Development Center: Vicksburg, MS, USA, 2013.
8. Lin, L.; Reed, C.W. Numerical modeling of mixed sediment transport in GIWW and West Galveston Bay, Texas. In Proceedings of the 8th International Symposium on Coastal Engineering and Science of Coastal Sediment Processes, Coastal Sediments'15, San Diego, CA, USA, 11–15 May 2015.
9. Lin, R.Q.; Lin, L. Wind input function. In Proceedings of the 8th International Workshop on Wave Hindcasting and Prediction, Oahu, HI, USA, 14–19 November 2004.

10. Lin, L.; Lin, R.Q. Wave breaking function. In Proceedings of the 8th International Workshop on Wave Hindcasting and Prediction, Oahu, HI, USA, 14–19 November 2004.

11. Lin, R.Q.; Lin, L. Wave breaking energy in coastal region. In Proceedings of the 9th International Workshop on Wave Hindcasting and Prediction, Victoria, BC, Canada, 24–29 September 2006.

12. Li, H.; Lin, L.; Burks-Copes, K. Numerical modeling of coastal inundation and sedimentation by storm surge, tides, and waves at Norfolk, Virginia, USA. In Proceedings of the 33rd International Conference on Coastal Engineering, Santander, Spain, 1–6 July 2012.

13. Demirbilek, Z.; Lin, L.; Ward, D.L.; King, D.B. *Modeling Study for Tangier Island Jetties, Tangier Island, VA*; ERDC/CHL TR-14-8; U.S. Army Corps of Engineers Research and Development Center: Vicksburg, MS, USA, 2015.

14. Lopez-Feliciano, O.L.; Herrington, T.O.; Miller, J.K. A Morphology change study using the CMS in a groin field during Hurricane Sandy. *J. Am. Shore Beach* **2015**, *83*, 19–26.

15. Lin, L.; Demirbilek, Z.; Mase, H.; Zheng, J.; Yamada, F. *A Nearshore Spectral Wave Processes Model for Coastal Inlets and Navigation Projects*; ERDC/CHL TR-08-13; U.S. Army Corps of Engineers Research and Development Center: Vicksburg, MS, USA, 2008.

16. Lin, L.; Demirbilek, Z. Recent capabilities of CMS-Wave: A coastal wave model for inlets and navigation projects. *J. Coast. Res.* **2011**. [CrossRef]

17. Lin, L.; Lin, R.Q.; Maa, J.P.Y. Numerical simulation of wind wave field. In Proceedings of the 9th International Workshop on Wave Hindcasting and Forecasting, Victoria, BC, Canada, 24–29 September 2006.

18. Vimswave-Vims Directional Wave Data. Available online: http://web.vims.edu/physical/research/VIMSWAVE/VIMSWAVE.htm (accessed on 8 September 2015).

19. Turning Operational Oceanographic Data into Meaningful Information for the Nation. Available online: https://tidesandcurrents.noaa.gov (accessed on 8 September 2015).

20. Demirbilek, Z.; Lin, L.; Bass, G.P. Prediction of storm-induced high water levels in Chesapeake Bay. In Proceedings of the Coastal Disaster 2005 Conference, ASCE, Charleston, SC, USA, 8–11 May 2005.

21. Demirbilek, Z.; Lin, L.; Mark, D.J. Numerical modeling of storm surges in Chesapeake Bay. *Int. J. Ecol. Dev.* **2008**, *10*, 24–39.

22. Basco, D.R.; Shin, C.S. *Design Wave Information for Chesapeake Bay and Major Tributaries in Virginia*; Technical Report, No. 93-1; Department of Civil Engineering, The Coastal Engineering Institute, Old Dominion University: Norfolk, VA, USA, 1993.

Journal of
*Marine Science
and Engineering*

MDPI

Article

Geometry of Wave-Formed Orbital Ripples in Coarse Sand

Gerben Ruessink *, Joost A. Brinkkemper and Maarten G. Kleinhans

Department of Physical Geography, Faculty of Geosciences, Utrecht University, P.O. Box 80.115, 3508 TC Utrecht, The Netherlands; E-Mails: j.a.brinkkemper@uu.nl (J.A.B.); m.g.kleinhans@uu.nl (M.G.K.)

* E-Mail: b.g.ruessink@uu.nl; Tel.: +31-30-2532780.

Academic Editor: Edward J. Anthony

Received: 3 November 2015 / Accepted: 14 December 2015 / Published: 21 December 2015

Abstract: Using new large-scale wave-flume experiments we examine the cross-section and planform geometry of wave-formed ripples in coarse sand (median grain size $D_{50} = 430$ μm) under high-energy shoaling and plunging random waves. We find that the ripples remain orbital for the full range of encountered conditions, even for wave forcing when in finer sand the ripple length λ_r is known to become independent of the near-bed orbital diameter d_s (anorbital ripples). The proportionality between λ_r and d_s is not constant, but decreases from about 0.55 for $d_s/D_{50} \approx 1400$ to about 0.27 for $d_s/D_{50} \approx 11,500$. Analogously, ripple height η_r increases with d_s, but the constant of proportionally decreases from about 0.08 for $d_s/D_{50} \approx 1400$ to about 0.02 for $d_s/D_{50} > 8000$. In contrast to earlier observations of coarse-grained two-dimensional wave ripples under mild wave conditions, the ripple planform changes with the wave Reynolds number from quasi two-dimensional vortex ripples, through oval mounds with ripples attached from different directions, to strongly subdued hummocky-type features. Finally, we combine our data with existing mild-wave coarse-grain ripple data to develop new equilibrium predictors for ripple length, height and steepness suitable for a wide range of wave conditions and a D_{50} larger than about 300 μm.

Keywords: orbital ripples; hummocks; flume experiment; empirical prediction

1. Introduction

Wave-formed ripples are ubiquitous small-scale bed forms in shelf to nearshore water depths with typical spacing (or, wave length) of $\mathcal{O}(0.1–1)$ m and height of $\mathcal{O}(0.01–0.1)$ m. Hydrodynamic and morphodynamic models often demand predictions of cross-section (*i.e.*, ripple spacing or wavelength, and height) and planform (orientation and along-crest regularity) ripple geometry, e.g., [1–3], because of the effect of ripples on waves, currents, and sediment suspension and transport. Accordingly, numerous empirical classification schemes and predictors have been proposed that relate ripple occurrence and equilibrium geometry to non-dimensional wave and sediment properties; for a recent overview, see [4]. While such schemes and predictors are now reaching considerable skill for sand with typical median diameters of 150–250 μm [4–6], there is considerable doubt on their applicability to ripples that form in coarser sand, especially for high-energy wave conditions [6–8].

One of the most commonly adopted wave-ripple classification schemes for 150–250 μm sand, due to Clifton [9], comprises orbital, suborbital and anorbital ripples and expresses wave forcing and sand characteristics as the ratio between orbital diameter and median grain size, d/D_{50}. For mild wave conditions ($d/D_{50} \leq 2000$) ripple length λ and height η scale linearly with d, with often-quoted constants of proportionality of about 0.65 and 0.10, e.g., [10,11], respectively. The steepness $\vartheta = \eta/\lambda$ of orbital ripples is thus near 0.15, implying them to be sufficiently steep to shed sand-laden vortices into the water column during flow reversal (vortex ripples; e.g., [12]). For energetic wave conditions

$(d/D_{50} \geq 5000; [11,13])$ the ripples are anorbital; that is, λ does not depend on d anymore. Instead, λ now relates to D_{50} as $535D_{50}$ on average, e.g., [11,14], implying that anorbital λ is substantially shorter than d. In addition, anorbital ripples are no longer vortex ripples; with an increase in d/D_{50} their steepness reduces rapidly to 0.01 [11], which is essentially flat bed. Suborbital ripples from a transitional type between orbital and anorbital ripples. Predictors that are, at least partly, based on the orbital-suborbital-anorbital scheme include Wiberg and Harris [11], Soulsby *et al.* [15] and Nelson *et al.* [4]. Often, anorbital ripples are superimposed on substantially longer, three-dimensional and also strongly subdued ripples, known as large wave ripples, mega-ripples or hummocks, e.g., [16–20], which these predictors do not consider. The combination of anorbital ripples and hummocks has been found in the field for sand with a D_{50} up to about 300 μm, e.g., [17,19,20].

Observations of wave ripples in coarser sediment are largely limited to mild wave conditions because of large water depths or low wave heights, e.g., [7,21–26]. They mostly show two-dimensional, steep (vortex) orbital or suborbital ripples with similar λ/d ratios as observed in finer sand. Limited laboratory experiments under stronger wave conditions ($d/D_{50} \approx 5000$–7500; [8,27]) do not show a transition to anorbital length scales or to large hummocky ripples. Instead, the ripples remain two-dimensional vortex ripples. Cummings *et al.* [8] found the ratio for λ/d to be lower (0.4) in 0.8-mm sand than in 0.12-mm sand (0.6) for the same d, while the experiments of Pedocchi and Garcia [27] suggest a negative dependence on the maximum orbital velocity. Whether the steep coarse-grained ripples remain orbital and develop into hummocks under even stronger wave forcing is not known. Interestingly, O'Donoghue *et al.* [28] postulated that ripples remain two-dimensional when D_{50} exceeds 300 μm and are three-dimensional for D_{50} is less than 220 μm, except when d is low. Other data [5,29] suggest that even coarse-grain ripples may become three-dimensional under strong wave forcing. This paper documents new coarse-sand, equilibrium ripple data collected for the $d/D_{50} \approx 1000$–20,000 range under high-energy shoaling and plunging random waves on a prototype laboratory beach. Our objectives are to investigate cross-section and planform ripple geometry and to derive a new coarse-sand equilibrium ripple predictor for which we combine our data with several existing coarse-grain ripple data sets collected under mild wave conditions.

2. Methods

2.1. Bardex II Experiment

The data analysed here were collected in the large-scale Delta flume facility of Deltares in Vollenhove, The Netherlands as part of the second Barrier Dynamics Experiment (Bardex II; [30]). The barrier, which filled the entire 5-m width of the flume, was constructed from coarse ($D_{50} = 430$ μm; mean grain size = 510 μm; $D_{16} = 280$ μm; $D_{84} = 830$ μm), moderately sorted (0.81ϕ) and coarse-skewed (-0.24ϕ) quartz sand that contained a small amount of gravel ($\approx$1%, >2000 μm) [30]. The median fall velocity w_s of this sand amounts to 0.061 m/s [31] and the Reynolds particle number $Re_p = \sqrt{gRD_{50}}D_{50}/v$ [5] to 35.9, where $g = 9.81$ m/s^2 is gravitational acceleration, $R = 1.65$ is the submerged specific density, and v is the kinematic viscosity of water, here set to 1×10^{-6} m^2/s. The seaward part of the barrier profile initially comprised a 0.5-m thick sand layer at cross-shore coordinates $x = 29$–49 m ($x = 0$ is at the wave paddle) and a 1:15 seaward-sloping section at $x = 49$–109 m (Figure 1) that ended at the 4.5-m high barrier crest.

J. Mar. Sci. Eng. **2015**, 3, 1568–1594

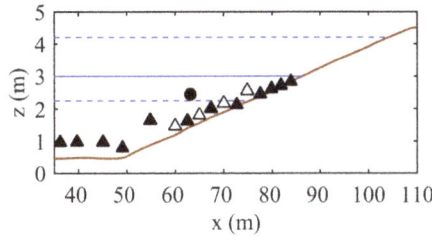

Figure 1. Initial bed elevation z versus cross-shore distance x from the wave paddle (brown line). The blue line is the default still water level h_s; the two dashed blue lines represent the lower and upper h_s, see Table 1. The 16 triangles represent the locations of pressure transducers. At the 4 open triangles the pressure transducer was co-located with an electromagnetic or acoustic current meter. The filled circle is the approximate location of the 3D Profiling Sonar.

The Bardex II test programme consisted of 19 distinct tests with different wave and water level conditions, grouped into 5 series that focused on surf-swash zone processes (series A–C), and barrier overwash and destruction (series D and E, respectively). The wave paddle steering signal in all tests was constructed from a JONSWAP spectrum using a target significant wave height H_{s0} and peak wave period T_{p0} with a peak-enhancement factor of 3.3. As can be seen in Table 1, H_{s0} was either 0.6 or 0.8 m, and T_{p0} varied from 4 to 12 s. The still water level h_s with respect to the concrete flume floor ranged between 2.25 and 4.2 m, with a default value of 3 m (Table 1). During series C, the barrier was subjected to a rising (C1) and falling tide (C2). The Automated Reflection Compensator was switched on during all tests to avoid seiching in the flume.

Table 1. Experimental conditions during Bardex II. H_s = significant wave height; T_p = peak wave period; h_s = still water level with respect to the concrete flume floor; T_{test} = test duration; and N_{runs} is number of wave runs.

Test	H_{s0} (m)	T_{p0} (s)	H_{s1} (m)	T_{p1} (s)	h_s (m)	T_{test} (min)	N_{runs}
A1	0.8	8	0.92	8.0	3	320	13
A2	0.8	8	0.90	8.0	3	200	5
A3	0.8	8	0.90	7.9	3	197	1
A4	0.8	8	0.90	8.0	3	200	5
A6	0.6	12	0.73	11.8	3	335	13
A7	0.6	12	0.77	12.6	3	213	5
A8	0.6	12	0.79	12.6	3	200	5
B1	0.8	8	0.91	8.3	3	165	5
B2	0.8	8	0.91	7.8	2.5	255	6
C1	0.8, 0.6	8	0.90, 0.57	7.3	2.25 → 3.65	330	11
C2	0.8, 0.6	8	0.91, 0.58	7.5	3.53 → 2.25	270	9
D1	0.8	4	0.79	4.0	3.15 → 4.2	160	8
D2	0.8	5	0.82	4.6	3.45 → 4.05	100	5
D3	0.8	6	0.86	6.0	3.45 → 3.9	80	4
D4	0.8	7	0.84	7.0	3.45 → 3.9	80	4
D5	0.8	8	0.86	7.7	3.45 → 3.75	60	3
D6	0.8	9	0.87	9.3	3.30 → 3.75	80	4
D7	0.8	10	0.93	10.0	3.15 → 3.6	80	4
E1	0.8	8	0.94	7.4	3.9	65	5

The H_{s0} and T_{p0} were target values at the wave paddle. The measured values at the most offshore pressure transducer (H_{s1} and T_{p1} at x = 36.2 m, Figure 1) differed from these target values and varied slightly between runs, see, for example, Figure 5d,j. The H_{s1} and T_{p1} listed here are values averaged over the runs. The final 17 min of A3 are labeled as A5 in Masselink et al. [30] and involved 8 short (2-min each) sequences of mono- and bi-chromatic wave runs.

51

J. Mar. Sci. Eng. **2015**, 3, 1568–1594

2.2. Measurements: Ripple Data

2.2.1. Profile Data

Each test was generally broken up in several wave runs (Table 1) that varied in duration from 10 min to 3 h. The center profile of the flume was surveyed after each wave run using a profiling wheel mounted on an overhead gantry. In total, 115 bed profiles were collected, each with a 0.01-m cross-shore resolution. A first inspection of the data revealed the presence of occasional spikes in most bed profiles, most likely induced by glitches in the profiling system. These spikes were removed by filtering each bed profile using a second-order loess interpolator [32] with a cross-shore scale parameter l_x of 0.05 m. This interpolator acts as a low-pass filter and removes variability with length scales less than $l_x/0.7$ (here, ≈ 0.07 m). Visual inspection of original and despiked bed profiles illustrated that this l_x was effective in removing the spikes while leaving the ripples unaffected. The despiked data are henceforth referred to as $z(x,t)$, where z is bed elevation defined positive upward from the concrete floor of the flume and t is time with $t = 0$ corresponding to the start of the first wave run in test A1.

Analysis of $z(x,t)$ [33] revealed that the waves in tests A1–A4 ($H_s = 0.8$ m, $T_p = 8$ s) reshaped the initially planar profile into a sandbar-trough system, with the sandbar crest at $x \approx 70$ m. The waves in subsequent A6–A8 tests ($H_s = 0.6$ m, $T_p = 12$ s) transported sand onshore, causing the decay of the sandbar and the generation of a pronounced berm in the upper swash zone. During series B and C the berm and the remains of the sandbar hardly changed, while during series D and E morphological change was most pronounced at the berm and the barrier crest [34].

To separate the large-scale sandbar-berm variability from the smaller scale wave ripples, $z(x,t)$ was low-pass filtered with $l_x = 3.5$ m yielding $z_{bb}(x,t)$, the data set that contains the sandbar-berm variability only. The residual series from this filtering step, i.e., $z_{bb}(x,t) - z(x,t)$, are zero-mean profiles with bed variability induced by wave ripples, $\tilde{z}_{wr}(x,t)$. Positive and negative $\tilde{z}_{wr}$ correspond to ripple crests and troughs, respectively. The cross-shore evolution in ripple length λ_r, ripple height η_r, and ripple steepness $\vartheta_r = \eta_r/\lambda_r$ were subsequently calculated every 0.5 m from overlapping (95%) 10-m wide, centered windows (subsets) of $z_{wr}(x)$. In each window the length and height of every individual ripple, defined with a zero-down-crossing technique, were determined. The window ripple length was taken as the mean of the individual lengths, and the window ripple height as the root-mean-square value of the individual heights. The center of the most seaward window was taken at $x = 35$ m, while the center of the most landward window was chosen at the location where the corresponding z_{bb} profile intersected the still water level h_s. This implies that this most landward window essentially encapsulated the swash zone. The use of 10-m windows was a compromise between having sufficient ripples within a window for robust statistics and quantifying cross-shore trends in λ_r and η_r. A first inspection of the length and height profiles illustrated that the window-to-window variability was considerable in the outer surf zone, where, as examined in detail below, ripple length was typically largest. As an example, this variability is illustrated in Figure 2a with λ_r determined after the fifth wave run of test A4. As can be seen, λ_r varied reasonably smoothly with x for $x < 70$ m, but fluctuated between 0.75 and 2.6 m for $x = 70$–75 m (waves started to break on the sandbar edge near $x = 68$ m, Figure 2b). To suppress these rapid and unrealistic fluctuations, all $\lambda_r(x)$ and $\eta_r(x)$ were low-pass filtered with $l_x = 3.5$ m (e.g., Figure 2a). In the following λ_r and η_r refer to these low-pass filtered values. The use of $l_x = 3.5$ m implies that λ_r and η_r vary on the same cross-shore scales as the sandbar-berm morphology. We stress that the main results presented below do not depend on this low-pass filtering step; the filtering primarily acted to suppress noise.

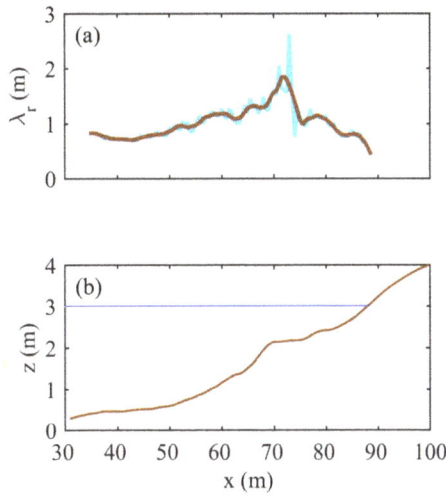

Figure 2. (**a**) Ripple length λ_r versus cross-shore distance x after wave run 5 in test A4. The blue line represents the original values based on the 10-m wide windows; the brown line is the $l_x = 3.5$ m smoothed version; (**b**) shows the bed profile at the end of this wave run for reference. The horizontal blue line is the still water level $h_s = 3$ m.

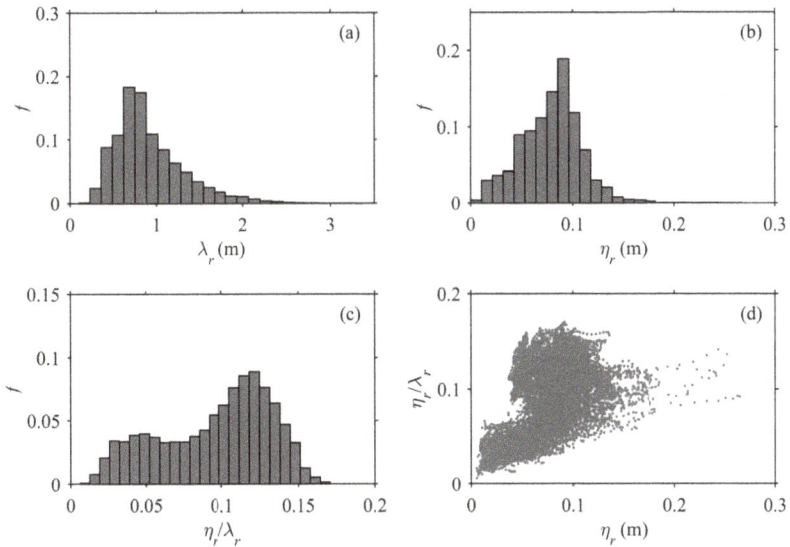

Figure 3. Probability f histograms of (**a**) ripple length λ_r, (**b**) ripple height η_r and (**c**) ripple steepness $\vartheta_r = \eta_r / \lambda_r$. For each parameter the full range was divided into 25 bins of equal width. Panel (**d**) is a scatter plot of ϑ_r versus η_r. The values shown are estimates from all available 0.5-m spaced, 10-m wide windows. The total number of observations amounts to 12,732.

Probability histograms of λ_r, η_r and ϑ_r are given in Figure 3a–c to illustrate the ripple characteristics in our data set. The central 99% intervals of λ_r and η_r range between 0.31 and 2.38 m, and 0.01 and 0.17 m, respectively, with median values of 0.81 and 0.07 m. As a consequence, ϑ_r spans the full range between values expected for vortex ripples (≈ 0.15) and for strongly subdued ripples (≤ 0.03), Figure 3c. This implies that, combining earlier terminology [8,16,23], all ripples in our data are large coarse-grained wave ripples and that these have substantially larger ranges in η_r and ϑ_r than their always low ($\eta_r \leq 0.05$ m) and subdued ($\vartheta_r \leq 0.03$) fine-grained counterparts [16,18], see Figure 3d. Interestingly, our data set does contain low and subdued ripples (Figure 3d), whereas most previous observations of ripples in coarse sand contained vortex ripples only [8,27,28]. The unimodality in the probability histogram of λ_r (Figure 3a) also suggests the absence of smaller-scale anorbital ripples (for $D_{50} = 430$ μm a length of about 0.2 m is expected) that in finer sand are often superimposed on the large wave ripples [16–20]. Visual inspection of all $\tilde{z}_{wr}(x)$ indeed confirmed a single ripple scale (no superimposed ripples), consistent with other coarse-sand ripple observations [8,27,28].

2.2.2. 3D Sonar Data

A 1.1 MHz 3D Profiling Sonar 2001 (Marine Electronics Ltd., Guernsey, UK) was mounted in a downward looking manner at $x = 63.1$ m, 1.65 m from the nearest flume wall. It was operated in two distinct modes to, firstly, obtain high-resolution circular elevation models of wave-ripple induced bed variability and hence planform geometry after a wave run and, secondly, to provide insight into cross-shore ripple migration during a wave run. In its first operation mode, the sonar was triggered manually after wave action had ceased to scan a 120° swath with a 0.9° resolution and then to rotate by 0.9° to capture the next swath until a complete circular area underneath the sonar was surveyed. Because the mounting height was about 1 m above the bed, the diameter of the surveyed circle was approximately 3.5 m. From each swath a bed profile was detected using a threshold algorithm. The detected bed points (typically, about 17,500) were interpolated on a regular horizontal grid with a 0.025 m spacing using $l_x = l_y = 0.15$ m, where l_y is an alongshore scale parameter. For consistency with the processing of the cross-shore profile data, we would have liked to use $l_x = l_y = 0.05$ m, but this resulted in rather gappy bed elevation models, especially at the edges of the scan. All resulting models of bed elevation were subsequently detrended using the $z_{bb}(x,t)$ beneath the sonar to yield zero-mean, approximately circular models of wave-ripple induced bed variability. As in the $\tilde{z}_{wr}(x)$ profiles, positive and negative perturbations correspond to ripple crests and troughs, respectively. The horizontal coordinates are relative to the sonar, with positive x_s onshore and negative y_s to the nearest flume wall. The ripple planform geometry was classified qualitatively for each circular elevation model as 2D, quasi-2D or 3D using definitions provided in [8]. This classification is not affected by the use of $l_x = l_y = 0.15$ m rather than $l_x = l_y = 0.05$ m.

The second mode of operation was applied during a wave run. Because we expected ripple location and/or planform geometry to change during the approximate 11-min duration of a complete circular scan, we essentially applied the sonar as a 2D (cross-shore) line scanner and triggered it manually every 2 to 5 min. All cross-shore swaths were produced into cross-shore profiles of bed variability with the same threshold algorithm as used to process a full no-wave scan and were subsequently demeaned using $l_x = 3.5$ m. The cross-shore ripple migration speed C_r was estimated using a cross-correlation of the time-separated, zero-mean, wave-ripple induced bed profiles, e.g., [7,19,20]. The cross-shore distance over which the ripples migrated shows up as the lag with the maximum correlation.

2.3. Measurements: Hydrodynamical Data

Estimates of well-established hydrodynamical parameters related to ripple characteristics and migration are available at up to 16 cross-shore locations (Figure 1). Pressure transducers were wall-mounted at all 16 locations and sampled near-bed pressure with frequencies of 4, 5 or 20 Hz depending on location. All pressure series were converted to water-surface elevation ζ series using linear wave theory, which were processed into the short-wave (0.05–2 Hz) (1) significant wave orbital diameter $d_s = H_s / \sinh(kh)$, where H_s is the local significant wave height, and k is the wave number estimated from linear theory using water depth h and the peak period T_{p1} at the most seaward pressure transducer ($x = 36.2$ m in Figure 1); (2) peak semi-orbital velocity $u_w = \pi d_s / T_{p1}$; (3) mobility number $\psi = u_w^2/(RgD_{50})$, (4) Shields parameter $\theta = 0.5f_w\psi$, where f_w is a friction factor for which we used Equations (60a) and (60b) in [35]; (5) wave Reynolds number $Re_w = 0.5u_wd_s/\nu$; (6) wave skewness $S_\zeta = \overline{\zeta^3}/\sigma_\zeta^3$, where the overbar represent a run average and σ_ζ is the standard deviation of ζ; and, (7) wave asymmetry $A_\zeta = \overline{\mathcal{H}(\zeta)^3}/\sigma_\zeta^3$, where $\mathcal{H}(\zeta)$ represents the Hilbert transform of ζ. Because the paddle motion was not repeated exactly in each run, the resulting wave height and period varied slightly from run to run [36]. We therefore preferred the use of the measured peak period at the most seaward sensor T_{p1} (see Table 1) over the target value T_{p0} in the computation of d_s and u_w. Both S_ζ and A_ζ are measures of wave non-linearity [37]. S_ζ is positive when waves have high, narrow crests and broad, shallow troughs and A_ζ is negative when waves are forward-leaning. At four locations the pressure transducer was co-located with a near-bed (typically, 0.11 m above the bed) electromagnetic or acoustic current meter (Figure 1), which sampled at 4 or 10 Hz, respectively. All instantaneous time-series of cross-shore velocity were processed into the run-average cross-shore velocity $\bar{u}$, with positive $\bar{u}$ directed landward.

Figure 4 visualizes the encountered forcing conditions. To put our ripple data in a broader perspective, the conditions of a number of other data sets are shown too. These include the field-laboratory data compiled by Goldstein *et al.* [6], the Sennen Beach data of Masselink *et al.* [7] ($D_{50} = 600$ μm), and the Georgia Shelf data of Nelson and Voulgaris [38] ($D_{50} = 388$ μm). The Bardex II data extends these earlier data sets with strong hydrodynamic conditions (large d_s and u_w) for given T_p and D_{50} (Figure 4a–e). The wave Reynolds number ranged between 1.1×10^5 and 6.4×10^6. When compared to the Re_w conditions for the coarse-sand ripple data compiled by Pedocchi and Garcia [5] (e.g., their Figure 11a indicates 1×10^3 to 1×10^6 for $Re_p \approx 40$), this range also reflects the strong hydrodynamic conditions in our data set. Our data were collected beneath shoaling waves, breaking (plunging) waves and bores. This is also reflected by the wide range in relative wave height H_s/h ($\approx$0.3–1.5), where, based on visual observations [39], $H_s/h \approx 0.7$ delineated the shoaling from the surf zone. The Shields parameter θ increased with H_s/h from $\approx$ 0.2 to 2 (Figure 4f). The orbital motion was mostly non-linear: S_u ranged from 0 to about 1.5, and was typically largest where waves started to break (Figure 4g), while A_ζ generally became non-zero in the wave shoaling zone and was largest ($\approx$−2) just seaward of the swash zone (Figure 4h). The mean current $\bar{u}$ also was mostly offshore directed and ranged from $\approx$ 0 m/s under non-breaking waves to about −0.2 m/s in the surf zone (Figure 4i).

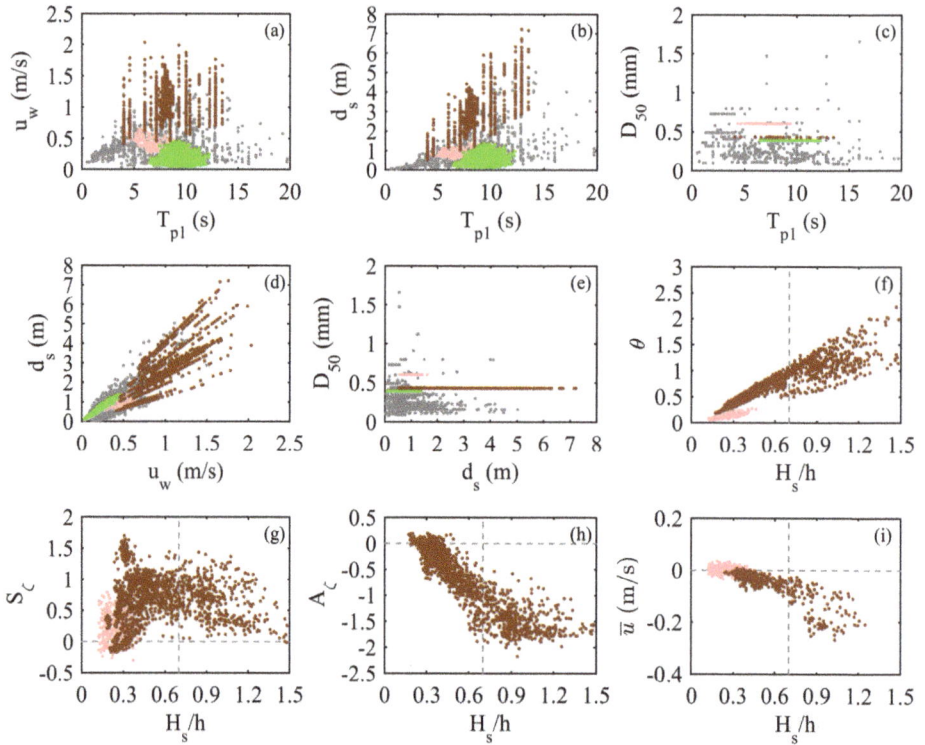

Figure 4. Visualization of the range in local forcing conditions, expressed as two-dimensional scatter plots involving the peak period T_{p1}, significant wave orbital diameter d_s, peak semi-orbital velocity u_w, median grain size D_{50}, relative wave height H_s/h, Shields parameter θ, skewness S_ζ, asymmetry A_ζ and the mean cross-shore flow $\bar{u}$. Brown dots: Bardex II data; grey dots: Goldstein *et al.* [6]; light-red dots: Masselink *et al.* [7]; green dots: Nelson and Voulgaris [38]. The vertical dashed line in (f)–(i) delineates the shoaling zone ($H_s/h < 0.7$) from the surf zone ($H_s/h \geq 0.7$).

Exploring a relationship between ripple geometry and hydrodynamical parameters demands the ripples to be in equilibrium with the hydrodynamical forcing. Based on full-scale flow tunnel experiments [40] established that the number of wave cycles to equilibrium, n_e, decreases exponentially with ψ as $n_e = \exp(-0.036\psi + 7.44)$ and depends little on the initial bed configuration. Most of our ψ (≈ 35–350) are well above the largest ψ in [40]'s data (≈ 55), suggesting that for our data n_e is $\mathcal{O}(10^2)$ or less. Given the wave periods deployed during the Bardex II tests, this implies that wave runs were mostly of sufficient duration for the ripples to reach equilibrium along the entire bed profile. To avoid any non-equilibrium conditions, the ripple data collected after the first three runs of A1 and of D1 were discarded.

3. Results

3.1. Cross-section Geometry

Examples of the cross-shore evolution of λ_r, η_r and ϑ_r are shown in Figure 5a–c and g–i for two tests (A4 and A7) with different H_{s0} and especially T_{p1} and hence d_s. During A4 waves started to break as plungers on the seaward edge of the sandbar ($x = 68$ m; Figure 5d), while during A7 plunging commenced slightly further landward ($x = 70 - 75$ m, Figure 5j). In both tests, λ_r and d_s/D_{50}

were approximately constant for $x < 60$ m, increased simultaneously to peak in the outer part of the surf zone, and then both decreased further onshore towards the beach face (compare Figure 5a,e to Figure 5g,k, respectively). Thus, the ripples were clearly orbital ripples, despite the fact that in both cases the d_s/D_{50} were well above the previously defined transition ($d_s/D_{50} = 2000$) from orbital into suborbital ripples and even above the lower limit ($d_s/D_{50} = 5000$) for anorbital ripples [11,13]. During both A4 and A7 η_r was approximately constant at 0.1 m seaward of the surf zone, but decreased rapidly to a few centimetres under the plunging breakers, especially in test A7. During all wave runs ϑ_r was approximately 0.15 over the deeper horizontal part of the flume, indicating the presence of vortex ripples, and slightly decreased to $\approx$0.1 at the landward edge of the shoaling zone. Inside the surf zone the ripples became much more subdued, with $\vartheta_r = 0.02$–0.03 in the most landward windows for which ripple geometry could be computed. Finally, we note that η_r and ϑ_r for a given d_s/D_{50} inside the surf zone were lower than for the same d_s/D_{50} outside the surf zone. For example, during A4 η_r and ϑ_r amounted to about 0.06 m and 0.05, respectively, at $x = 82$ m ($d_s/D_{50} \approx 6,500$), while the ripples were higher ($\eta_r \approx 0.09$ m) and steeper ($\vartheta_r \approx 0.10$) at $x = 62.5$ m despite similar d_s/D_{50}. This difference seems to be substantially less for λ_r.

The temporal evolution of λ_r, η_r and ϑ_r is shown in Figure 6a–c for two example locations ($x = 55$ and 80 m), together with the local values of H_s/h, d_s/D_{50} and θ (Figure 6d–f, respectively). The seaward ($x = 55$ m) location was near the seaward end of the steep sloping profile (Figure 1) and always experienced non-breaking wave conditions (Figure 6d, $H_s/h < 0.7$), while waves at the shallower landward location ($x = 80$ m) waves could either be shoaling or breaking depending on H_{s0}, T_{p1}, h_s and the sandbar morphology. Temporal variability in H_s/h, d_s/D_{50} and θ mostly reflected changes in H_{s0}, T_{p1} or h_s. For example, d_s/D_{50} and θ increased step-wise between tests in series D because of an increase in peak period T_{p1} and reduced slightly within a test because of an increase in the still water level h_s (Table 1). Also note the "tidal" signal in the wave conditions during tests C1 and C2. The gradual reduction in H_s/h, d_s/D_{50} and θ during A6 at $x = 80$ m was induced by the increase in local water depth associated with onshore sandbar migration. The wider range in wave conditions at $x = 80$ m is also reflected in a wider range in the cross-section ripple characteristics. Especially series C and D showed a clear positive dependence of λ_r on d_s/D_{50} (compare Figure 6a and Figure 6e). This is consistent with the suggestion based on Figure 5 that the ripples were orbital, even though again d_s/D_{50} extended well into the previously defined d_s/D_{50}-space of anorbital ripples. Under non-breaking conditions η_r also increased with d_s/D_{50}, resulting in an approximately constant ϑ_r of 0.1–0.15 (vortex ripples; Figure 6c). Clearly, the ripples were substantially less pronounced under breaking waves, with a clear reduction in ϑ_r with θ (compare Figure 6c and Figure 6f).

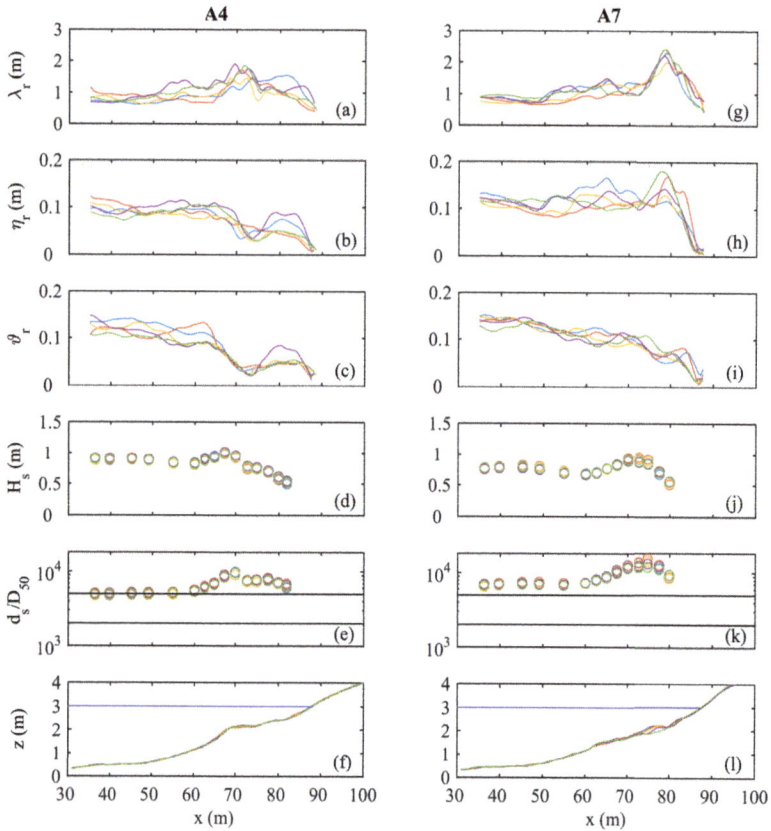

Figure 5. Cross-shore profiles of (a) ripple length λ_r; (b) ripple height η_r; (c) ripple steepness ϑ_r; (d) significant wave height H_s and (e) normalized orbital diameter d_s/D_{50} for all five runs in test A4. Panel (f) shows the corresponding bed profiles. Panels (g)–(l) are the same as (a)–(f) but for test A7. The two horizontal lines in (e) and (k) are previously defined values ($d_s/D_{50} = 2000$ and 5000; [11,13]) to delineate the orbital, suborbital and anorbital ripple regimes. The horizontal blue line in (f) and (l) is the still water level $h_s = 3$ m. The colors in all panels represent the different runs.

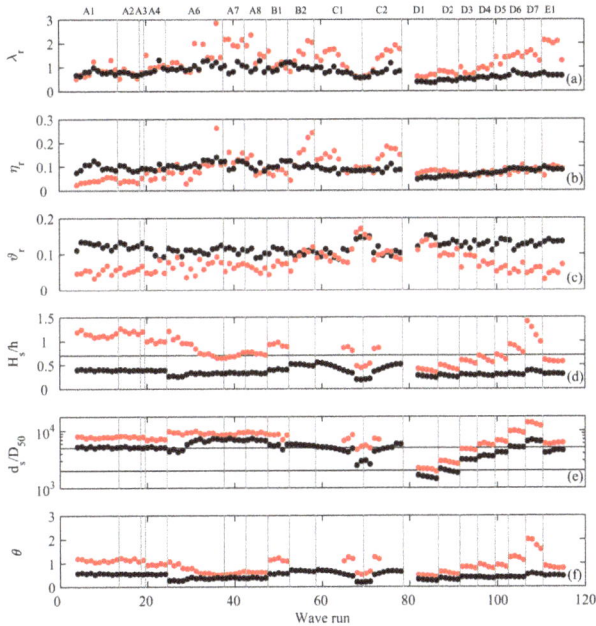

Figure 6. Temporal evolution of (**a**) ripple length λ_r; (**b**) ripple height η_r; (**c**) ripple steepness ϑ_r; (**d**) relative wave height H_s/h; (**e**) normalized orbital diameter d_s/D_{50} and (**f**) Shields parameter at $x = 55$ (black dots) and 80 m (red dots). The cross-section ripple data in (**a**)–(**c**) are based on a 10-m wide window centered at each x. The vertical grey lines in each panel mark the transition between tests. Note that the horizontal axis corresponds to the cumulative number of wave runs. Because runs were of different duration, the horizontal axis is not equidistant with time. The horizontal line in (**d**) marks the approximate transition between non-breaking and breaking conditions ($H_s/h = 0.7$). The two horizontal lines in (**e**) are previously defined values ($d_s/D_{50} = 2000$ and 5000; [11,13]) to delineate the orbital, suborbital and anorbital ripple regimes. No values at $x = 80$ m are shown in (**d**)–(**f**) for B2 and for several C1 and C2 wave runs because the instrument was not submerged continuously during these low h_s runs.

Figure 7a illustrates that λ_r normalized by D_{50} did indeed not follow the orbital-suborbital-anorbital trend with d_s/D_{50} found in fine-grained sand, where here this trend is indicated by the empirical predictor of Nelson *et al.* [4]. Instead, the trend is a growth in λ/D_{50} with d_s/D_{50} over the entire d_s/D_{50} range in the data, implying all our large coarse-grained wave ripples to be orbital ripples. It is, however, also obvious from Figure 7a that the ratio of λ_r to d_s was less than 0.65, a typical value quoted for orbital ripples. What is more, λ_r/d_s reduced with d_s from about 0.55 for $d_s/D_{50} \approx 1400$ to about 0.27 for $d_s/D_{50} \approx 11,500$, in contrast to earlier laboratory observations [8] that indicated λ_r/d_s to be constant for given D_{50}. The λ_r/d_s values quoted here were obtained by averaging λ/D_{50} in 0.1 wide $\log_{10}(d_s/D_{50})$ bins. Unsurprisingly, η_r/D_{50} did not follow the anorbital trend with d_s/D_{50} for $d_s/D_{50} > 5000$ (Figure 7b). As for λ_r, the proportionality between η_r and d_s decreased with d_s, from about 0.08 for $d_s/D_{50} \approx 1400$ to about 0.02 for $d_s/D_{50} > 8000$. The scatter in the observations near $d_s/D_{50} \approx 6000$–8000 is substantial. This is caused by the occurrence of these d_s/D_{50} values both inside and outside the surf zone with, as illustrated with Figure 5, substantially different η_r. Figure 7c demonstrates the expected change from vortex ripples ($\vartheta_r \approx 0.1 - 0.15$) for low d_s/D_{50} (here, $d_s/D_{50} \leq 5000$) to more subdued ripples at high d_s/D_{50}, although the scatter is again appreciable. We compare measured λ_r, η_r and ϑ_r to empirical predictors of ripple geometry designed for orbital ripples over the full d_s/D_{50} range in Section 4.

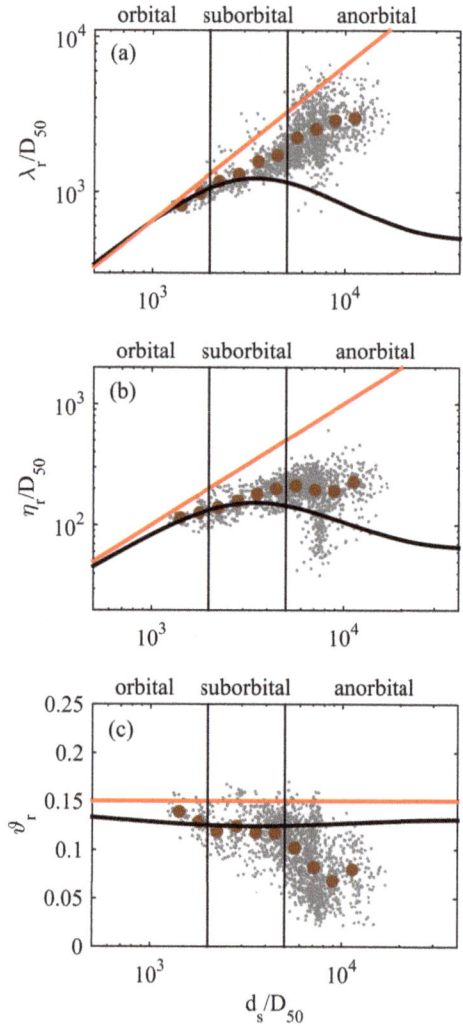

Figure 7. Measured (a) ripple length λ_r and (b) ripple height η_r, normalized by the median grain size D_{50}, and (c) ripple steepness ϑ_r versus normalized orbital diameter d_s/D_{50}, based on all observations at the 16 instrumented locations (1687 observations in total). The two vertical lines in all three panels are previously defined values ($d_s/D_{50} = 2000$ and 5000; [11,13]) to delineate the orbital, suborbital and anorbital ripple regimes. The curved black line in each plot represents the empirical predictor of Nelson *et al.* [4] to illustrate expected orbital-suborbital-anorbital trends in λ_r/D_{50}, η_r/D_{50} and ϑ_r, respectively. The red sloping lines in (a)–(c) are $\lambda_r/D_{50} = 0.65$, $\eta_r/D_{50} = 0.10$ and $\vartheta_r = 0.15$, values often quoted for orbital vortex ripples. The brown dots in each panel are average values computed from the measurements for 0.1-wide $\log_{10}(d_s/D_{50})$ bins.

3.2. Planform Geometry

None of the bed elevation models collected at $x = 63.1$ m revealed two-dimensional ripples with straight, uniform crests perpendicular to the wave direction that are characteristic of the coarse-grained wave ripples in existing field, e.g., [21–23,26], and several laboratory settings [8,28]. Instead, all models showed quasi two-dimensional to highly three-dimensional planform geometries. We illustrate this for series D and test C1 in Figures 8 and 9, respectively. In both figures the flow conditions are expressed as $0.06Re_w^{0.5}$, because [5] illustrated that ripples become two-dimensional for $Re_p > 0.06Re_w^{0.5}$ and are three-dimensional otherwise. At the end of test D1, when $0.06Re_w^{0.5}$ was below Re_p and at its lowest at $x = 63.1$ m for the entire Bardex II experiment, the ripples were quasi two-dimensional: ripple crests were reasonably continuous (sometimes up to several metres) and the crest-to-crest distance along two adjacent ripples was, at least visually, fairly constant, but the crests varied in orientation and some had notably variations in height. Also, some ripples bifurcated and several defects can be seen. With an increase in $0.06Re_w^{0.5}$ to about 65 (*i.e.*, well above Re_p), this planform changed gradually into oval mounds to which ripples with remarkably different orientations were attached (D6 and D7). This planform geometry was also observed during test C1 when $0.06Re_w^{0.5}$ was about 75 (*i.e.*, runs 8 and 9 in Figure 9). For $0.06Re_w^{0.5} > 80$ in test C1 (runs 4, 5 and 6 in Figure 9; the highest values observed at $x = 63.1$ m), the bed contained large, three-dimensional and gentle ($\vartheta_r \approx 0.07$) highs and lows only, closely resembling hummocky bed forms in fine-sand laboratory experiments [8,41–43] and field conditions [44]. For smaller $0.06Re_w^{0.5}$ in test C1 (runs 10 and 11 in Figure 9), the ripples became quasi two-dimensional, similar in appearance to that observed in tests D1 and D2 (Figure 8). The hydrodynamic forcing never became sufficiently energetic at $x = 63.1$ m to reach flat bed conditions. Cross-shore profiles of ϑ_r suggest that such conditions were reached only in the swash zone on the beach face (Figure 5c,i). Although we do not have bed elevation models for $0.06Re_w^{0.5}$ well below Re_p, our data confirm Pedocchi and Garcia [5]'s findings that ripple planform geometry is related to the wave Reynolds number and that $0.06Re_w^{0.5} = Re_p$ is a reasonable threshold above which ripples are three-dimensional.

The succession of equilibrium planform geometry for coarse sand that thus follows from Figures 8 and 9 for $0.06Re_w^{0.5} > 25$ is a transition from quasi two-dimensional ripples, through oval mounds with ripples attached from different directions, to three-dimensional hummocky bed forms. This change in planform is also associated with an increase in λ_r and hence, as deduced from Figure 7a, a reduction in λ_r/d_s. In other words, the strongly three-dimensional ripples in our data set generally had lower λ_r/d_s than the more moderately three-dimensional or quasi two-dimensional ripples. As an illustration, λ_r/d_s was near 0.3 for the strongly three-dimensional ripples in D7 and D8, and about 0.5 for the quasi two-dimensional ripples in D1. The dependence of λ_r/d_s on ripple planform observed here is qualitatively consistent with previous field [24] and laboratory [28] observations.

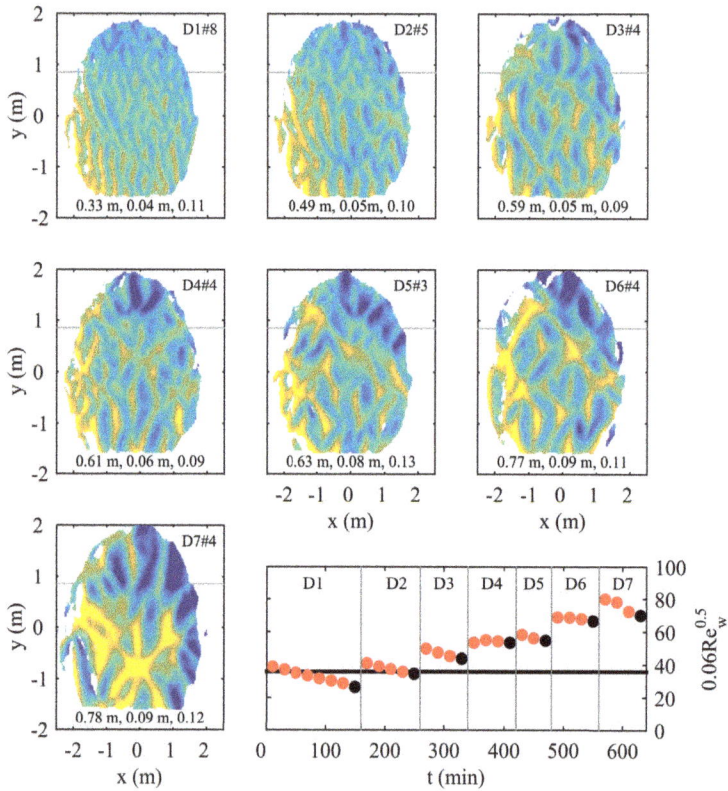

Figure 8. Bed elevation models for selected tests and runs during series D. Warm (yellow) colours are ripple crests, cold (blue) colours are ripple troughs. The colors range from −0.08 to 0.08 m. DX#Y stands for run Y in test DX. The local x and y coordinates are relative to the sonar, with $(x, y) = (0, 0)$ vertically below the sonar, and with x positive onshore. The local $x = 0$ m corresponds to $x = 63.1$ m in Figure 1. The nearest flume wall is at $y = -1.65$ m. The 3 numbers at the bottom of each panel are the ripple wavelength λ_r, height η_r and steepness ϑ_r based on the profile data. The center line of the flume is indicated by the gray line. The forcing conditions are expressed as a time series of $0.06 Re_w^{0.5}$; time $t = 0$ is the start of series D. The horizontal black line is the particle Reynolds number Re_p. The black dots represent the times of the shown bed elevation models.

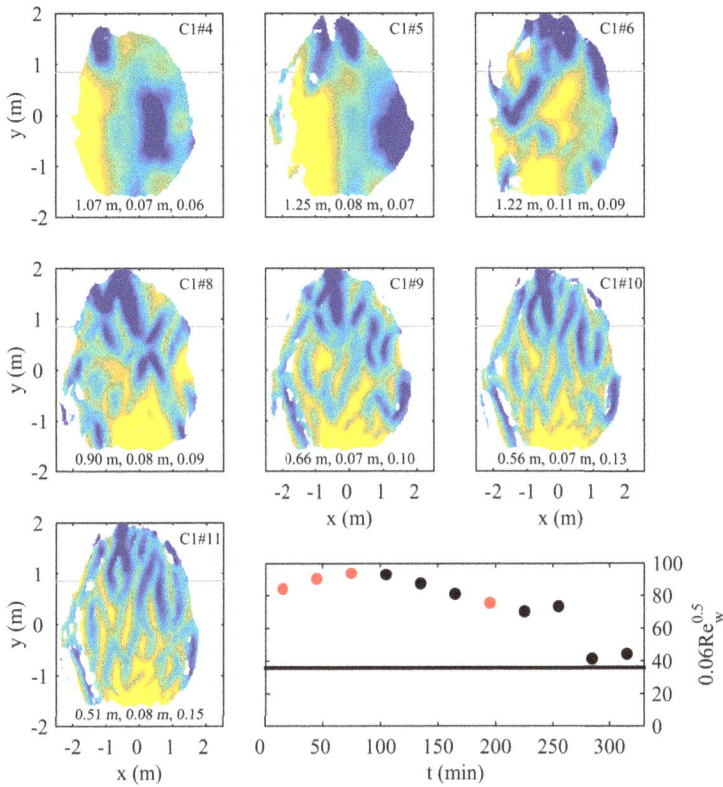

Figure 9. Bed elevation models for selected runs in test C1. For additional explanation, see the caption of Figure 8. Here, time $t = 0$ is the start of test C1. During test C1 $0.06Re_w^{0.5}$ reduced by a decrease in the still water level h_s and in the last two runs by a decrease in the significant wave height H_{s0} (Table 1).

The five consecutive bed elevation models in test A4 (Figure 10) clearly document that the three-dimensional ripples shifted and changed perpetually under approximately constant hydrodynamical forcing (Figure 5d,e). Yet, their average size (λ_r and η_r) remained fairly constant, see also Figure 5a–c. Ripple migration, separation and amalgamation seem to have taken place continuously, although it is difficult to tell how individual bed forms actually evolved within a run. Comparable dynamics were also observed during all other tests in series A and B (not shown). Similar to fine-grained three-dimensional ripples, e.g., [27,42,45], the coarse-grained three-dimensional ripples in our data thus exhibited dynamic-equilibrium behaviour.

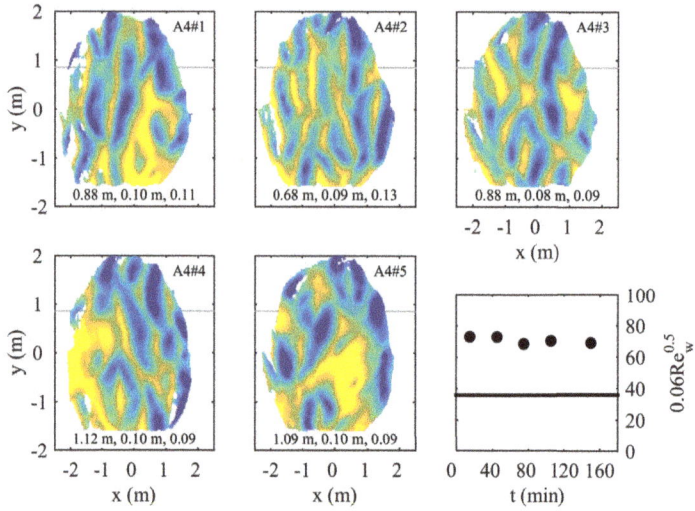

Figure 10. Bed elevation models for all five runs in test A4. For additional explanation, see the caption of Figure 8. Here, time $t = 0$ is the start of test A4.

3.3. Ripple Migration

Spatially and temporally coherent ripple migration was discernable at the sonar location only when the planform geometry was classified as quasi two-dimensional. This was, as illustrated in Figure 11a, the case for tests D1–D3. Ripple migration was found to be offshore directed and to amount to about 0.01 m/min. During these tests S_ζ and A_ζ were both close to 0 (Figure 11c,d) and $\bar{u}$ was weak (-0.03 m/s or less, see Figure 11e), representative of conditions well seaward of the surf zone (Figures 4 and 11b). This would indicate that neither wave non-linearity nor the current are driving the ripple migration. Instead, the offshore ripple migration presumably reflects downslope gravity-induced bedload transport. This would corroborate numerical sand transport computations [33,46] that demonstrate downslope gravity-induced transport to be a significant contributor to the total sand transport on the present steep profile under non-breaking wave conditions.

For the more energetic D4–D7 tests the individual ripples could simultaneously migrate landward or seaward, or not migrate at all (Figure 11a). In a few cases a ripple changed migration direction (e.g., D7, the ripple between $x \approx 0.5$ and 1.0 m) or split into two parts (e.g., near the end of D6 at $x \approx -0.75$ m) during a test. In addition, two ripples could merge (e.g., near the end of D6 at $x \approx -0.25$ m). Similar observations were made during all tests in which the ripple planform resembled oval mounds with ripples attached from different directions (not shown). The now rather large (and positive) S_ζ (Figure 11c) is obviously not associated with coherent onshore bedform migration as found previously for two-dimensional orbital wave ripples [24,25] and for anorbital [47] and subdued large wave ripples [19] in finer sand. The present spatially and temporally incoherent pattern expresses the dynamic-equilibrium behaviour of the ripples, including ripple separation and amalgamation.

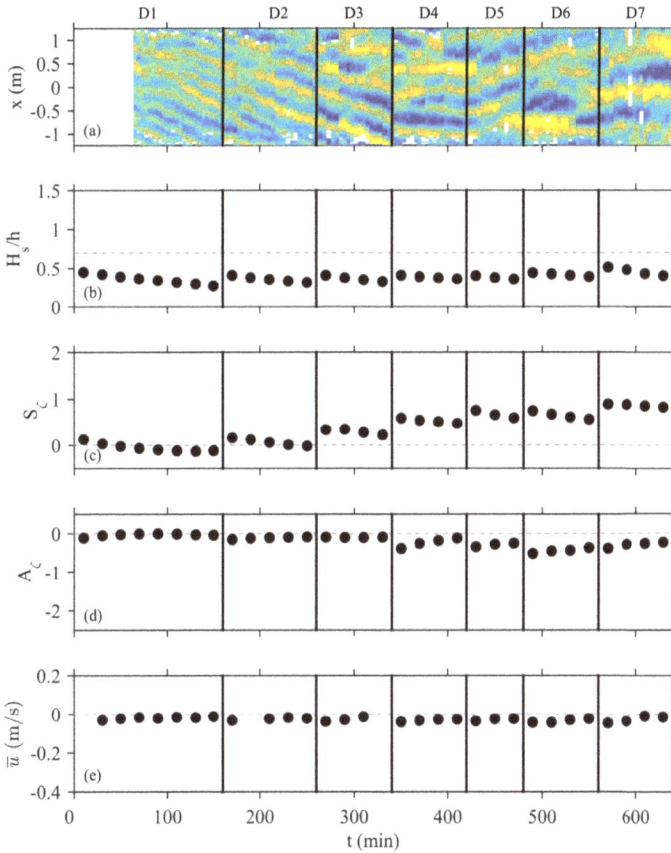

Figure 11. (a) Time-space diagram of wave-ripple induced bed variability beneath the sonar during series D. Warm (yellow) colours are ripple crests, cold (blue) colours are ripple troughs. The colors range from -0.05 to 0.05 m. The cross-shore coordinate is relative to the sonar, with $x = 0$ vertically below the sonar, and with positive x onshore. The local $x = 0$ m corresponds to $x = 63.1$ m in Figure 1. Time series of (b) the relative wave height H_s/h, (c) skewness S_ζ, (d) asymmetry A_ζ and (e) the mean cross-shore flow $\bar{u}$. In (b)–(e) the limits of the vertical axis were set to equal to the range in the entire data set, see Figure 4. H_s/h, S_ζ and A_ζ were measured at $x = 62.5$ m, $\bar{u}$ at $x = 65$ m. The dashed line in (b) marks the approximate H_s/h transition between non-breaking and breaking waves.

4. Empirical Prediction

The persistent orbital-nature of coarse-sand wave ripples precludes the use of empirical predictors that produce orbital-suborbital-anorbital trends in λ_r and η_r, e.g., [4,11]. Therefore, we now test the orbital Goldstein *et al.* [6] (henceforth GCM) and the coarse-sand Pedocchi and Garcia [5] (henceforth PG) predictors against our data (Figure 12). The GCM predictors are based on a compilation of laboratory and field data sets from which suborbital and anorbital ripples were discarded when they were superimposed on large wave ripples, and retained anorbital ripples were additionally removed for $u_w/w_s \geq 25$. Using genetic programming Goldstein *et al.* [6] obtained the following predictors

$$\lambda_{\text{GCM}}/d_s = \frac{1}{1.12 + 2.18D_{50}} \tag{1}$$

$$\eta_{\text{GCM}}/d_s = \frac{0.313 D_{50}}{1.12 + 2.18 D_{50}} \qquad (2)$$

$$\vartheta_{\text{GCM}} = \frac{3.42}{22 + (\lambda_{\text{GCM}}/D_{50})^2} \qquad (3)$$

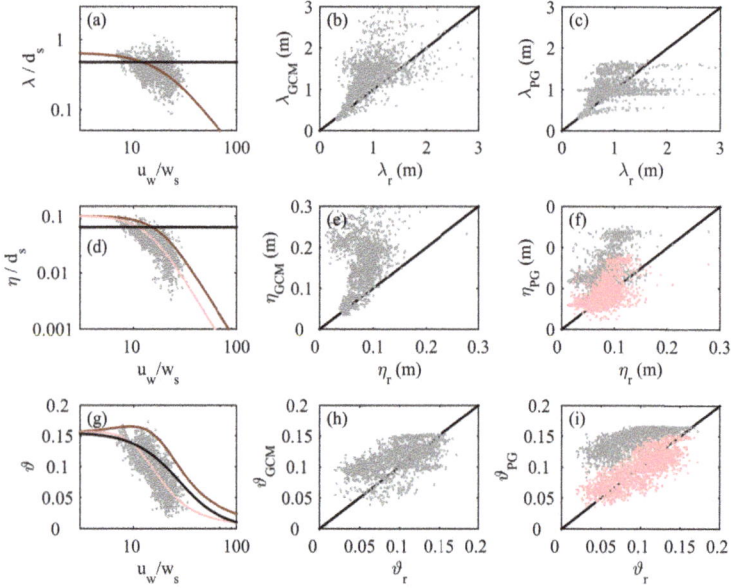

Figure 12. Measured (**a**) normalized ripple length λ_r/d_s, (**d**) normalized ripple height η_r/d_s and (**g**) ripple steepness ϑ_r versus the ratio of the peak semi-orbital velocity u_w over the sediment fall velocity w_s (gray dots). The black and brown lines in (**a**), (**e**) and (**g**) are the GCM and PG predictors, respectively; in (**g**), $T_p = 8$ s was used to relate u_w to d_s in the GCM predictor. Other panels are scatter plots of predicted against measured ripple parameters together with the 1:1 line. The light-red lines in (**d**) and (**g**), and the light-red dots in (**f**) and (**i**) correspond to the modified PG η/d_s predictor, see text for further explanation.

For a given D_{50} λ_{GCM} and η_{GCM} thus scale linearly with d_s (*i.e.*, orbital ripples). The present $D_{50} = 430$ μm results in $\lambda_{\text{GCM}}/d_s = 0.49$ and $\eta_{\text{GCM}}/d_s = 0.065$ (Figure 12a,d, respectively). The PG predictors are also based on laboratory and field data sets, from which PG discarded the characteristics of the large wave ripples when they were overlain by suborbital or anorbital ripples. PG related ripple dimensions to the ratio u_w/w_s as

$$\lambda_{\text{PG}}/d_s = 0.65 \left[(0.050 u_w/w_s)^2 + 1 \right]^{-1} \qquad (4)$$

$$\eta_{\text{PG}}/d_s = 0.1 \left[(0.055 u_w/w_s)^3 + 1 \right]^{-1} \qquad (5)$$

and $\vartheta_{\text{PG}} = \eta_{\text{PG}}/\lambda_{\text{PG}}$. In contrast to the GCM predictors, the ratios λ_{PG}/d_s and η_{PG}/d_s are not constant for a given D_{50}, but decrease with u_w (Figure 12a,d). For $u_w/w_s < 10$, ϑ_{PG} is about constant near 0.16 and then decreases to < 0.05 for $u_w/w_s > 50$ (Figure 12g). Three error measures were computed to quantify the performance of the two predictors for λ_r, η_r and ϑ_r: the bias b, the root-mean-square error ϵ_{rms}, and the correlation-coefficient squared r^2 of the best-fit linear line between a predicted and measured ripple parameter. Both b and ϵ_{rms} were normalized by the range of the observed values for each parameter predicted:

$$b = \frac{\frac{1}{N} \sum (X_p - X_m)}{\max(X_m) - \min(X_m)} \qquad (6)$$

and

$$\epsilon_{rms} = \frac{\sqrt{\frac{1}{N} \sum (X_p - X_m)^2}}{\max(X_m) - \min(X_m)} \qquad (7)$$

where X is the evaluated parameter, the subscripts p and m denote predicted and measured values, respectively, and N is the total number of observations. All error statistics are listed in Table 2.

Table 2. Equilibrium predictor error statistics.

	GCM			PG			RBK		
	λ_r	η_r	ϑ_r	λ_r	η_r	ϑ_r	λ_r	η_r	ϑ_r
b	0.11	0.34	0.10	0.02	0.23	0.32	0.00	0.02	0.01
ϵ_{rms}	0.17	0.43	0.20	0.12	0.27	0.35	0.13	0.11	0.13
r^2	0.36	0.07	0.32	0.21	0.30	0.58	0.33	0.29	0.61

The constant values for the ratios λ_{GCM}/d_s and η_{GCM}/d_s are inconsistent with our data (Figure 7a,b, and Figure 12a,d), and the GCM predictors overestimate λ_r, η_r, and ϑ_r substantially (Figure 12b,e,h), with bias values of 0.11, 0.33 and 0.10, respectively. While the PG predictor produces λ/d_s values that are roughly accurate, Figure 12a casts substantial doubt on the suitability of u_w/w_s to predict the trend in λ/d_s. For $u_w/w_s < 20$, most λ_r/d_s are lower than predicted, while the opposite is true for larger u_w/w_s. As a consequence, the overall agreement between λ_{PG} and λ_r is low (Figure 12c), with $r^2 = 0.21$ only. The η_r/d_s does decrease with u_w/w_s, but predicted values are substantially larger (Figure 12d). As a consequence, the bias values for η_{PG} and ϑ_{PG} are large (0.23 and 0.32, respectively). This systematic difference can be removed largely by modifying Equation (5) into

$$\eta_{PG}/d_s = 0.1 \left[(0.075 u_w/w_s)^3 + 1 \right]^{-1} \qquad (8)$$

This modified predictor is shown with the light-red lines in Figure 12d,g, and corresponding dots in Figure 12f,i. Although this change results in near-zero bias, the r^2 for η_{PG} remains low (0.27); for ϑ_{PG} the r^2 is substantially higher (0.61).

The overall poor performance of the GCM and PG predictors motivated us to design alternative predictors for the equilibrium length, height and steepness of coarse-grained wave ripples. To this end, we combined our Bardex II data with all equilibrium ripple data from the Goldstein *et al.* [6] database with $D_{50} \geq 300$ μm and the Sennen Beach data of Masselink *et al.* [7]. A scatter plot of observed λ/d_s versus the Shields parameter θ (Figure 13a) suggest that $\log_{10}(\lambda/d_s) = a_0 + a_1 \log_{10}(\theta)$, where a_0 and a_1 are fit parameters, is a meaningful predictor. A least-squares fit resulted in $a_0 = -0.471 \pm 0.008$ and $a_1 = -0.163 \pm 0.014$, where the ±value provides the 95% confidence range. This fit can be rewritten into

$$\lambda_{RBK}/d_s = 0.338 \theta^{-0.163} \qquad (9)$$

Our fit, indicated by the subscript RBK, thus results in a reduction of λ/d_s from 0.55 at $\theta = 0.05$ to 0.3 at $\theta = 2$. The dependence of ϑ on θ (Figure 13e) suggests a predictor of the form $\vartheta = a_2 - a_2 \tanh(a_3 \theta^{a_4})$, where a_2 to a_4 are fit parameters. A least-squares fit resulted in $a_2 = 0.164 \pm 0.004$, $a_3 = 0.630 \pm 0.020$ and $a_4 = 1.038 \pm 0.080$. Thus,

$$\vartheta_{RBK} = 0.164 - 0.164 \tanh(0.630\theta) \qquad (10)$$

considering that the 95% confidence band of a_4 encompasses 1. This fit produces $\vartheta \approx 0.15$ for $\theta <\approx 0.2$ and a subsequent reduction in ϑ to 0.02 for $\theta = 2$. We note that an earlier θ-based predictor for ripple steepness was proposed in Nielsen [48], $\vartheta = 0.342 - 0.34\sqrt[4]{\theta}$. It is obvious from Figure 13e that this fit is not a good approximation of the present combined data set. Finally, the normalized ripple height, η_{RBK}/d_s, can be computed for a given θ as the product of $\lambda_{RBK}(\theta)$ and $\vartheta_{RBK}(\theta)$ (Figure 13c). This results in $\eta_{RBK}/d_s \approx 0.1$ for low θ and a reduction in η_{RBK}/d_s with θ to $\eta_{RBK}/d_s < 0.01$ for $\theta > 1.5$.

The application of our fits to the Bardex II ripple data only (Figure 13b,d,f) results in improved error statistics compared to the GCM and PG predictors (Table 2), with near-zero bias and reduced root-mean-square error, the latter in particular for η_r and ϑ_r. Figure 13b illustrates that the differences between λ_{RBK} and λ_r are most pronounced for large λ_r and can amount to 1 m or more. A closer inspection of these differences revealed that they are largest for the $H_s/h = 0.6 - 1.0$ range, that is, in the outer surf zone where the waves broke as plungers. Vortices in plunging breakers are known to penetrate into the water column, e.g., [49,50] and, upon impact with the bed, to forcefully lift sand into suspension, e.g., [51–53]. It is feasible that this modifies ripple geometry and explains, at least partly, the generally poor agreement between predicted and observed ripple lengths. It would also explain why non-filtered ripple length estimates scatter most on the sandbar (Figure 2). For lower H_s/h the waves are non-breaking, while for larger H_s/h the plunging waves have evolved into bores in which breaking-induced turbulence near the bed is less intense and wave ripples are presumably again determined largely by the near-bed orbital flow.

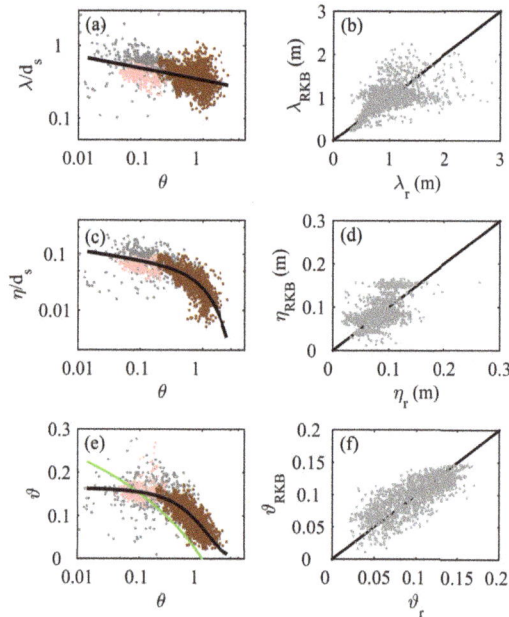

Figure 13. Measured (**a**) normalized ripple length λ/d_s, (**c**) normalized ripple height η/d_s and (**e**) ripple steepness ϑ versus the Shields parameter θ. Brown dots: Bardex II data; grey dots: Goldstein *et al.* [6]; light-red dots: Masselink *et al.* [7]. The black lines are our least-squares fits based on all data. The green line in (**e**) is based on Nielsen [48]. Panels (**b**), (**d**) and (**f**) are scatter plots of predicted versus measured ripple parameters (Bardex II data only) together with the 1:1 line.

5. Discussion

In this paper we have extended earlier work on coarse-grain wave ripples by exploring their characteristics under full-scale, irregular waves with large orbital motion (Figure 4). The data illustrated that the ripples are orbital for the entire range of d_s/D_{50} encountered ($\approx$1000–20,000), with a reduction in the ratios λ_r/d_s and η_r/d_s with increasing d_s/D_{50} (Figure 7a,b). Simultaneously, the ripple steepness reduced from 0.1–0.15 (vortex ripples) to <0.05 (Figure 7c), and the planform geometry changed from quasi two-dimensional ripples to strongly three-dimensional hummocky bed forms under the most energetic conditions (Figures 8 and 9). In other words, coarse-grain ripples can become three-dimensional when the wave forcing is sufficiently strong and wave-formed hummocks are not restricted to fine sands.

We realize that by deriving Equations (9) and (10) we have added yet another predictor to the existing plethora of equilibrium ripple predictors. For mild wave conditions ($d_s/D_{50} < 2000$) most predictors, including ours, produce orbital vortex ripples with $\lambda_r/d_s \approx 0.65$, $\eta_r/d_s \approx 0.1$ and $\vartheta_r \approx 0.15$. For more energetic conditions our predictor produces cross-section ripple characteristics that deviate considerably from suborbital-anorbital trends. Field data sets, e.g., [17,19,20], have shown suborbital or anorbital ripples up to about $D_{50} = 300$ µm. Accordingly, we propose that our predictor should be used only when D_{50} exceeds 300 µm. This is larger than Pedocchi and Garcia [5]'s definition of coarse sand; their $Re_p \geq 13$ corresponds to $\geq$220 µm (quartz sand) at 20 °C. For finer sand we recommend the use of the Nelson *et al.* [4] predictor as it is based on a vast amount of data and outperforms many other predictors. This implies that, as [5], we advocate the use of grain size dependent predictors. This has the disadvantage of potential spatial discontinuities in predictions of cross-section ripple geometry when spatially explicit grain size maps are used as input in hydrodynamic and morphodynamic models. We do not know whether there is a need to include a third, intermediate grain size predictor to, for example, minimize the discontinuities. This will depend on the width of the grain size range into which ripple type changes with increasing orbital flow from orbital-suborbital-anorbital into orbital only. New laboratory experiments, for instance in a large oscillating water tunnel, using sands with D_{50} ranging from $\approx$250 µm to $\approx$350 µm under mild, intermediate and high orbital flow may shed further light on this issue and will aid in providing a physical explanation for the change in ripple behaviour with flow conditions near 300 µm.

6. Conclusions

Wave-formed ripples with equilibrium length $\lambda_r = 0.31 - 2.38$ m, height $\eta_r = 0.01 - 0.17$ m and steepness $\vartheta = 0.01 - 0.16$ were observed in the shoaling and surf zone of a coarse sand, prototype laboratory beach under a range of wave conditions and water depths. Our data confirm findings from earlier limited laboratory data that coarse-grained wave ripples remain orbital, even when the ratio of orbital diameter to median grain size d_s/D_{50} is in the part of parameter space where in fine sand anorbital ripples form. The ratio of λ_r to d_s is not constant, but decreases from about 0.55 for $d_s/D_{50} \approx 1400$ to about 0.27 for $d_s/D_{50} \approx 11,500$. Analogously, ripple height η_r increases with d_s, but the proportionally decreases from about 0.08 for $d_s/D_{50} \approx 1400$ to about 0.02 for $d_s/D_{50} > 8000$. Ripple planform geometry changes with increasing wave Reynolds number from vortex ripples with wavy crests, through oval mounds with ripples attached from different directions, to strongly subdued hummocky features. Our data thus indicate that coarse-grained wave ripples can be three-dimensional if the orbital flow is sufficiently strong and that wave-formed hummocks are not restricted to fine sands. The three-dimensional ripples show dynamic-equilibrium behaviour with ripple amalgamation and separation, but without clear onshore migration even though the orbital motion is positively (onshore) skewed and mean currents are weak. Finally, we propose new empirical equilibrium ripple predictors for $D_{50} > 300$ µm, in which λ_r/d_s, η_r/d_s and ϑ_r are a function of the Shields parameter θ. For finer sand we recommend a predictor that follows the orbital-suborbital-anorbital trend in cross-section geometry, such as that of Nelson *et al.* [4].

J. Mar. Sci. Eng. **2015**, *3*, 1568–1594

Acknowledgments: Bardex II was supported by the European Community's 7th Framework Programme through the grant to the budget of the Integrating Activity HYDRALAB IV, contract No. 261520. The academic lead of the project was Gerd Masselink and the Deltares coordinator was Guido Wolters. Winnie de Winter, Daan Wesselman and Florent Grasso contributed significantly to the data collection, and Marcel van Maarseveen, Henk Markies and all Deltares Delta flume staff provided excellent technical support. Gerd Masselink and Evan Goldstein provided the additional ripple data sets. G.R. and J.A.B. were funded by the Dutch Technology Foundation STW, which is part of the Netherlands Organisation for Scientific Research (NWO), and which is partly funded by the Ministry of Economic Affairs (project number 12397). All data are available from the first author upon request.

Author Contributions: The authors contributed in the following proportions to design and data collection, data processing, analysis and conclusions, and manuscript preparation, respectively: G.R. (100%, 50%, 60%, 60%), J.A.B. (0%, 50%, 30%, 20%) and M.G.K. (0%, 0%, 10%, 20%).

Conflicts of Interest: The authors declare no conflict of interest.

References

1. Van Rijn, L.C.; Walstra, D.J.R.; Grasmeijer, B.; Sutherland, J.; Pan, S.; Sierra, J.P. The predictability of cross-shore bed evolution of sandy beaches at the time scale of storms and seasons using process-based profile models. *Coast. Eng.* **2003**, *47*, 295–327.

2. Van Rijn, L.C.; Walstra, D.J.R.; van Ormondt, M. Unified view of sediment transport by currents and waves. IV: Application of morphodynamic model. *J. Hydraul. Eng.* **2007**, *133*, 776–793.

3. Ganju, N.K.; Sherwood, C.R. Effect of roughness formulation on the performance of a coupled wave, hydrodynamic, and sediment transport model. *Ocean Model.* **2010**, *33*, 299–313.

4. Nelson, T.R.; Voulgaris, G.; Traykovski, P. Predicting wave-induced ripple equilibrium geometry. *J. Geophys. Res.* **2013**, *118*, 3202–3220.

5. Pedocchi, F.; Garcia, M.H. Ripple morphology under oscillatory flow: 1. Prediction. *J. Geophys. Res.* **2009**, *114*, C12014, doi:10.1029/2009JC005354.

6. Goldstein, E.B.; Coco, G.; Murray, A.B. Prediction of wave ripple characteristics using genetic programming. *Cont. Shelf Res.* **2013**, *71*, doi: 10.1016/j.csr.2013.09.020.

7. Masselink, G.; Austin, M.J.; O'Hare, T.J.; Russell, P.E. Geometry and dynamics of wave ripples in the nearshore zone of a coarse sandy beach. *J. Geophys. Res.* **2007**, *112*, C10022, doi:10.1029/2006JC003839.

8. Cummings, D.I.; Dumas, S.; Dalrymple, R.W. Fine-grained versus coarse-grained wave ripples generated experimentally under large-scale oscillatory flow. *J. Sediment. Res.* **2009**, *79*, 83–93.

9. Clifton, H.E. Wave-formed sedimentary structures: A conceptual model. In *Beach and Nearshore Sedimentation*, Special Publication 24; Society for Sedimentary Geology, Tulsa, OK, USA, 1976; pp. 126–148.

10. Miller, M.C.; Komar, P.D. Oscillation sand ripples generated by laboratory apparatus. *J. Sediment. Res.* **1980**, *50*, 173–182.

11. Wiberg, P.L.; Harris, C.K. Ripple geometry in wave-dominated environments. *J. Geophys. Res.* **1994**, *99*, 775–789.

12. Van der Werf, J.J.; Doucette, J.S.; O'Donoghue, T.; Ribberink, J.S. Detailed measurements of velocities and suspended sand concentrations over full-scale ripples in regular oscillatory flow. *J. Geophys. Res.* **2007**, *112*, F02012, doi:10.1029/2006JF000614.

13. Clifton, H.E.; Dingler, J.R. Wave-formed structures and paleoenvironmental reconstruction. *Mar. Geol.* **1984**, *60*, 165–198.

14. Maier, I.; Hay, A.E. Occurrence and orientation of anorbital ripples in near-shore sands. *J. Geophys. Res.* **2009**, *114*, doi: 10.1029/2008JF001126.

15. Soulsby, R.L.; Whitehouse, R.J.S.; Marten, K.V. Prediction of time-evolving sand ripples in shelf seas. *Cont. Shelf Res.* **2012**, *38*, 47–62.

16. Hanes, D.M.; Alymov, V.; Chang, Y.S.; Jette, C. Wave-formed sand ripples at Duck, North Carolina. *J. Geophys. Res.* **2001**, *106*, 22575–22592.

17. Grasmeijer, B.T.; Kleinhans, M.G. Observed and predicted bed forms and their effect on suspended sand concentrations. *Coast. Eng.* **2004**, *51*, 351–371.

18. Williams, J.J.; Bell, P.S.; Thorne, P.D. Unifying large and small wave-generated ripples. *J. Geophys. Res.* **2005**, *110*, C02008, doi:10.1029/2004JC002513.

19. Miles, J.R.; Thorpe, A.; Russell, P.; Masselink, G. Observations of bedforms on a dissipative macrotidal beach. *Ocean Dyn.* **2014**, *64*, 225–239.
20. Larsen, S.M.; Greenwood, B.; Aagaard, T. Obervations of megaripples in the surf zone. *Mar. Geol.* **2015**, *364*, doi: 10.1016/j.margeo.2015.03.003.
21. Forbes, D.L.; Boyd, R. Gravel ripples on the inner Scotian shelf. *J. Sediment. Petrol.* **1987**, *57*, 46–54.
22. Hunter, R.E.; Dingler, J.R.; Anima, R.J.; Richmond, B.M. Coarse-sediment bands on the inner shelf of Southern Monterey Bay, California. *Mar. Geol.* **1988**, *80*, 81–98.
23. Leckie, D. Wave-formed, coarse-grained ripples and their relationship to hummocky cross-stratification. *J. Sediment. Petrol.* **1988**, *58*, 607–622.
24. Traykovski, P.; Hay, A.E.; Irish, J.D.; Lynch, J.F. Geometry, migration, and evolution of wave orbital ripples at LEO-15. *J. Geophys. Res.* **1999**, *104*, 1505–1524.
25. Doucette, J.S. Bedform migration and sediment dynamics in the nearshore of a low-energy sandy beach in southwestern Australia. *J. Coast. Res.* **2002**, *18*, 576–591.
26. Yoshikawa, S.; Nemoto, K. The role of summer monsoon-typhoons in the formation of nearshore coarse-grained ripples, depression, and sand-ridge systems along the Shimizu coast, Suruga Bay facing the Pacific Ocean, Japan. *Mar. Geol.* **2014**, *353*, 84–98.
27. Pedocchi, F.; Garcia, M.H. Ripple morphology under oscillatory flow: 2. Experiments. *J. Geophys. Res.* **2009**, *114*, C12015, doi:10.1029/2009JC005356.
28. O'Donoghue, T.; Doucette, J.S.; Van der Werf, J.J.; Ribberink, J.S. The dimensions of sand ripples in full-scale oscillatory flows. *Coast. Eng.* **2006**, *53*, 997–1012.
29. Kleinhans, M.G. Large bedforms on the shoreface and upper shelf, Noordwijk, The Netherlands. In *SANDPIT: Sand Transport and Morphology of Offshore Sand Mining Pits*; van Rijn, L.C., Soulsby, R.L., Hoekstra, P., Davies, A.G., Eds.; Aqua Publications: Amsterdam, The Netherlands, 2005.
30. Masselink, G.; Ruju, A.; Conley, D.; Turner, I.; Ruessink, G.; Matias, A.; Thompson, C.; Castelle, B.; Puleo, J.; Citerone, V.; et al. Large-scale Barrier Dynamics Experiment II (BARDEX II): Experimental design, instrumentation, test programme, and data set. *Coast. Eng.* **2015**, in press.
31. Ferguson, R.I.; Church, M. A simple universal equation for grain settling velocity. *J. Sediment. Res.* **2004**, *74*, 933–937.
32. Plant, N.G.; Holland, K.T.; Puleo, J.A. Analysis of the scale of errors in nearshore bathymetric data. *Mar. Geol.* **2002**, *191*, 71–86.
33. Ruessink, B.G.; Blenkinsopp, C.; Brinkkemper, J.A.; Castelle, B.; Dubarbier, B.; Grasso, F.; Puleo, J.A.; Lanckriet, T. Sandbar and beachface evolution on a prototype coarse-grained sandy barrier. *Coast. Eng.* **2016**, in press.
34. Matias, A.; Masselink, G.; Castelle, B.; Blenkinsopp, C.; Kroon, A. Measurements of morphodynamic and hydrodynamic overwash processes in a large-scale wave flume. *Coast. Eng.* **2015**, doi:10.1016/j.coastaleng.2015.08.005.
35. Soulsby, R.L. *Dynamics of Marine Sands*; Thomas Telford: London, UK, 1997.
36. De Winter, W.; Wesselman, D.; Grasso, F.; Ruessink, G. Large-scale laboratory observations of beach morphodynamics and turbulence beneath shoaling and breaking waves. *J. Coast. Res.* **2013**, *65*, 1515–1520.
37. Elgar, S. Relationships involving third moments and bispectra of a harmonic process. *IEEE Trans. Acoust. Speech Signal Proc.* **1987**, *35*, 1725–1726.
38. Nelson, T.R.; Voulgaris, G. Temporal and spatial evolution of wave-induced ripple geometry: Regular versus irregular ripples. *J. Geophys. Res. Oceans* **2014**, *119*, 664–688.
39. Brinkkemper, J.A.; Lanckriet, T.; Grasso, F.; Puleo, J.A.; Ruessink, B.G. Observations of turbulence within the surf and swash zones of a field-scale sandy laboratory beach. *Coast. Eng.* **2015**, in press.
40. Doucette, J.S.; O'Donoghue, T. Response of sand ripples to change in oscillatory flow. *Sedimentology* **2006**, *53*, 581–596.
41. Arnott, R.W.C.; Southard, J.B. Exploratory flow-duct experiments on combined-flow bed configurations, and some implications for interpreting storm-event stratification. *J. Sediment. Petrol.* **1990**, *60*, 211–219.
42. Southard, J.B.; Lambie, J.M.; Frederico, D.C.; Pile, H.T.; Weidman, C.R. Experiments on bed configurations in fine sands under bidirectional purely oscillatory flow, and the origin of hummocky cross-stratification. *J. Sediment. Petrol.* **1990**, *60*, 1–17.

J. Mar. Sci. Eng. **2015**, *3*, 1568–1594

43. Dumas, S.; Arnott, R.W.C. Origin of hummocky and swaley cross-stratification—The controlling influence of unidirectional current strength and aggradation rate. *Geology* **2006**, *34*, 1073–1076.

44. Osborne, P.D.; Vincent, C.E. Dynamics of large and small scale bedforms on a macrotidal shoreface under shoaling and breaking waves. *Mar. Geol.* **1993**, *1993*, 207–226.

45. Perillo, M.M.; Best, J.L.; Yokokawa, M.; Sekiguchi, T.; Takagawa, T.; Garcia, M.H. A unified model for bedform developement and equilibrium under unidirectional, oscillatory and combined-flows. *Sedimentology* **2014**, *61*, 2063–2085.

46. Dubarbier, B.; Castelle, B.; Marieu, V.; Ruessink, B.G. On the modeling of sandbar formation over a steep beach profile during the large-scale wave flume experiment BARDEX II. In Proceedings of the Coastal Sediments Conference 2015, San Diego, CA, USA, 11–15 May 2015.

47. Crawford, A.M.; Hay, A.E. Linear transition ripple migration and wave orbital velocity skewness: Observations. *J. Geophys. Res.* **2001**, *106*, 14113–14128.

48. Nielsen, P. Dynamics and geometry of wave-generated ripples. *J. Geophys. Res.* **1981**, *86*, 6467–6472.

49. Nadaoka, K.; Hino, M.; Koyano, Y. Structure of the turbulent flow field under breaking waves in the surf zone. *J. Fluid Mech.* **1989**, *204*, 359–387.

50. Kimmoun, O.; Branger, H. A particle image velocimetry investigation on laboratory surf-zone breaking waves over a sloping beach. *J. Fluid Mech.* **2007**, *588*, 353–397.

51. Miller, R.L. Role of vortices in surf zone prediction: Sedimentation and wave forces. In *Beach and Nearshore Sedimentation*; Davis, R.A., Ethington, R.L., Eds.; SEPM: Tulsa, OK, USA, 1976; pp. 92–114.

52. Nadaoka, K.; Ueno, S.; Igarashi, T. Sediment suspension due to large scale eddies in the surf zone. In Proceedings of the 21st International Conference on Coastal Engineering, Torremolinos, Spain, 20–25 June 1988; ASCE: Reston, VA, USA, 1988; pp. 1646–1660.

53. Sato, S.; Homma, K.; Shibayama, T. Laboratory study on sand suspension due to breaking waves. *Coast. Eng. Japan* **1990**, *33*, 219–231.

Journal of
Marine Science and Engineering

MDPI

Article

Natural and Human-Induced Dynamics on Big Hickory Island, Florida

Tiffany M. Roberts Briggs [1,*] and Nicole Elko [2]

[1] Department of Geosciences, Florida Atlantic University, 777 Glades Road, Boca Raton, FL 33431, USA
[2] Elko Coastal Consulting, Inc., P.O. Box 1451, Folly Beach, SC 29439, USA; nelko@elkocoastal.com
* Correspondence: briggst@fau.edu; Tel.: +1-561-297-4669

Academic Editor: Gerben Ruessink
Received: 30 December 2015; Accepted: 2 February 2016; Published: 22 February 2016

Abstract: Big Hickory Island, located in Lee County along the mixed-energy west Florida coast, experiences high long-term rates of shoreline recession, with much of the erosion concentrated along the central and southern portions of the island. In 2013, approximately 86,300 cubic meters of sand from an adjacent tidal inlet to the north were placed along 457 m to restore the beach and dune system. In an effort to combat erosion, seven concrete king-pile groins with adjustable panels were constructed subsequent to the completion of the beach nourishment. Natural and human-induced dynamics of Big Hickory Island are discussed through analysis of shoreline and morphologic change using historic aerial photographs and topographic and bathymetric field surveys of the recent beach erosion mitigation project. Although much of the long-term anomalously high rates of erosion for the area are related to natural interchanges between the sand resources of the barrier islands and adjacent ebb tidal shoals, additional reduction in sand supply is a result of human-interventions updrift of Big Hickory over the last several decades. The coupled natural and anthropogenic influences are driving the coastal processes toward a different morphodynamic state than would have occurred under natural processes alone.

Keywords: shoreline change; beach erosion; beach-inlet interactions; groin stabilization

---

1. Introduction

Chronic erosion plagues many developed beachfront communities in the U.S. [1–4]. Maintaining some minimum dry-beach width is critical for storm protection and sustainability of coastal environments [5–7]. A number of engineering approaches have been used to counteract the effect of erosion by stabilizing or restoring beaches [8]. Recent studies have shown that the implementation of groins designed specifically to retain beach fill material or stabilize the shoreline have proven effective in reducing erosion and mitigating downdrift impacts [6,9–12]. Improved understanding of the influence of anthropogenic modifications on the morphodynamics of the coastal system [13–15] is critical as human impacts on these environments increase concomitantly with sea level and storminess [16,17].

Many communities have reduced long-term erosion rates with beach nourishment alone or nourishment combined with erosion control structures [16,18–20]. Recent coastal management challenges such as fewer sediment sources, higher dredging costs, and environmental impacts on nearby habitats constrain engineers to use less sand with more cost effective beach management projects [21]. For example, most communities offer public access making them eligible for public funding assistance. Privately held beaches are often not eligible for public funds, so a number of small, private U.S. communities are financing long-term beach erosion mitigation projects with minimal to no government assistance [22,23]. Cost-effectiveness is a primary goal in these projects. In addition, from an engineering/science perspective, the relatively small shoreline frontage resulting in short community-scale beach erosion mitigation projects (sometimes bordered by non-engineered beaches and influenced by nearby tidal inlets, as in this example) creates a shoreline planform in disequilibrium with adjacent beaches.

The Pelican Landing Community Association owns approximately 700 m of shoreline along Big Hickory Island, FL with the remaining shoreline frontage to the north and south owned by Lee County. Big Hickory Island is a short barrier island (<1,300 m long) in the Gulf of Mexico, bordered to the north by New Pass and to the south by Big Hickory Pass (Figure 1). North and south of Big Hickory Island are Lover's Key and Little Hickory Island, respectively. Regional longshore sediment transport is north to south [24,25]. Big Hickory Island is subjected to large shoreline fluctuations due to its short length and closely spaced adjacent tidal inlets. As a result, the island has experienced chronic long-term erosion, ranging between 2.3 to 2.9 meters per year over the last century [25].

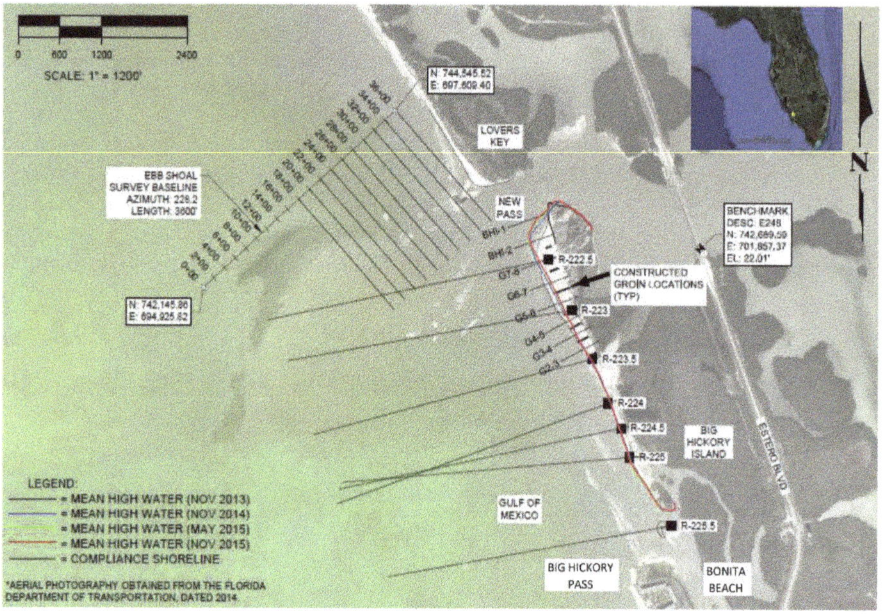

Figure 1. Big Hickory Island, FL survey plan with Google Earth image inset of Big Hickory Island location along the west Florida coast (yellow star).

As a result of beach erosion that threatened community facilities (club house and beach pavilions), the Pelican Landing Beach Restoration and Groins Project ("beach erosion mitigation project") was constructed along the central portion of the barrier island from May to October 2013. Approximately 86,300 cubic meters of sand from New Pass were placed between R-222.5 and R-224 in Lee County (Figure 1). This was followed by the construction of seven (7) concrete king-pile groins [26] with adjustable concrete panels that fit between concrete I-beam pilings (Figure 2). An advantage to king-pile groins is the ability to remove or add panels to control or tune the amount of sediment trapped by the structures. The groins are numbered 8 to 2 from north to south, shown as black shore-perpendicular lines on Figure 1.

Figure 2. Photos of the king-pile groins on Big Hickory Island, FL.

The effect of the community's decision to restore the beach using nourishment and erosion control structures is driving the island toward a different morphodynamic state than what was occurring under natural processes alone. These natural and human-induced dynamics are discussed through an analysis of shoreline and morphologic change using historic aerial photographs and recent topographic and bathymetric field surveys of the beach erosion mitigation project. Evaluation of the barrier island dynamics at multiple temporal scales is critical to elucidate both the near-term (decades to half-centuries) and short-term (events, seasons, years) patterns of morphologic change and the natural and anthropogenic drivers within the system [27,28]. It is important to understand and include the role of barrier island processes and inlet dynamics in the modeling, cost-benefit analysis, and design of similar beach preservation projects incorporating erosion control structures [3,8,29].

The objective of this paper is to evaluate the natural and human-induced dynamics of shoreline and morphologic change on Big Hickory Island in southwest Florida. Near-term changes (1944 through 2012) are analyzed using historic aerial photographs and short-term changes (2012 through 2015) using recent topographic and bathymetric field surveys.

2. Experimental Section

2.1. Near-Term Shoreline Change Analysis Methods

Near-term morphodynamics are established through an analysis of aerial photography (obtained from Lee County and the University of Florida Libraries) and available literature on historical shoreline change on Big Hickory Island [25,30] from 1944 through 2012. Historical aerial photographs (*i.e.*, prior to 1996 in this study), were georeferenced using Geographic Information System (GIS) software and identifying control points from a rectified (NAD83 State Plane FL West) 1996 image [31] prior to shoreline digitization.

The Big Hickory Island shoreline is manually determined in GIS utilizing the common proxy-indicator of the visibly discernable coastal feature of the high-water line (HWL) from aerial photography [32,33]. The HWL is identified based on the change in color tone along a sandy beach (e.g., water-saturated area due to total wave runup at the time of the flight) [32,33]. It is recognized that uncertainties and error are attributed to utilizing the visually determined proxy-based shoreline indicator of the HWL, however for determining the general trends of the morphologic state of the barrier island, this method for qualitative assessment (as opposed to quantifying shoreline recession/advance or volume change) is common and adequate [32–37].

2.2. Short-Term Shoreline Change Analysis Methods

Short-term project performance is analyzed through the following survey monitoring plan, illustrated in Figure 1:

1. MHW shoreline survey along the northern portion of Big Hickory Island,
2. Topographic and hydrographic beach profiles from R-222.5 to R-225.5 to the depth of closure,
3. Wading depth (topographic only) profiles at the centerline of each groin cell, and
4. Volumetric changes were also calculated to supplement the short-term dynamics analysis.

The survey plan utilized advances in survey-grade Global Positioning Systems (GPS) technology, coupled with Real-Time Kinematic baseline processing (RTK-GPS). Greater spatial coverage can be achieved by collecting data continuously along the shoreline (number 1 above) in conjunction with a traditional beach profile cross-section survey (number 2 above).

MHW shoreline data and topographic beach profile data were surveyed using hand-held Trimble RTK-GPS rovers. The MHW data were collected by walking along the HWL, described in Section 2.1. Topographic and bathymetric surveys were collected at six beach profiles (R-222.5 through R225.5) and wading depth (topographic only) surveys were collected at the center line of each groin cell and at two additional locations to the north of the groin field (Figure 1). The bathymetric portion of the survey was collected using a boat equipped with Trimble R8, Hypack 2014, and a 456 Innerspace single beam echo sounder with a side-mounted transducer. The wading depth profiles (topographic portion only) within and north of the groin field extended approximately 1.2 to 1.5 meters water depth NAVD (or about wading depth during surveying).

The profiles were measured along the same azimuth and commenced at a Florida Department of Environmental Protection (FDEP) R-monument extending seaward to the short-term depth of closure (NAD83 State Plane FL West). The topographic portion of the beach profile extended seaward to a point overlapping the bathymetric component of the survey. Topographic elevation measurements were collected at 7.6 meter (25 feet) intervals or at significant changes in beach slope. The data were collected according to the state monitoring standards for beach erosion control projects [38].

The analysis of data collected through this survey plan relies primarily on shoreline position data for two reasons: (1) for comparison to the near-term shoreline change analysis described above; and (2) to capture high spatial resolution alongshore variability within the project area. Complex spatial changes, such as erosional hot-spots [5] and beach response adjacent to engineering structures, are often not captured in a series of widely-spaced (e.g., 300-m) beach profiles [6].

The volumetric changes were determined by the average end area method [39]: $V = L\left(\dfrac{A_1 + A_2}{2}\right)$, where V is the volume, A_1 and A_2 are areas of cross-section (assuming each station is a trapezoid) and L is the distance between stations. Volume changes represent the difference in quantity of sand measured between the FDEP R-monument (generally landward of the dunes) and the short-term depth of closure. The short-term depth of closure is defined as the seaward limit of active sediment transport across a beach profile, beyond which negligible sediment transport is presumed to occur [40]. All volumetric changes are in cubic meters.

3. Results

The following sections present data evaluating the natural and human-induced dynamics of shoreline and morphologic change on Big Hickory Island in southwest Florida. Near-term changes (1944 through 2012) are analyzed through shoreline change using historic aerial photographs and short-term changes (2012 through 2015) are analyzed through both shoreline and volumetric change using recent topographic and bathymetric field surveys.

3.1. Near-Term Shoreline Change

The long-term evolution of Big Hickory Island suggests that the island has been highly migratory since the late 1800s [30]. A major change in the overall barrier island morphology occurred between 1885 and 1927, when the island shortened and widened as the inlet to the north (Little Carlos Pass, approximately 1.8 km north of present-day New Pass [41] substantially migrated south and Big Hickory Pass (to the south) migrated north. In addition to larger-scale drivers of change, such as the global acceleration of sea level rise [42], the natural hydrodynamic interactions between the barrier and its bounding inlets and the dynamics of the adjacent barrier islands were the primary localized drivers of morphologic change at that time (*i.e.*, little to no human-induced change).

Evaluation of decadal trends in the near-term evolution of Big Hickory Island reveals an island continuing to exhibit unstable shoreline conditions and large morphologic variations throughout the 1900s and into the 2000s. In 1944, Big Hickory Island was an elongated barrier with a well-developed channel separating the barrier from the vegetated island to the east (landward of the barrier), similar to the general morphologic conditions of 1927. Between 1944 and 1958, overall landward migration of the barrier island, likely through overwash processes associated with the passage of several hurricanes [41], closed the channel passage along the bayside of the barrier and connected it to the vegetated landmass to the east (Figure 3). New Pass, bounding Big Hickory Island to the north, had become a well-developed and dominant inlet [41]. Shoreline recession at the south tip contributed to an overall shortening of the barrier. However, the most substantial change occurred along the northern end of the barrier island, which not only retreated south but also recurved landward closing the northern extent of the 1944 backbarrier channel. By 1958, development had begun on the barrier island to the south (Little Hickory Island). Any unvegetated areas or sediment shoals to the north of the island in 1944 disappeared by 1958. In addition, significant sediment accumulation along the northern tip of Little Hickory Island appears to have occurred during this time. This period denotes the introduction of human-induced changes to the natural barrier island system.

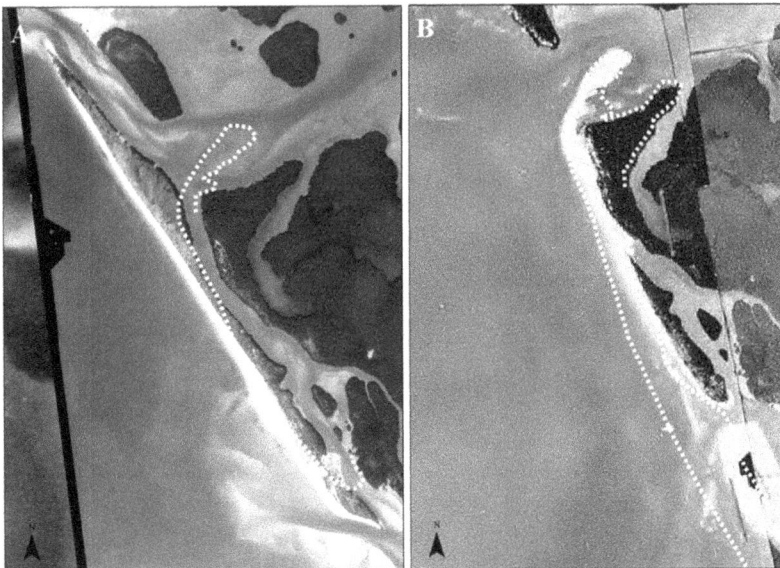

Figure 3. Morphologic changes of Big Hickory Island. (**A**) 1944 aerial photograph with the approximate 1958 shoreline shown as a white dashed line; (**B**) 1958 aerial photograph with the approximate 1980 shoreline shown as a white dashed line.

Between 1958 and 1980, continued morphologic change occurred in conjunction with significant anthropogenic activities within the area. Between 1958 and 1965, a coastal causeway was constructed connecting Estero Island (to the north) to Little Hickory Island, influencing the hydrodynamics of several tidal channels within the area [31]. Northward sediment transport resulted in the closure of Big Hickory Pass, consequentially connecting Big Hickory Island and Little Hickory Island. Despite anthropogenically reopening the inlet in 1976 [30], by 1980 the inlet was again infilled by northward longshore sediment transport (Figure 4). Remnants of the dredged 1976 inlet are apparent from the 1980 aerial image. The closure of this inlet allowed for significant quantities of northward transported sediment to naturally supplement the beaches along Big Hickory Island. The apparent northward longshore sediment transport represents a localized reversal in the regional north to south longshore sediment transport patterns [24,25]. Figure 4 shows a comparison of a 1970 aerial photograph to the 1980 shoreline illustrating the widening of the beach as a result of the closure of the southern inlet [25].

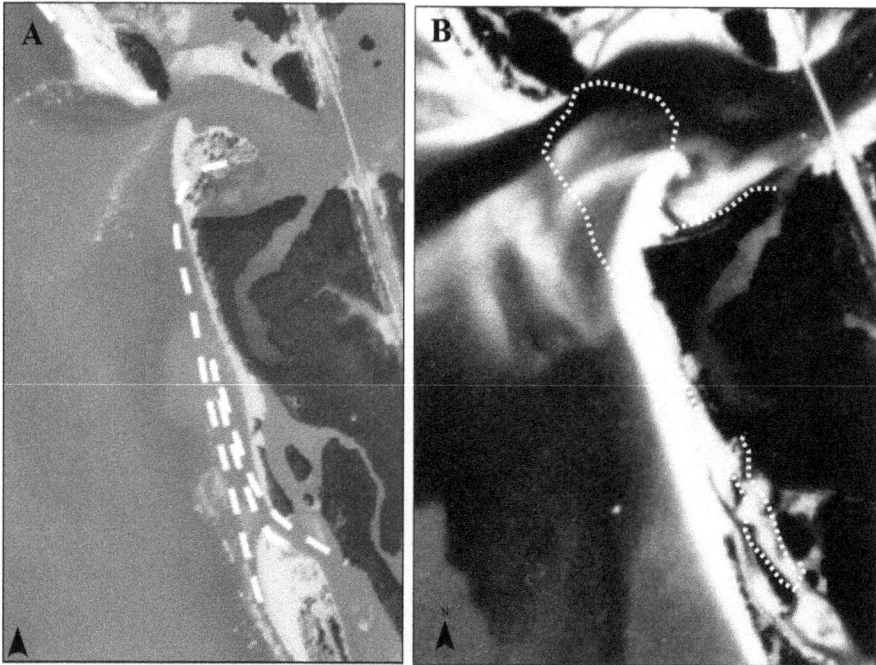

Figure 4. Morphologic changes of Big Hickory Island. (**A**) 1970 aerial photograph with the approximate 1980 shoreline shown as a white dashed line (*from* [25] with permission from Pelican Landing Community Association); (**B**) 1980 aerial photograph with the approximate 1996 shoreline shown as a white dashed line.

In November 1995, Big Hickory Pass (south channel-side) was stabilized in an open configuration with two terminal rock groins at the north end of Little Hickory Island [43] to prevent infilling by northward transported sediment. Due to the predominant northward longshore sediment transport, sediment was depleted from the southern portion of Big Hickory Island, resulting in shoreline recession along the southern extent of the island and northward shoreline advance from sediment accumulation along the northern tip of the island (Figure 4). It is evident that the groin structures along Bonita Beach (south of Big Hickory Island) had a significant impact on the morphology and sediment supply of Big Hickory Island. It also appears that when Big Hickory Pass is open, Big Hickory Island will erode due to a reduced sediment supply.

Following the stabilization of Little Hickory Pass south of Big Hickory Island, sediment supply to the island was diminished with little mechanism for sediment by-passing. As a result, the barrier island began to migrate landward, with shoreline retreat observed between 1996 and 2005 (Figure 5). During this time, community facilities were permitted and constructed on Big Hickory Island. The trend of shoreline recession continued through 2012, with rapid shoreline retreat along the northern portion of Big Hickory Island (Figure 5). The south inlet's northern ebb tidal delta appears to have equilibrated after groin construction and started contributing sediment to the southern portion of Big Hickory Island, evident from the slight shoreline advance along this section of the island. However, overall the island appears to be in a state of severe sediment depletion as evidenced by the erosive trends exhibited leading up to the 2012 morphologic state.

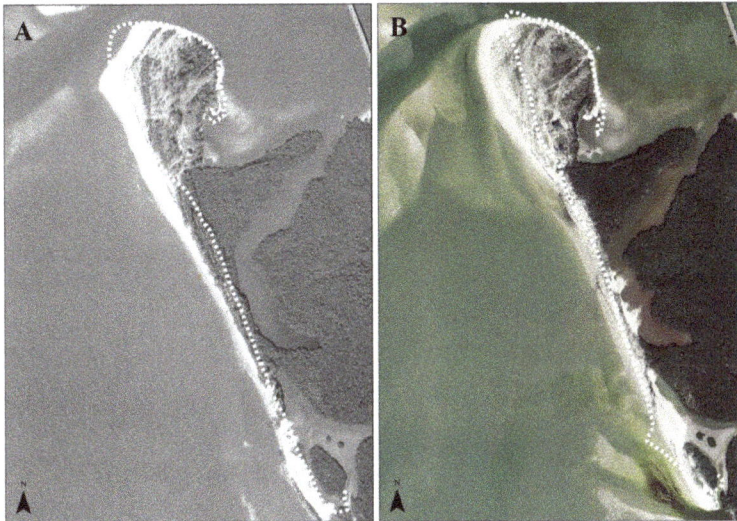

Figure 5. Morphologic changes of Big Hickory Island. (**A**) 1996 aerial photograph with the approximate 2005 shoreline shown as a white dashed line; (**B**) 2005 aerial photograph with the approximate 2012 shoreline shown as a white dashed line.

Near-term evaluation of aerial photographs illustrates the dramatic morphologic changes of Big Hickory Island between 1944 and 2012. Natural processes associated with the hydrodynamic fluctuations of nearby inlets and event-driven changes resulting from the passages of storms dominated the morphodynamics of Big Hickory Island until the late 1950s. The 1960 and 1970s represent the temporal shift from natural processes to human-induced changes dominating the barrier island system. Throughout the last two decades, shoreline-stabilization structures updrift (south) of Big Hickory Island (and removal of sediment for nearby beach nourishment projects [25]) resulted in a significant deficit of sediment input onto the barrier island, causing marked barrier island retrogradation by 2012.

3.2. Short-Term Shoreline Change

Between 2012 and early 2013, the north end of Big Hickory Island continued to retreat landward, as illustrated by the Mean High Water (MHW) change (Figure 6, red and orange lines). Beach nourishment and groin construction were implemented in mid- to late-2013 in response to the rapid erosion occurring on Big Hickory Island (Figure 6, yellow line). Nourishment sediment spreading is evident with advancement of the island shoreline to the north during the two years post-construction (Figure 6, purple line). By late 2015, a new equilibrium shoreline location is emerging along the groin field. Detailed evaluation of the MHW and volumetric changes across Big Hickory Island between 2012 and 2015 provides information on the short-term morphodynamics on Big Hickory Island in response to the most recent anthropogenic influences (nourishment and groin placement), suggesting a trend toward a new barrier island dynamic equilibrium state [44] that is more consistent with the 2005 state (Figure 5B).

Tabulated annual shoreline change from construction completion (November 2013) to November 2015 at the FDEP R-monuments is given in Table 1. The average shoreline change during the first two years after project construction, from November 2013 to November 2015, was a landward movement of 5.4 ($\pm$ 12.2) meters. The large standard deviation (σ) implies significant alongshore variability. The greatest shoreline change in the project area occurred at R223, which represents roughly the center of the beach nourishment perturbation. Note that no standard R-monuments exist north of the project area in the volatile region adjacent to New Pass. Shoreline change immediately south of the project area was negative (representing landward change or erosion); whereas, change along southern Big Hickory Island, adjacent to Big Hickory Pass was positive or accretional (Figure 6).

Shoreline change within the groin field since construction (November 2013 to November 2015) was on average 15.7 ($\pm$ 4.2) meters landward. This change is visualized in Figure 6, illustrating that despite the substantial shoreline retreat, the 2015 shoreline position is seaward of the pre-nourishment shoreline position.

When shoreline change is calculated for all wading depth profiles, including BHI-1 and BHI-2, the total shoreline change averaged only 2.8 ($\pm$ 8.9) meters landward between November 2013 and November 2015. As suggested by the high σ, the shoreline location moved 47.9 m seaward at BHI-2 during this time period (Table 2), representing significant spreading of nourished sediment to the north. As expected [45], the greatest landward shoreline movement occurred in the G5-6 groin cell, which is located close to the center of the beach erosion mitigation project.

Figure 6. Morphologic changes of Big Hickory Island between 2012 and 2015, represented by the Mean High Water (MHW) line surveyed bi-annually, shown on a 2012 aerial photograph.

Table 1. MHW shoreline positions measured at the Florida Department of Environmental Protection (FDEP) R-monuments and changes from construction completion (November 2013) to November 2015. Shaded rows represent the project area.

MON	MHW Position	MHW Position	MHW Position	Total Change
	2013-Nov (m)	2014-Nov (m)	2015-Nov (m)	Nov 2013–Nov 2015 (m)
R222.5	46.0	54.1	41.9	−4.1
R223	58.4	39.3	29.8	−28.6
R223.5	7.4	12.4	7.9	0.5
R224	−7.1	−10.1	−11.7	−4.6
R224.5	8.6	8.1	5.7	−2.9
R225	−1.6	-3.0	5.9	7.5
Average (Standard Deviation, σ)				−5.4 (12.2)

MHW shoreline changes for both the R-monument beach profile surveys (Table 1) and the groin profile surveys (Table 2) are summarized in Figure 7. Note that Figure 7 does not represent shoreline position (*i.e.*, not a planform or a map). Overall shoreline change after construction followed a typical planform spreading signature [45] of landward shoreline movement in the center of the nourished area and shoreline advancement to the north and south, with considerably more advancement to the north, the direction of longshore sediment transport. Shoreline change stabilized (*i.e.*, near zero change) in the vicinity of groins 2, 3, 4, and 5 during the second year after construction; whereas, the pattern of spreading continued along the northern project area with spreading to the north.

Table 2. MHW shoreline positions* measured at within each groin cell and changes from construction completion (November 2013) to November 2015. *MHW positions measured from the MHW survey plan view (Figure 1).

Groin Line	MHW Position	MHW Position	MHW Position	Total Change
	2013-Nov (m)	2014-Nov (m)	2015-Nov (m)	Nov 2013–Nov 2015 (m)
G2-3	68.8	70.4	69.9	1.1
G3-4	79.6	73.0	72.0	−7.6
G4-5	88.8	70.0	69.4	−19.4
G5-6	110.0	87.9	77.0	−33.0
G6-7	117.3	99.2	88.3	−29.0
G7-8	104.4	109.7	98.4	−6.1
BHI-1*	−9.2	−4.8	27.0	36.2
BHI-2*	61.1	107.7	109.0	47.9
Average change within the groin field (σ)				−15.7 (4.2)
Average change including north end (σ)				−2.8 (8.9)

Based on the shoreline change performance in the vicinity of R223.5 to G4-5, the groins have stabilized shoreline changes two years after project construction. The data suggest that the groins will serve to stabilize shoreline changes north of G4-5 to R222.5 once the nourished sediment is distributed outside of the project area. Provided periodic renourishment, the island should reach a new dynamic equilibrium shoreline position controlled by the groins that is farther seaward than the pre-nourishment shoreline position. Without periodic renourishment, the groin field may have adverse impacts on the downdrift shoreline located to the north of the project area.

3.3. Short-Term Volumetric Change

Tabulated volumetric changes calculated from construction completion (November 2013) to November 2015 are given in Table 3. The volumetric analysis is limited to the R-monument surveys because they extend to the depth of closure and capture all volume change across the profile. However, these monument surveys are spaced at roughly 150-m alongshore; therefore, high-resolution changes within the groin field are not analyzed in detail in this section. The volumetric change analysis supports the findings in sections 3.1 and 3.2.

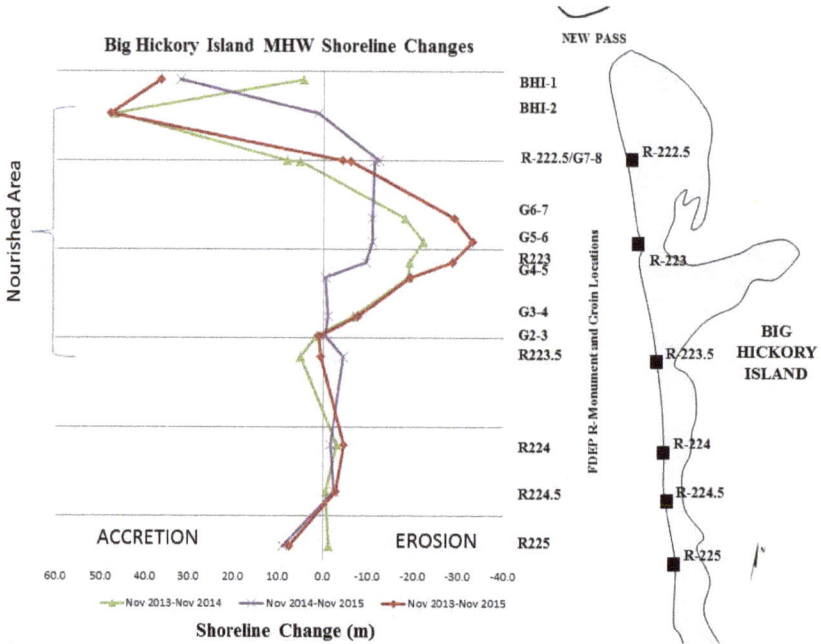

Figure 7. Time-series Mean High Water (MHW) shoreline changes at FDEP R-monuments and groin locations (Nov 2013–Nov 2014; Nov 2014–Nov 2015; Nov 2013–Nov 2015). Figure 6 illustrates shoreline position, whereas this figure quantifies the change.

A total of 7132 yd^3 of sediment volume change (gain) was measured across the Big Hickory Island study area from R222.5 to R225.5 from November 2013 to November 2015. Based on R-monument calculations, the nourished area was erosional, while the area located to the south (from R224 to R225) was accretional. This corresponds to the planform spreading pattern noted in the previous section.

Relatively low shoreline change statistics south of the project area (Figure 7) and high volume change data in this area suggest that much of the sediment has accumulated below mean high water. The volumetric change data indicate that the groins have considerably reduced post-nourishment sediment volume losses.

Table 3. Volumetric change measured at FDEP R-monuments from Nov 2013–Nov 2014, Nov 2014–Nov 2015, and Total (*i.e.*, Nov 2013–Nov 2015).

MON	Volume (m^3)		
	Nov 2013–Nov 2014	Nov 2014–Nov 2015	Total
R222.5			
	−2085	−5778	−7863
R223			
	1926	−6603	−4678
R223.5			
	9568	−2027	7542
R224			
	4395	96	4491
R224.5			
	4251	1710	5960
R225			

Volumetric changes for the R-monument surveys (Tables 1 and 3) are summarized in Figure 8. Volume loss was measured within the project area and accretion was observed to the south. Photographic and field observations, as well as shoreline change measurements, indicate substantial accretion to the north of the project area. As noted above, this is the expected post-nourishment volume change response. The positive volume change statistics suggest good beach erosion mitigation project performance.

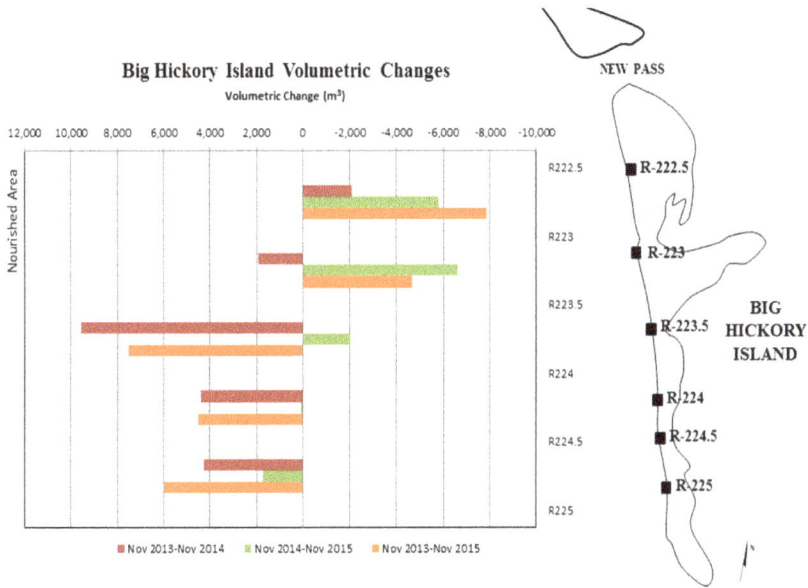

Figure 8. Time-series volumetric changes on Big Hickory Island, corresponding to FDEP R-monuments (Nov 2013–Nov 2014, Nov 2014–Nov 2015, and Nov 2013–Nov 2015).

4. Discussion

The near-term and short-term analysis of the shoreline and volume change (representing overall barrier island morphology) of Big Hickory Island suggest that coupled natural and anthropogenic influences are driving the coastal processes toward a different morphodynamic state than would have occurred under natural processes alone. The initial shift from a naturally influenced barrier island to a combined natural-human influenced barrier island system occurred in the mid-1900s with infrastructure development commencing on nearby islands. Closure of Big Hickory Pass (to the south) by 1980 allowed the island to begin morphologic recovery as natural sediment bypassing resumed. However, subsequent erosion mitigation efforts on Little Hickory Island to structurally maintain Big Hickory Pass (in an open position) resulted in severe erosion on Big Hickory Island, similar to the eroded conditions observed in the 1970s. Continued shoreline retreat was observed through the 1990s and 2000s.

Construction of community infrastructure in the mid-2000s was the first direct anthropogenic activity on Big Hickory Island. Subsequent impacts to the island were from human influences on nearby barriers and inlets. In 2013, severe erosion prompted the private landowners (Pelican Landing Community Association) to implement and self-finance a beach erosion mitigation project, consisting of the construction of king-pile groins combined with beach nourishment.

The groin field was constructed within the central region of the barrier island with undeveloped (and unmanaged) shoreline on either side of the project. To date, the groins have stabilized the central shoreline of Big Hickory Island in a more seaward position, similar to the 2005 shoreline location (Figure 9). A conceptual model illustrates the observed lateral spreading of the substantial sediment volume added by the beach nourishment, which has resulted in shoreline progradation of the northern tip of Big Hickory Island and volume gain to the south. However, the volume gain to the south has not resulted in shoreline progradation (Figure 9). The 2005 aerial photo used in Figure 9 illustrates that the project is stabilizing the shoreline to near-2005 conditions. The project is functioning as designed, thus Figure 9 represents both realized and idealized project performance.

Figure 9. Conceptual model of the post-project generalized directions of sediment movement along Big Hickory Island, suggested from observations in this study.

Big Hickory Island is approaching a new dynamic equilibrium state, in response to the recent direct anthropogenic erosion mitigation efforts on the barrier island and natural coastal processes within the region. It is recognized that any new significant perturbation to the system (e.g., human-implemented coastal construction or natural storm impacts) will disrupt the current trajectory of barrier island equilibration.

As noted, little sediment has been transported to Big Hickory Island naturally since the installation of the terminal groins at Big Hickory Pass to the south (at Bonita Beach). Thus, periodic renourishment is essential. Provided continued renourishment, the barrier island should reach a new dynamic equilibrium controlled by the groins. Without periodic renourishment, the groin field may have adverse impacts on the beach located to the north of the project area. Continued monitoring will determine the effects that the newly placed groin field will have on the northern extent of the island; however to date, no adverse impacts have been observed. Results from continued monitoring will help the planning, cost-benefit analysis, and design of similar projects. This privately-funded project, in the middle of a barrier island in a relatively low-energy setting, provides coastal managers and engineers an opportunity to evaluate alternative erosion mitigation strategies.

With continued direct placement of sediment to supplement the shoreline stabilization efforts of the king pile groins, Big Hickory Island may reach a more stable morphodynamic state as compared to the last several decades of severe erosion and retrogradation due to diminished sediment input. Because of limited sediment transport to the barrier through natural processes of inlet bypassing (at Big Hickory Pass) due to the groin structures on the updrift adjacent barrier island (Little Hickory Island), anthropogenically-introduced sediment input into the barrier island system is critical to the longevity of the shoreline stability of Big Hickory Island. However, as sea-level rise [42] potentially couples with increased storminess [46], amplified rates of coastal erosion will likely require a reevaluation of the amount of sediment needed to maintain the stability not only of Big Hickory Island, but barrier islands worldwide.

Acknowledgments: Funding was provided by the Pelican Landing Community Association (PLCA). Special thanks to Marie Martel, PLCA General Manager, and Tom Schemenaur. The authors wish to thank Dr. Robert Dean for his insightful observations on island morphodynamics and Doug Mann for his engineering expertise and contributions to the construction of the beach erosion mitigation project.

Author Contributions: N.E. conceived and implemented the fieldwork involved in the project. N.E. and T.R.B. conducted data analysis and interpretation. T.R.B. and N.E. contributed to the manuscript preparations and revisions.

Conflicts of Interest: The authors declare no conflict of interest.

References

1. Bird, E.C. The modern prevalence of beach erosion. *Mar. Pollut. Bull.* **1987**, *18*, 151–157. [CrossRef]
2. Morton, R.A.; McKenna, K.K. Analysis and projection of erosion hazard areas in Brazoria and Galveston Counties, Texas. *J. Coast. Res.* **1999**, *28*, 106–120.
3. Kraus, N.C.; Galgano, F.A. *Beach Erosional Hot Spots: Types, Causes, Solutions*; Engineer Research and Development Center: Vicksburg, MI, USA, 2001; pp. 1–17. Available online: http://citeseerx.ist.psu.edu/viewdoc/download?doi=10.1.1.492.7258&rep=rep1&type=pdf (accessed on 14 February 2016).
4. Elko, N.A.; Mann, D.W. Implementation of geotextile t-groins in Pinellas County, Florida. *Shore Beach* **2007**, *75*, 2–10.
5. CEM. *Coastal Engineering Manual, Part V, Chapter 3*; U.S. Army Corps of Engineers: Washington, DC, USA, 2002.
6. Basco, D.R.; Pope, J. Groin functional design guidance from the Coastal Engineering Manual. *J. Coast. Res.* **2004**, *33*, 121–130.
7. Van Rijn, L. Coastal erosion and control. *Ocean Coast. Manag.* **2011**, *54*, 867–887. [CrossRef]
8. National Research Council. *Beach Nourishment and Protection, Committee on Beach Nourishment and Protection*; National Academies Press: Washington, DC, USA, 1995; p. 352. Available online: http://www.nap.edu/catalog/4984.html (accessed on 27 January 2016).
9. Kraus, N.C.; Hanson, H.; Blomgren, S. Modern functional design of groins systems. In Proceedings 24th Coastal Engineering Conference; ASCE: New York, NY, USA, 1994; pp. 1327–1342.
10. Donohue, K.A.; Bocamazo, L.M.; Dvorak, D. Experience with groin notching along the Northern New Jersey Coast. *J. Coast. Res.* **2004**, *33*, 198–214.

11. Galgano, F.A. Long-term effectiveness of a groin and beach fill system: A case study using shoreline change maps. *J. Coast. Res.* **2004**, *33*, 3–18.
12. Traynum, S.B.; Kana, T.W.; Simms, D.R. Construction and performance of six template groins at Hunting Island, South Carolina. *Shore Beach* **2010**, *78*, 21–32.
13. Nordstrom, K.F. Beaches and dunes of human-altered coasts. *Progr. Phys. Geogr.* **1994**, *18*, 497–516. [CrossRef]
14. Peterson, C.H.; Bishop, M.J. Assessing the Environmental Impacts of Beach Nourishment. *BioScience* **2005**, *55*, 887–896. [CrossRef]
15. Roberts, T.; Wang, P. Four-year performance and associated controlling factors of several beach nourishment projects along three adjacent barrier islands, west-central Florida, USA. *Coast. Eng.* **2012**, *70*, 21–39. [CrossRef]
16. Rahmstorf, S. A semi-empirical approach to projecting future sea-level rise. *Science* **2007**, *315*, 368–370. [CrossRef] [PubMed]
17. Houston, J.; Dean, R. Erosional impacts of modified inlets, beach encroachment, and beach nourishment on the east coast of Florida. *J. Coast. Res.* **2015**. in press. [CrossRef]
18. Browder, A.E.; Dean, R.G. Monitoring and comparison to predictive models of the Perdido Key beach nourishment project, Florida, USA. *Coast. Eng.* **2000**, *39*, 173–191. [CrossRef]
19. Davis, R.J.; Wang, P.; Silverman, B. Comparison of the performance of the three adjacent and differently constructed beach nourishment projects on the gulf peninsula of Florida. *J. Coast. Res.* **2000**, *16*, 396–407.
20. Elko, N.A.; Holman, R.A.; Gelfenbaum, G. Quantifying the rapid evolution of a nourishment project with video imagery. *J. Coast. Res.* **2005**, *21*, 633–645. [CrossRef]
21. American Shore and Beach Preservation Association (ASBPA). Reintroducing coastal structures for erosion control on the open coasts of America. *Shore Beach* **2011**, *79*, 62–67.
22. Giannio, S.; Stevens, R.; Watts, G. Local financing for beach nourishment at Captiva Island, Florida. In *Coastal Zone*; Magoon, O., Converse, H., Miner, D., Clark, D., Tobin, L., Eds.; American Society of Civil Engineers: New York, NY, USA, 1985; Volume 85, pp. 2154–2170.
23. Rudeen, K. *Analysis of Proposed DeBordieu Groin and Beach Nourishment Project*; Western Carolina University: Cullowhee, NC, USA, 2011; pp. 1–14.
24. Walton, T.L. *Littoral Drift Estimates along the Coastline of Florida*; Sea Grant Report No. 13; Coastal and Oceanographic Engineering Laboratory, University of Florida: Gainesville, FL, USA, 1976.
25. Dean, R.G. *Investigation of Erosional Causes and Option for Big Hickory Island*; Robert G. Dean: Gainesville, FL, USA, 2008.
26. Dean, R.G. Compatibility of borrow material for beach fills. In Proceedings of the 14th Coastal Engineering Conference; American Society of Civil Engineers: New York, NY, USA, 1975; pp. 1319–1333.
27. Morton, R.A.; Miller, T.; Moore, L. Historical shoreline changes along the US Gulf of Mexico: A summary of recent shoreline comparisons and analyses. *J. Coast. Res.* **2005**, *21*, 704–709. [CrossRef]
28. Lentz, E.E.; Hapke, C.J.; Stockdon, H.F.; Hehre, R.E. Improving understanding of near-term barrier island evolution through multi-decadal assessment of morphologic change. *Mar. Geol.* **2013**, *337*, 125–139. [CrossRef]
29. Weathers, H.D.; Voulgaris, G. Evaluation of beach nourishment evolution models using data from two South Carolina, USA beaches: Folly Beach and Hunting Island. *J. Coast. Res.* **2013**, *69*, 84–89. [CrossRef]
30. Davis, R.A. Barriers of the Florida Gulf Peninsula. In *Geology of the Holocene Barrier Island Systems*; Davis, R.A., Ed.; Springer-Verlag: Berlin, Germany, 1994; p. 456.
31. Hiland, M.; Byrnes, M.; McBride, R.; Jones, F. Change analysis and spatial information management for coastal environments. *MicroStation Manag.* **1993**, *3*, 58–61.
32. Gorman, L.; Morang, A.; Larson, R. Monitoring the coastal environment; Part IV: Mapping, shoreline changes, and bathymetric analysis. *J. Coast. Res.* **1998**, *14*, 61–92.
33. Boak, E.H.; Turner, I.L. Shoreline definition and detection: A review. *J. Coast. Res.* **2005**, *21*, 688–703. [CrossRef]
34. Smith, G.L.; Zarillo, G.A. Calculating long-term shoreline recession rates using aerial photographic and beach profiling techniques. *J. Coast. Res.* **1990**, *6*, 111–120.
35. Crowell, M.; Leatherman, S.; Buckley, M. Historical shoreline change: Error analysis and mapping accuracy. *J. Coast. Res.* **1991**, *7*, 839–852.
36. Moore, L.J. Shoreline mapping techniques. *J. Coast. Res.* **2000**, *16*, 111–124.

37. Moore, L.J.; Ruggiero, P.; List, J.H. Comparing mean high water and high water line shorelines: Should proxy-datum offsets be incorporated into shoreline change analysis? *J. Coast. Res.* **2006**, *22*, 894–905. [CrossRef]

38. Florida Department of Environmental Protection (FDEP). Monitoring Standards for Beach Erosion Control Projects. 2014; Retrieved from Division of Water Resources, FDEP. Available online: http://www.dep.state.fl.us/beaches/publications/pdf/PhysicalMonitoringStandards.pdf (accessed on 19 December 2015).

39. Dean, R.G.; Dalrymple, R.A. Field techniques and analysis. In *Coastal Processes: With Engineering Applications*; Davis, R.G., Dalrymple, R.A., Eds.; Cambridge University Press: Cambridge, UK, 2002; p. 47.

40. Kraus, N.C.; Larson, M.; Wise, R.A. *Depth of Closure in Beach-Fill Design, Coastal Engineering Technical Note CHTN II-40*; U.S. Army Engineer Waterways Experiment Station, Coastal and Hydraulics Laboratory: Vicksburg, MI, USA, 1998.

41. Jones, C. *Big Hickory Pass, New Pass, and Big Carlos Pass Glossary of Inlets*; Report No.8; Department of Coastal and Oceanographic Engineering, University of Florida: Gainesville, FL, USA, 1980.

42. Church, J.A.; White, N.J. Sea-level rise from the late 19th to the early 21st Century. *Surv. Geophys.* **2011**, *32*, 585–602. [CrossRef]

43. Olsen Associates Inc. *Bonita Beach—Beach Restoration Project, Six-Months Post Construction Monitoring Report*; Bureau of Beaches and Coastal Systems, Florida Department of Environmental Protection, and the Lee County Board of County Commissioners: Jacksonville, FL, USA, 1996.

44. Dean, R. Cross-shore sediment transport processes. In *Advances in Coastal and Ocean Engineering*; Philip Liu, L., Ed.; World Scientific: Singapore, Singapore, 1995; pp. 159–220.

45. Dean, R.G. Beach nourishment, theory and practice. In *Advanced Series on Ocean Engineering*; World Scientific: Singapore, Singapore, 2002; Volume 18, p. 399.

46. Slott, J.M.; Murray, A.B.; Ashton, A.D.; Crowley, T.J. Coastline responses to changing storm patterns. *Geophys. Res. Lett.* **2006**, *33*, L18404. [CrossRef]

Journal of
Marine Science and Engineering

MDPI

Article

On the Intersite Variability in Inter-Annual Nearshore Sandbar Cycles

Dirk-Jan R. Walstra [1,2,*], Daan A. Wesselman [3,†], Eveline C. van der Deijl [3,†] and Gerben Ruessink [3,†]

1   Marine and Coastal Systems, Deltares, PO Box 177, 2600 MH, Delft, The Netherlands
2   Hydraulic Engineering Section, Faculty of Civil Engineering and Geosciences,
     Delft University of Technology, PO Box 5048, 2600 GA, Delft, The Netherlands
3   Department of Physical Geography, Faculty of Geosciences, Institute for Marine and Atmospheric Research,
     Utrecht University, PO Box 80115, 3508 TC, Utrecht, The Netherlands; D.A.Wesselman@uu.nl (D.A.W.);
     E.C.vanderDeijl@uu.nl (E.C.D.); B.G.Ruessink@uu.nl (G.R.)
*   Correspondence: DirkJan.Walstra@Deltares.nl; Tel.: +31-88-335-8287; Fax: +31-88-335-8582
†   These authors contributed equally to this work.

Academic Editor: Dong-Sheng Jeng
Received: 4 December 2015; Accepted: 3 February 2016; Published: 25 February 2016

Abstract: Inter-annual bar dynamics may vary considerably across sites with very similar environmental settings. In particular, the variability of the bar cycle return period (T_r) may differ by a factor of 3 to 4. To date, data studies are only partially successful in explaining differences in T_r, establishing at best weak correlations to local environmental characteristics. Here, we use a process-based forward model to investigate the non-linear interactions between the hydrodynamic forcing and the morphodynamic profile response for two sites along the Dutch coast (Noordwijk and Egmond) that despite strong similarity in environmental conditions exhibit distinctly different T_r values. Our exploratory modeling enables a consistent investigation of the role of specific parameters at a level of detail that cannot be achieved from observations alone, and provides insights into the mechanisms that govern T_r. The results reveal that the bed slope in the barred zone is the most important parameter governing T_r. As a bar migrates further offshore, a steeper slope results in a stronger relative increase in the water depth above the bar crest which reduces wave breaking and in turn reduces the offshore migration rate. The deceleration of the offshore migration rate as the bar moves to deeper water—the morphodynamic feedback loop—contrasts with the initial enhanced offshore migration behavior of the bar. The initial behavior is determined by the intense wave breaking associated with the steeper profile slope. This explains the counter-intuitive observations at Egmond where T_r is significantly longer than at Noordwijk despite Egmond having the more energetic wave climate which typically reduces T_r.

Keywords: morphodynamic feedback loop; Egmond; Noordwijk; inter-annual bar dynamics; process based modeling; Unibest-TC; sandbars; bar switch; morphodynamic modeling; cyclic bar behavior; Jarkus

1. Introduction

Alongshore sand bars are common features in shallow nearshore coastal environments (water depth typically less than 10 m) with a striking variability in the cross-shore and longshore geometry (e.g., [1–4]). Bars are the net result of cross-shore sediment accumulation resulting from the highly non-linear morphological feedback between the bed profile and nearshore hydrodynamics (e.g., [2,5]). As bars may also influence upper beach morphology [6–8] and are often altered by shoreface nourishments (e.g., [9–11]), their relevance for coastal managers is evident.

The behavior of (multiple) bar systems has been studied extensively over the past decades. These studies focused on bar behavior at time scales ranging from hours, days and weeks (e.g., [7,12,13]), via months and seasons (e.g., [14–16]) to years and decades (e.g., [5,8,17–21]). Common findings are that bars mostly have a multi-annual lifetime and that up to five bars can occur simultaneously in the cross-shore. As the most seaward (outer) bar limits the amount of wave energy by enforcing waves to break, it controls the evolution of the shoreward located (inner) bars [13,21,22]. Decay of the outer bar typically initiates a cascaded response in which the next (shoreward) bar experiences amplitude growth and net seaward migration. This in turn creates accommodation space for its shoreward neighbor and so on, eventually resulting in the generation of a new bar near the shoreline. This offshore directed cyclic character is typically measured by the period between two bar decay events, referred to as the bar cycle return period (T_r).

This T_r can vary markedly at a site and between sites, but the underlying reasons and environmental controls are not well understood [3,20,23–25]. Intra-site differences in T_r are typically related to (quasi) persistent three-dimensional bar behavior referred to as bar switching (e.g., [5,18,26]). It is defined as bars being alongshore discontinuous, either in a different phase of the bar cycle [5] or with a completely different T_r [18,23]. For the latter case, intra-site differences in T_r can be substantial (exceeding a factor 4) and appears to be continuously present in time [18], here referred to as a persistent bar switch. Bar switches that separate sections with similar T_r are usually less persistent as alongshore interactions cause bar switches to disappear when the adjacent bars temporarily are in a similar phase [5], here referred to as a non-persistent bar switch.

Wijnberg and Terwindt [18] were among the first to study the inter-site differences in T_r. To that end they introduced the concept of a large-scale coastal behavior (LSCB) region. It is defined as an area in which the sandbars show similar cross-shore migration (*i.e.*, approximately constant T_r) and exhibit comparable changes in bar morphology over several decades. For the Holland coast (Figure 1) the annual surveys of the coastal profiles (Jarkus database) revealed that the transitions between LSCB regions were primarily persistent bar switches. In general, the transitions between LSCB regions were relatively distinct and of limited alongshore length (about 2 km). One of the most prominent differences in T_r was found between the area northward of the IJmuiden harbor moles to the Petten Seawall and the area southward of IJmuiden to the harbor moles of Scheveningen (see Figure 1). The overall inter-annual bar cycle characteristics are similar for both areas. However, the T_r differ significantly: in the southern area the return period is much smaller (about 4 *versus* 15 years for the area northwards of IJmuiden). In addition, the alongshore coherence in offshore bar movement seems to be larger in the southern region [18], that is, there are less non-persistent bar switches.

For the Holland coast, Wijnberg [24] found that changes in decadal coastal behavior were primarily coupled to large man-made structures and alongshore changes in the offshore bathymetry (ebb delta and shoreface terrace). No link could be established with any other investigated environmental variables, such as the sediment composition and wave forcing. A similar change across a manmade structure was also observed at Duck, NC (USA), where a factor 2 difference in T_r in the areas just north and south of a pier was observed [23]. Wijnberg [24] hypothesized that structures inhibit the alongshore interaction between the intersected coastal sections causing an independent evolution that ultimately results in different equilibrium states originating from, for example, small differences in the local wave climate or bed slopes.

The nearshore bar response is sensitive to initial perturbations in the bed profile and is dominated by the morphologic feedback to the wave and current fields (e.g., [5,15,23,27]). The inter-annual bar amplitude response is primarily governed by the water depth above the bar crest, h_{Xb}, and the incident wave angle, θ [16,21]. As a consequence, the morphological developments do not only depend on the instantaneous small-scale processes; they also incorporate some degree of time history in profile configuration. Using a process-based profile model (*i.e.*, assuming alongshore uniformity), Walstra *et al.* [5] showed that specific initial profile and wave forcing combinations could affect the bar characteristics over the entire inter-annual cycle period. This is qualitatively in line with [18,24] who hypothesized that regions with different large-scale coastal behavior are controlled by the combined effects of different hydrodynamic forcing, sedimentological constraints (viz. grain size, stratigraphy) and/or morphological constraints (viz. shoreline orientation, shoreface morphology, surf zone morphology). To the best of our knowledge, all comprehensive data analysis studies were unable to further detail the (relative) contribution of these parameters and to identify the dominant physical processes that govern the bar cycle return period in different LSCB regions or sites.

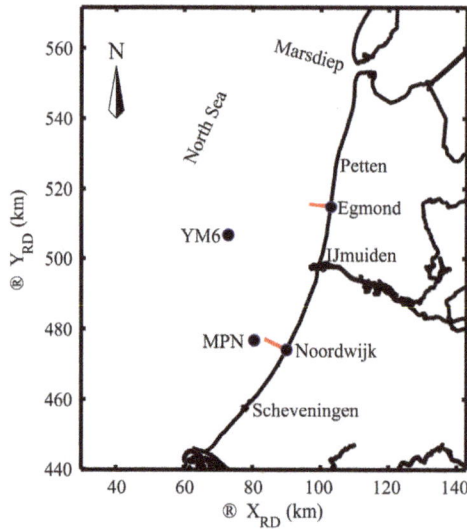

Figure 1. The Holland Coast with the sites at Egmond and Noordwijk indicated, as well as the location of the wave buoys YM6 (IJmuiden Munitie Stortplaats) and MPN (MeetPost Noordwijk). Red lines indicate the considered profiles at Noordwijk and Egmond, X_{RD} and Y_{RD} are the "Rijksdriehoek" coordinates.

Therefore, the present study utilizes a process-based forward model to identify the dominant environmental variables and the associated mechanisms that govern T_r. To that end, the profile model developed in [5,21] is applied at two locations 42 km apart (Noordwijk and Egmond, located at RSP 38 km and 80 km, respectively; RSP (RijksStrandPalen) is the Dutch alongshore beach pole numbering system). The sites are located in the LSCB regions just South and North of the IJmuiden harbor moles (Figure 1) with distinctly different bar cycle return periods. The model is utilized to investigate the influence of various environmental parameters on T_r. To that end, a range of model simulations are evaluated by comparing the predicted bar cycle return periods for various combinations of environmental variables from the Noordwijk and Egmond sites. The considered variables comprise the wave forcing (viz. wave height and incident wave angle), sediment size, and various geometric profile properties (viz. bar size, bar location and profile steepness). Subsequently, the underlying processes that predominantly govern T_r are identified. We finalize the paper with a discussion on the main findings and with the conclusions.

2. Environmental Settings

Both Noordwijk and Egmond are located along the Holland coast which is enclosed by the Marsdiep inlet in the north and the Rotterdam harbor moles in the south (Figure 1). The Holland coast is characterized by sandy beaches and multiple barred near-shore zones [28]. The entire Holland coast is an inlet free, sandy and wave dominated coast, with relatively small alongshore variations in offshore wave height and tide [24]. Due to the concave shape of the Holland Coast, the coastline orientation at Egmond (277 °N) and Noordwijk (298 °N) differs by about 21°. Furthermore, the sediment at Egmond is markedly coarser than at Noordwijk (see Table 1).

Table 1. Sediment diameters for Egmond and Noordwijk expressed as the 50 and 90 percentile, d_{ss} is the estimated d_{50} of the sediment in suspension, as applied in the model, small cross-shore variations in grain size are ignored.

Grain size	Noordwijk (µm) [14]	Egmond (µm) [28]
d_{50}	180	265
d_{90}	280	380
d_{ss}	170	240

2.1. Cross-Shore Bed Profile Characteristics

First, in order to exclude the bar morphology, the time-averaged cross-shore bed profile characteristics are analyzed for both sites. The time-averaged profiles were derived for Noordwijk and Egmond based on the annual profile surveys of the Jarkus database [18] for the period 1965 to 1998. Data from 1999 onwards were excluded because both sites were regularly nourished since that time, e.g., [9,10]. The shoreface (between −18 m and 0 m NAP (Normaal Amsterdams Peil); NAP is the Dutch datum at approximately mean sea level) is sub-divided into four sections, for each of which we compare the mean slopes in Figure 2: the beach section (Section 1) comprises the beachface between the dune foot (3 m NAP) and the mean water level (0 m); the upper shoreface (Section 2) the profile between 0 and −8 m; the middle shoreface (Section 3) is enclosed by the −8 m and −15 m depth contour and the lower shoreface (Section 4) is the part of the profile between −15 m and −18 m. The boundary between the upper and middle shoreface is defined at –8 m, because it is the edge of the near-shore zone [28]. Sandbars, and accordingly the temporal variability in sea bed elevation, are significantly reduced [29] and bars do not occur beyond this depth. The seaward limit of the analyzed profiles is set to −18 m, which corresponds to the water depth at the location of the wave observations at Noordwijk (MPN). As indicated in Figure 2, the beach and lower shoreface have similar slopes, whereas the upper and middle shoreface are notably steeper at Egmond.

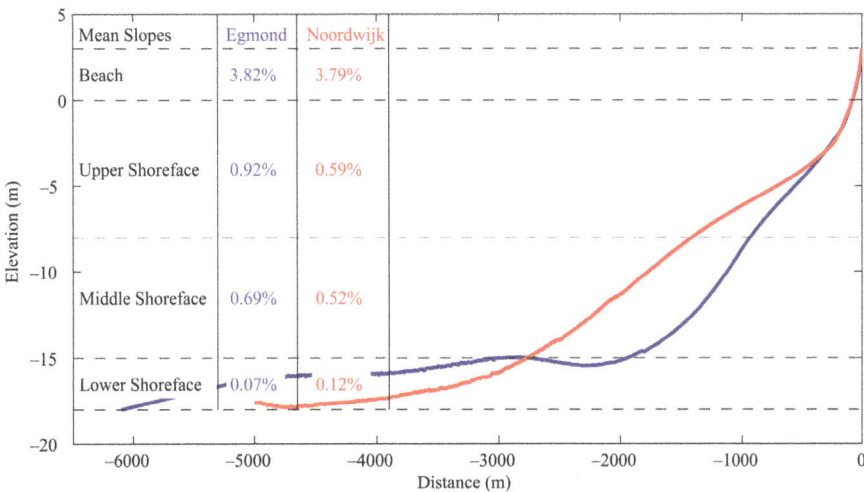

Figure 2. Time-averaged profiles for Noordwijk and Egmond on the same cross-shore axis with the origin for both at NAP 0 m.

2.2. Sandbar Characteristics

The sandbars are studied by subtracting the time averaged profile (Figure 2) from the actual bed profiles; especially at the upper and middle shoreface the resulting profile perturbations result primarily from the bar morphology. Figure 3 shows the profile perturbations for Egmond and Noordwijk for the part of the cross-shore profile at which the bars are prevalent.

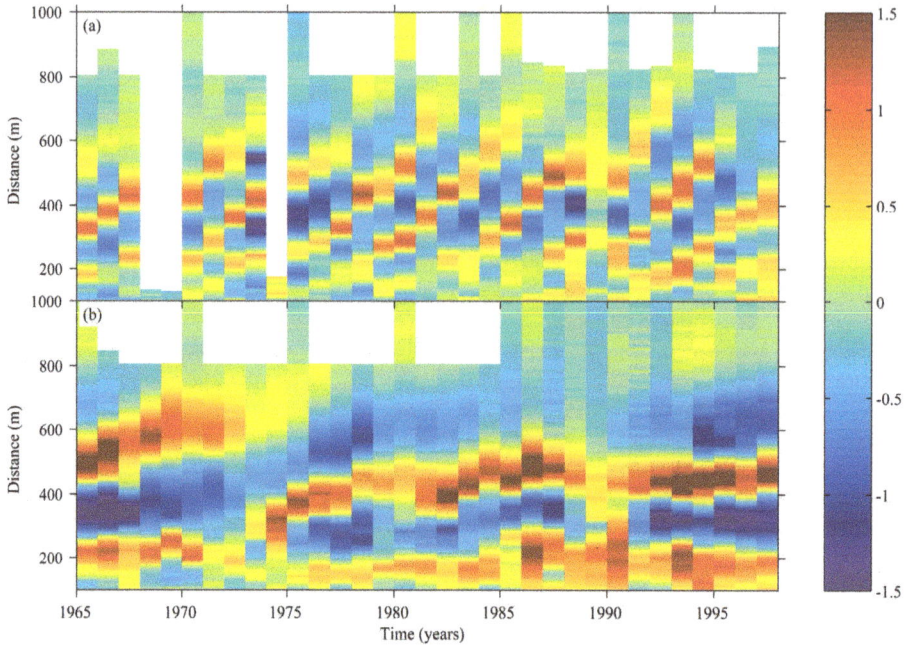

Figure 3. Profile perturbations of the time averaged near-shore profile are shown for (**a**) Noordwijk (RSP 80 km) and (**b**) Egmond (RSP 38 km).

Both at Egmond and Noordwijk mostly three bars are present [18,30]. The positive and negative perturbations indicate the bar and trough regions, respectively. The time stack plots (Figure 3a,b) clearly reveal the inter-annual cyclic bar characteristics. That is, bar initiation in the inter-tidal region, gradual offshore migration and amplitude growth and finally gradual decay at the seaward limits of the surf zone. However, the difference in bar cycle return period between both sites is striking. Estimates of T_r, derived earlier with a complex EOF method are 3.9 and 15.1 years for Noordwijk and Egmond, respectively [3]. Furthermore, the bars at Egmond are noticeably wider and higher.

2.3. Wave and Tidal Characteristics

We considered the period from 1 January 1990 to 31 December 1999 for which detailed hourly and three-hourly wave observations (root-mean-square wave height H_{rms}, peak wave period T_p and wave direction θ) were available for Noordwijk (Meetpost Noordwijk, MPN; see Figure 1) and IJmuiden (about 17 km south of Egmond, Munitie stortplaats, YM6; see Figure 1), respectively. To ensure a consistent comparison at the same water depth, the wave conditions at YM6 were converted to the water depth at MPN (from −21 m to −18 m) using Snell's law.

Figure 4a compares the time-mean H_{rms} of Noordwijk and Egmond as a function of θ. Apart from the waves from the southwestern direction, the wave height at Egmond is larger. Especially for the northwestern direction this difference increases as Egmond is more exposed to the North Sea.

Differences in the time-mean wave period are relatively small (Figure 4b). Storms ($H_{rms} > 1.5$ m) are predominantly obliquely incident (Figure 5) and occur throughout the year, although the fall and winter are usually more energetic than spring and summer [14]. This gives rise to a weak seasonality in H_{rms} [24]. In addition, there is some year-to-year variability in the wave climate [5]. At Noordwijk, for example, the annual cumulative wave energy can be up to 30% higher or lower than the multi-annual mean, although the differences are usually substantially smaller [5]. In addition, there is no periodicity in the year-to-year variability.

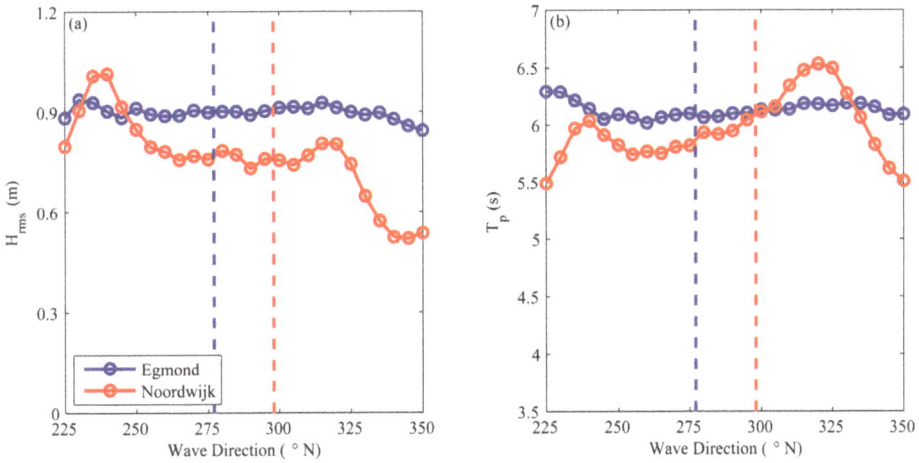

Figure 4. Comparison of the time-mean H_{rms} wave height (**a**) and the time-mean peak wave period (**b**) at Noordwijk and Egmond as a function of the incident wave direction. The vertical lines indicate the shore normal orientation for both sites.

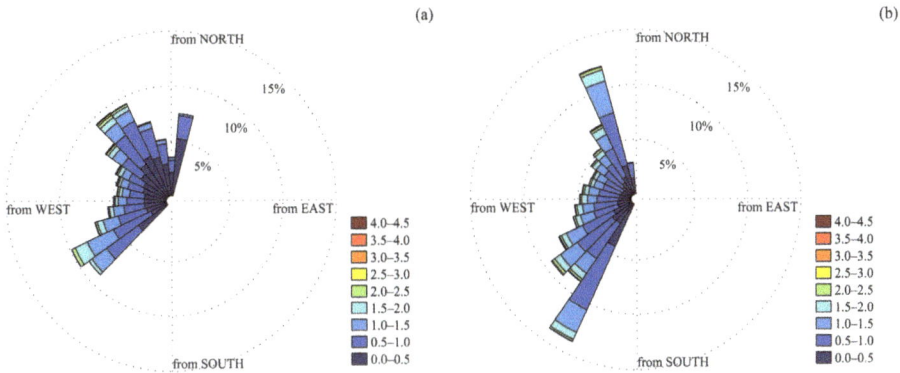

Figure 5. Wave roses of the imposed wave time series at Noordwijk (**a**) and Egmond (**b**).

The tide along the Holland coast is micro-tidal, with a mean tidal range of about 1.6 m. The tidal range decreases slightly in northward direction, which results in a tidal range that is on average about 0.1 m smaller at Egmond than at Noordwijk [24]. Tidal currents are generally lower than 1 m/s with little alongshore variations.

3. Approach

The main objective is to identify which environmental parameters and processes primarily govern the bar cycle duration. To that end, we apply the calibrated Noordwijk model [21] to a profile at Egmond as well. Although profile models typically require a site-specific calibration [13], we maintain the Noordwijk model settings in the application at the Egmond site. Only the site specific environmental variables from Egmond are used (*i.e.*, profile, d_{50} and time series of the waves and water levels). It is not our aim to achieve an optimal performance at Egmond (*i.e.*, best agreement with the observed inter-annual profile evolution) as long as the model is able to predict a significant difference in T_r between both sites. That will allow us to generate consistent predictions for both sites in which, for example, one specific (known) variable is modified. This approach allows us to identify the influence of the main environmental parameters such as wave height, near shore profile shape and sediment size on T_r. A comparison of two separately calibrated models would hamper such a comparison. Although different model settings will not influence the overall characteristics of the simulated bar morphology (*i.e.*, the net offshore directed cycle), it will affect the magnitude of the morphodynamic response. This will influence the subtle interdependencies between the hydrodynamic forcing and the morphodynamic response, which, in turn, will convolute the analysis of the predictions at both sites. However, as stated earlier, the primary concern is to verify that the predicted T_r at Egmond differs sufficiently (*i.e.*, larger) than at Noordwijk in the reference simulations. Therefore, as a first step, the predictions for both sites are evaluated. Next, the main environmental variables will be interchanged to identify the relative contribution of the wave climates, profiles and sediment size to changes in the bar cycle return period (e.g., the Egmond wave climate is combined with the Noordwijk profile and *vice versa*). The results of these hindcast simulations and the overall effects of the Egmond and Noordwijk wave climates, profiles and sediment sizes on T_r are discussed in detail in Section 4. In Section 5, these overall effects are further examined in order to identify the mechanisms and processes that govern T_r. For this, detailed schematic simulations are conducted and analyzed in which, for example, the influence of the profile slope on T_r is quantified.

This section continues with a brief description of the model in Section 3.1, followed by a description of the hindcast simulations in Section 3.2. Finally, the adopted analysis method is briefly discussed in Section 3.3.

3.1. Model Description

Unibest-TC is a cross-shore profile model and comprises coupled, wave-averaged equations of hydrodynamics (waves and mean currents), sediment transport, and bed level evolution. Straight, parallel depth contours are assumed. Starting with an initial, measured cross-shore depth profile and boundary conditions offshore, the cross-shore distribution of the hydrodynamics and sediment transport are computed. Transport divergence yields bathymetric changes, which feedback to the hydrodynamic model at the subsequent time step, forming a coupled model for bed level evolution. The phase-averaged wave model is based on [31] extended with the roller model according to [32] and the breaker delay concept [33] to have an accurate cross-shore distribution of the wave forcing. The cross-shore varying wave height to depth ratio, γ, of [34] was used in the breaking wave dissipation formulation as it results in more accurate estimates of the wave height across bar-trough systems than a cross-shore constant γ. The vertical distribution of the flow velocities is determined with the 1DV current-model of [35]. Based on the local wave forcing, mass flux, tide and wind forcing a vertical distribution of the longshore and cross-shore wave-averaged horizontal velocities are calculated. These advective currents are combined with the instantaneous oscillatory wave motion in such a way that the resulting velocity signal has the same characteristics of short-wave velocity skewness, amplitude modulation, bound infragravity waves, and mean flow as a natural random wave field [36]. The transport formulations distinguish between bed load and suspended load transport. The bed load formulations [37] are driven by the instantaneous velocity signal. The suspended transports are based on the integration over the water column of the sediment flux. The wave-averaged near-bed sediment concentration is prescribed according to [38], which among other factors, is driven by a time-averaged bed shear stress based on the instantaneous velocity signal. A detailed description of the Unibest-TC model can be found in [13,21].

3.2. Hindcast Model Simulations

The simulations are based on the settings according to the Noordwijk model calibrated for 1980 to 1984 period (*i.e.*, one bar cycle period, see [21]). As the calibrated model was shown to be valid for other periods at Noordwijk as well [5] and the primary focus of the present study is to investigate the difference between the two sites, we did not perform additional calibration or validation simulations for the Noordwijk and the Egmond model application.

The hindcast simulations have a net duration of about 9.5 years (1990–1999) and were forced with the locally observed (MPN and YM6 stations, see Figure 1) hydrodynamic forcing time series for this period for both sites (water levels and wave characteristics). The initial bed profiles were derived from the measured 1990 Jarkus transects (see Figure 6) and the sediment characteristics are according to Table 1.

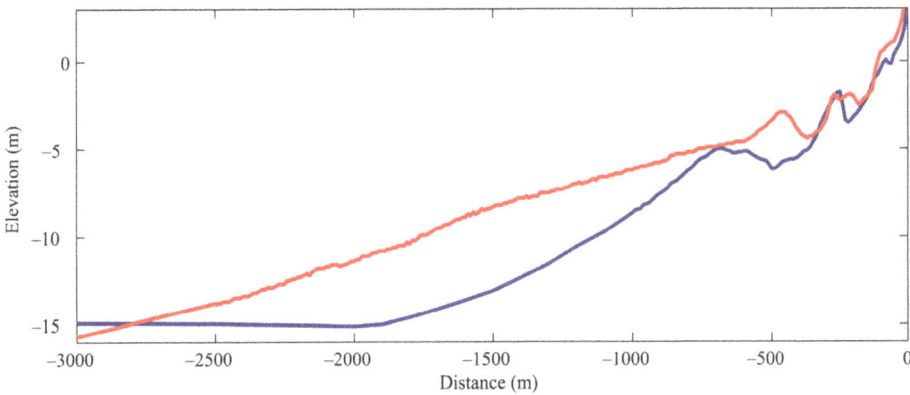

Figure 6. The nearshore part of the initial profiles for Noordwijk (red) and Egmond (blue), the offshore boundary of the model is at *x* = –6500 m.

Next, model simulations were performed in which the profile (and sediment diameter), wave climate (wave height, period and angle) for Noordwijk and Egmond were interchanged. Since the sediment size and the profile slope are correlated (e.g., [39]), we did not consider these separately. This implies that four combinations of wave time series and profile/d_{50} could be evaluated (Table 2).

Table 2. Hindcast simulations for Noordwijk and Egmond with interchanged wave forcing and profiles sediment diameter.

Scenario	Profile and Sediment	Wave Time Series
NN	Noordwijk	Noordwijk
EN	Egmond	Noordwijk
NE	Noordwijk	Egmond
EE	Egmond	Egmond

To investigate whether specific profile characteristics influenced the bar cycle period, we constructed synthetic profiles in which parts of the Noordwijk and Egmond (time-averaged) profiles and bars were combined. These profiles were subsequently used to perform hindcast simulations forced with the wave climates of both sites. We considered combinations of the upper shoreface (upper profile up to 8 m water depth), the middle shoreface (profile between 8 and 15 m water depth) and the lower shoreface (profile deeper than 15 m water depth) from both sites (see Table 3 and Figure 7). As the sediment size is assumed to be cross-shore constant in the model, it cannot be varied together with the profile sections. The choice of sediment size was therefore associated with the upper shoreface profile as in test simulations it was found that especially these required to be correlated to avoid an unstable or unrealistic profile evolution.

Table 3. Definition of the profiles constructed from parts of the Egmond and Noordwijk profiles.

Profile Code	Bar	Shoreface		
		Upper/ d_{50}	Middle	Lower
1 (ENNN)	Egmond	Noordwijk	Noordwijk	Noordwijk
2 (NENN)	Noordwijk	Egmond	Noordwijk	Noordwijk
3 (EENN)	Egmond	Egmond	Noordwijk	Noordwijk
4 (NNEN)	Noordwijk	Noordwijk	Egmond	Noordwijk
5 (NNNE)	Noordwijk	Noordwijk	Noordwijk	Egmond

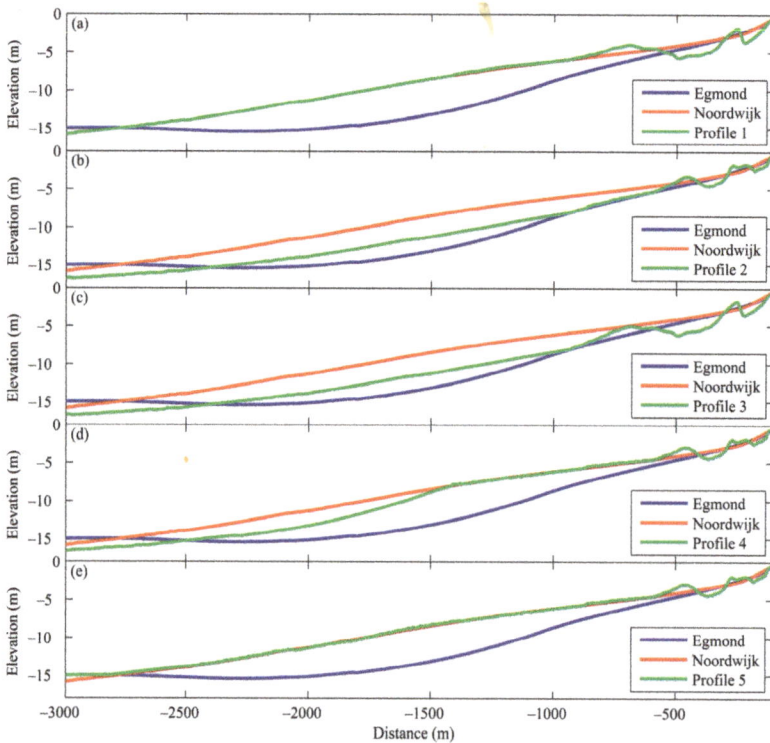

Figure 7. Constructed profiles from part of the Egmond and Noordwijk profiles. See Table 3 for profile composition details shown in plots **a–e**.

3.3. Analysis Method

The bar cycle return period T_r was determined by the time it takes a bar to be at the same cross-shore position as its predecessor. Ruessink *et al.* [3] showed that the complex EOF analysis is a robust method to derive T_r and it is therefore also used in this study. Complex EOF was preferred over classic EOF because it can capture the migrating sandbar pattern in a single (complex) mode and, as such, allows for a straightforward quantification of spatial and temporal sandbar characteristics (see [3]). Classic EOF is restricted to the description of standing patterns and thus requires two modes that contain approximately equal variance to describe migrating sandbars (see [18]). While these two modes can be combined into a complex pair, the technique that produces the complex mode inherently was preferred. An extensive description of complex EOF can be found in [3,40].

4. Model Results

First the reference cases for Noordwijk and Egmond are presented. Subsequently, the results of the modified model set ups described in Section 3 are discussed by comparing these to the reference case predictions.

4.1. The Reference Cases (Scenarios NN and EE)

From the comparison of the predicted profile development (Figure 8), the difference in bar cycle duration stands out immediately. The bar cycle period for Noordwijk (Scenario NN) is 4.8 years, which compares well to that derived from the observations for the same period (T_r = 3.9 years). For Egmond (Scenario EE), the predicted T_r of 8.7 years is significantly larger. However, it is still a significant under-estimation of the value derived from the profile surveys (T_r = 15.1 years). Ruessink *et al.* [13] showed that the model required a site specific calibration effort on weekly time scales. Given the multi-annual time scales considered in the present study, relatively larger model errors are to be expected as the model was not calibrated to the Egmond site. Since we are primarily interested in identifying the causes for the difference in the bar cycle period, we consider the model performance at Egmond to be adequate since the model predicts a significant difference in T_r between both sites. Furthermore, the short-term response to periods of increased or reduced wave energy is relatively stronger for Noordwijk (*i.e.*, short-term variations around the annual trend are larger at Noordwijk). The difference in T_r primarily originates from the combined effects of a larger annual offshore migration at Noordwijk (averaged offshore migration rate is approximately 55 m/year compared to 40 m/year for Egmond) and an approximately 200 m narrower cross-shore bar zone because the bars decay at a relatively shallow water depth.

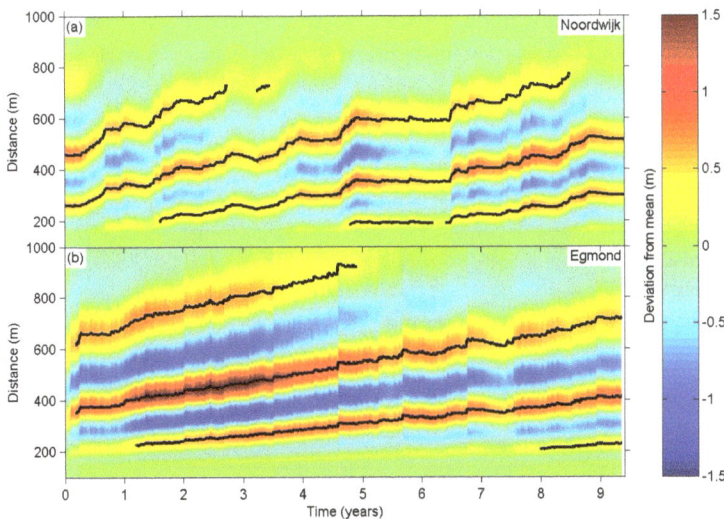

Figure 8. Predicted profile perturbations for (**a**) Noordwijk (Scenario NN) and (**b**) Egmond (Scenario EE).

4.2. Effects of Wave Climate vs. Sediment Size and Profile (Scenarios EN and NE)

The initial profile and wave climate have a profound impact on the resulting profile evolution (Figure 9a,b). Imposing the slightly more energetic Egmond wave climate on the Noordwijk profile (Scenario NE, see Figure 9a) results in a 50% reduction of the bar cycle period compared to the Noordwijk reference (Scenario NN, see Figure 8a). The opposite occurs when subjecting the Egmond profile to the Noordwijk wave climate (Scenario EN, see Figure 9b): the bar cycle period is almost doubled to 14.6 years. Although the Egmond wave climate reduced T_r, the wave climate increases the bar zone width by about 200 m and also results in slightly increased maximum bar amplitude. Due to the increased T_r, the bar zone width is difficult to determine for Scenario EN, but the results seem to suggest that it decreases by at least 100 m. Furthermore, the maximum bar amplitude in this scenario is about 0.5 m less compared to the Egmond reference case (Scenario EE, see Figure 8b).

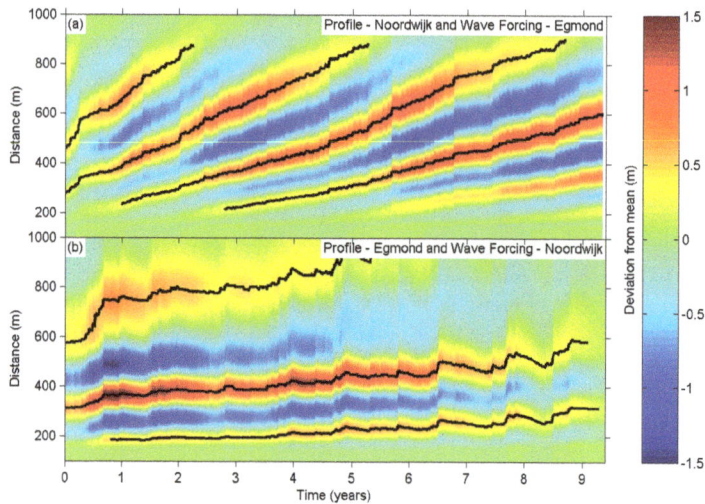

Figure 9. Predicted profile perturbations for scenarios with swapped wave forcing: (**a**) Noordwijk profile with wave forcing from Egmond (Scenario NE) and (**b**) *vice versa* (Scenario EN).

Consistent with [3], the energy level of the wave climate appears to influence T_r significantly. However, the effect of the initial profile and bar morphology has an even larger influence. Comparing T_r for the four scenarios (summarized in Table 4), an indication of the relative importance of the initial profiles and wave climates can be obtained. The interchange of wave climates results in a change of T_r of about 200% (compare scenarios NN, NE, EE, and EN). The influence of the initial profile, bar morphology and sediment size results in a variation T_r of about 300%. For example, the Egmond climate on the Noordwijk profile results in a T_r of 2.4 years compared to $T_r = 8.7$ years for the Egmond profile.

Table 4. Hindcast simulations for Noordwijk and Egmond with interchanged wave forcing and profiles (and d_{50}).

Scenario	Profile/Sediment	Wave Conditions	Cycle Period (years)
NN	Noordwijk	Noordwijk	4.8
EN	Egmond	Noordwijk	14.6
NE	Noordwijk	Egmond	2.4
EE	Egmond	Egmond	8.7

4.3. Effects of Profile Slope and Bar Characteristics

The various profile compositions as summarized in Section 3.2 are used as the starting point for 10 year morphodynamic simulations using the wave and water level time series of both Noordwijk and Egmond as boundary conditions. The predicted return periods are collected in Table 5. The table shows the return periods for the composite profiles forced with the Noordwijk and Egmond wave climates as well as the relative change compared to the appropriate hindcast simulations.

Table 5. Bar cycle periods and relative change to reference simulations for the different profile compositions subjected resulting from 10 year simulations for both the Noordwijk and Egmond wave time series. Scenarios between the brackets in columns 4 and 5 are according to Table 4. Profile codes in first column according to Table 3, indicating the origin of (from left to right): the bar, the upper shoreface (and sediment), middle shoreface and lower shoreface.

Profile Code	Bar return period, T_r (years) Wave Time Series		Relative change in T_r (–) Wave Time Series	
	Noordwijk	Egmond	Noordwijk	Egmond
1 (ENNN)	6.5	2.8	1.36 (NN)	1.17 (NE)
2 (NENN)	*7.0	6.1	*1.46 (NN)	2.55 (NE)
3 (EENN)	12.9	7.0	0.89/2.69 (EN/NN)	0.80/2.91 (EE/NE)
4 (NNEN)	4.6	2.2	0.95 (NN)	0.90 (NE)
5 (NNNE)	5.1	2.6	1.05 (NN)	1.10 (NE)

* indicates simulation for which bar cycle period could not be determined reliably).

Combining the Egmond bars with the Noordwijk profile (profile 1—ENNN) clearly causes an increased T_r for both wave climates (*i.e.*, compare T_r values for profile 1 in Table 5). Compared to the original Noordwijk profile the increase is about twice as large for the Noordwijk wave climate compared to the Egmond wave climate (1.36 *vs.* 1.17). However, incorporating the Egmond upper shoreface in the Noordwijk profile (*i.e.*, bar zone; profile 2—NENN) has a larger impact. Profile 2 combined with the Noordwijk climate results in a somewhat unrealistic profile evolution for which only a visual estimate of the bar cycle period could be made; however, a clear substantial increase in T_r was present (7 years). For the Egmond wave climate, the relatively steep slope of the Egmond upper shoreface results in a major (2.55) relative increase in T_r.

The comparison of profile 3 (*i.e.*, Egmond bar and upper shoreface combined with the middle and lower shoreface of Noordwijk; EENN) with the original Noordwijk profile simulations shows significantly increased T_r for both wave forcing time series (changes in T_r for profile 3 are 2.69 and 2.91 compared original Noordwijk profile, see Table 5). This implies that the combined effect of the upper shoreface slope and bar volume (and sediment size) has the largest effect on T_r of all the considered scenarios by far. The influence of the bed slope of the upper shoreface is especially clear for the Egmond wave forcing (*i.e.*, for NENN—only upper shoreface is taken from Egmond—T_r is 2.55 larger than for the complete Noordwijk profile, using the Egmond bar results in an T_r of 2.91). For the Noordwijk wave forcing this is less obvious (T_r respectively 1.46 and 2.69 larger). This is probably due to the unrealistic predictions starting from profile 2 subjected to the Noordwijk wave forcing.

The return periods for profile 3 were reduced by only 10% to 20% relative to original Egmond profile simulations. This implies the effect of the middle and lower shoreface are relatively limited. This is also reflected by Profiles 4 and 5. Interestingly, comparison of the perturbation time stacks revealed that the slope of the upper shoreface also influenced the bar amplitude. This was especially clear for the simulations with Profile 2 in which the bar amplitude rapidly increased to similar values as observed at Egmond (not shown).

In the simulations with the composite profiles the upper shoreface and bar volume appear to contribute about 80% to 90% of the profile induced changes on T_r. The Egmond wave climate reduces T_r by about a factor 2–2.5 and is approximately similar for most composite profiles (except for profile 2).

The relative influence of the profile and wave climate on T_r are therefore similar as found for the reference simulations (Sections 4.1 and 4.2).

5. The Relative Influence of Environmental Parameters on T_r

5.1. Introduction

From the evaluation in the previous section it is apparent that the wave climate, profile geometry and sediment size all have a significant effect on T_r. Increased sediment size causes a decrease in sediment transport and T_r (and *vice versa*). A relatively energetic wave climate results in an enhanced net bar offshore migration and consequently reduces T_r, whereas relatively large bars and steeper upper shoreface bed slopes have the opposite effect. Of the latter two, it was found in the previous section that especially the upper shoreface bed slope has a major influence on T_r. At first sight this is somewhat counter-intuitive as a steeper slope typically results in more intense wave breaking and consequently enhanced undertow and offshore sediment transport at the bar crest. This is addressed in Section 5.2 by comparing outcomes from morphostatic simulations (*i.e.*, no bed updating) for profiles with identical bars in the inner surf zone, but different profile slopes. This approach is extended in Section 5.3 to investigate the influence of the water depth at the bar crest (h_{Xb}) on T_r by considering sets of simulations in which a bar with constant shape is placed at 21 equidistant locations across the barred zone.

5.2. Effect of the Profile Slope on the Bar Migration Rate in the Inner Surf Zone

The effect of the profile slope was further investigated by considering morphostatic simulations starting from schematic profiles in which identical bars (with the crest at identical water depth) are combined with bed slopes representative for Egmond and Noordwijk (Figure 10) which were subjected to the full 9.5 year Noordwijk wave and water level time series. Detailed comparisons of wave height, undertow and sediment transport at the crest of the bars (location indicated in Figure 10) clearly confirmed that, despite the identical wave height at the top of the bar (Figure 11a), the undertow (depth-averaged return flow) is indeed larger due to more intense wave breaking at the bar crest for the steeper Egmond profile (Figure 11b). The enhanced turbulence levels due to the wave breaking and the increased return flow velocities consequently enhance the offshore sediment transports (Figure 11c). Potentially, this would induce an enhanced offshore bar migration.

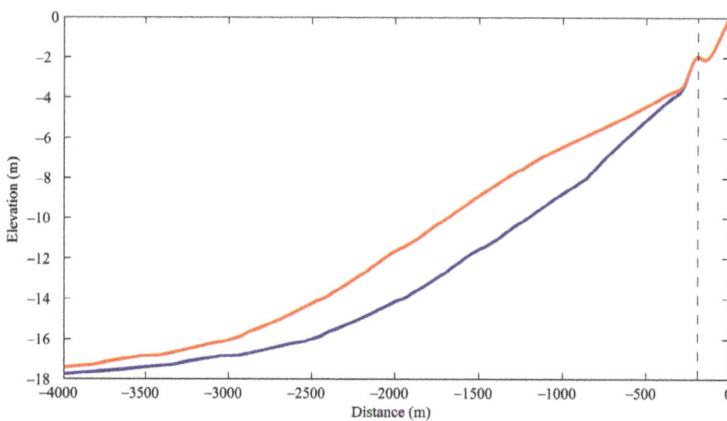

Figure 10. Schematic upper shoreface profiles combined with the middle and lower shoreface profiles for Noordwijk (red) and Egmond (blue) with the same water depth at the bar crest. Vertical dashed line indicates bar crest location at which model predictions are compared in Figure 9.

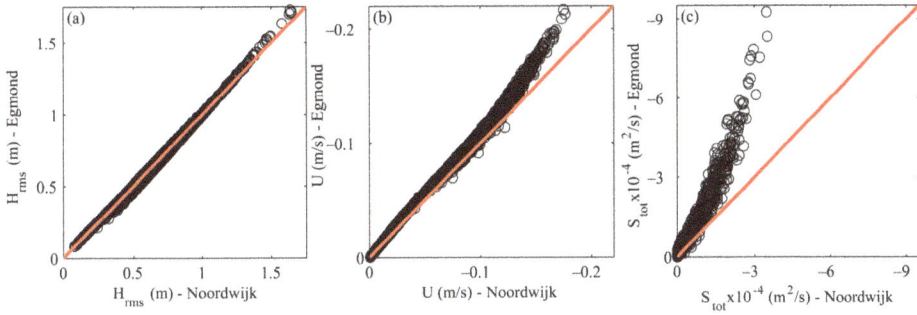

Figure 11. Comparison of the root-mean-square wave height H_{rms} (**a**), depth-averaged return flow U (**b**) and total sediment transport S_{tot} (**c**) at the top of the bar crest Noordwijk *vs.* Egmond (location shown in Figure 8). Red line indicates equality between Egmond and Noordwijk.

5.3. Identification of the Effects of H_{rms}, θ and d_{50} on T_r

In the hindcast simulations the initial response described above apparently does not result in an increased T_r. Therefore, it is assumed that the cumulative effect of the morphodynamic feedback between the barred profile and the wave forcing primarily governs T_r. In [21], the water depth above the bar crest (h_{Xb}) was identified to be a crucial parameter. Therefore, we need to investigate how h_{Xb} and the morphodynamic feedback loop affects T_r. In other words, how is the offshore migration rate affected as the bar migrates offshore and can we quantify the impact on T_r? To estimate T_r we conduct a set of one-day simulations starting from plane profiles in which a bar is placed at 21 equidistant locations across the bar zone. In order to exclude the effect of the transient bar amplitude response (*i.e.*, the change from growth to decay as the bar migrates across the surf zone) we considered a bar with a constant shape. For each simulation the daily migration rate and bar amplitude response are determined by considering the change in the horizontal and vertical bar crest position. Subsequently, the daily migration rates are integrated over the set of 21 simulations to estimate the time it takes for a bar to migrate across the bar zone as a proxy for T_r.

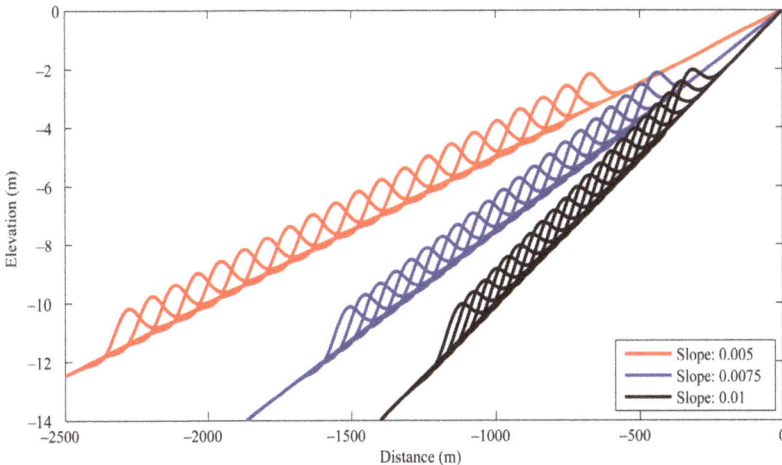

Figure 12. Plane profiles with the 21 schematic bars for 3 of the 10 considered profile slopes. Each bar was subjected to a one-day simulation with $H_{rms} = 1.7$ m, $T_p = 8$ s and $\theta = 20°$, and various additional scenarios.

By modifying a single environmental variable in each considered set we are able to isolate its influence on T_r. We considered 10 profile slopes ranging from 0.5% to 1% (see Figure 12). The same single wave condition as also used in [21] ($H_{rms} = 1.7$ m, $T_p = 8$ s, $\theta = 20°$) was applied. Normally a single wave condition is not sufficient to represent the full wave climate [41]. However, since we are primarily interested in the relative changes in T_r, the full wave climate is not required. In addition to the profile slope, the wave height and wave direction were also varied with ranges that are representative of the difference in these parameters between Egmond and Noordwijk. The relevant Noordwijk environmental variables were used as a reference. Since in this approach T_r is derived from the initial profile response, it will also allow us to isolate the effect of the sediment size (this was not possible in the morphodynamic simulations as unrealistic profiles or instabilities resulted if the upper profile and bar zone were inconsistent with the sediment size).

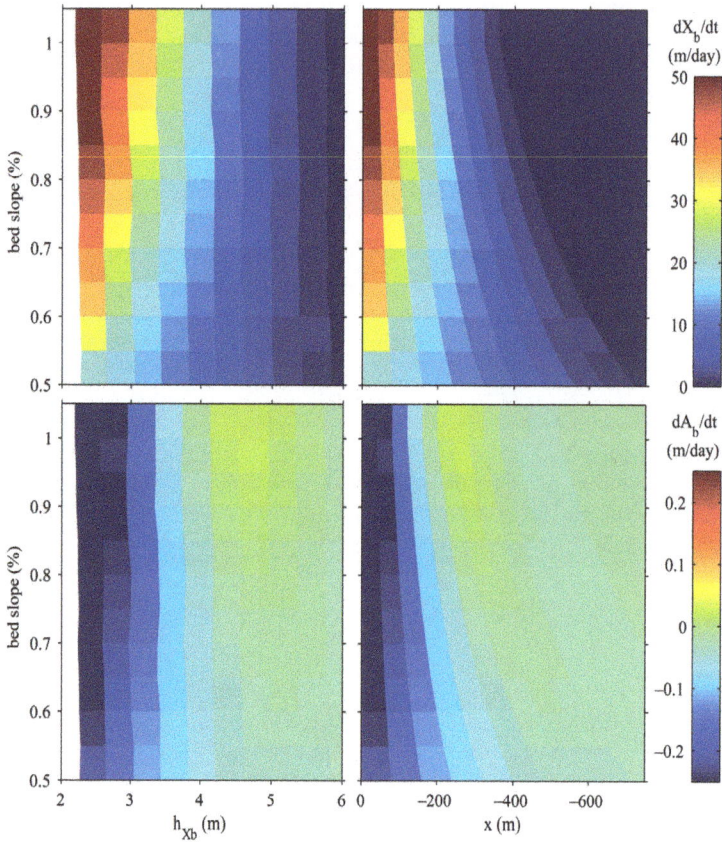

Figure 13. The migration rates, *dXb/dt* (**a,b**) and bar amplitude response, *dAb/dt* (**c,d**) for the reference case as a function of the bed slope plotted with h_{Xb} (**a,c**) and *x* (**b,d**).

The migration rate (dX_b/dt) and bar amplitude response (dA_b/dt) as derived for the set of reference simulations as a function of the bed slope are shown in Figure 13 for both h_{Xb} and *x*. The influence of the bed slope on both dX_b/dt and dA_b/dt is striking. A steeper profile clearly results in an offshore migration of the bar into larger water depths, but in a narrower cross-shore region (compare Figure 13a,b). It clearly illustrates the importance of h_{Xb}: steeper slopes initially induce an increased offshore migration but it quickly reduces as the bar migrates to deeper water. As a result, the cross-shore region at which this offshore migration occurs is also narrower. The bar amplitude growth is significantly larger for steeper profile slopes, extends into larger water depths, and also occurs in a relatively narrow region (Figure 13c,d). The integrated positive (*i.e.*, offshore) migration rates across the surf zone are used as a proxy for T_r. In this way the varying width of the barred zone (see Figure 13b) is included in the analysis.

The predicted T_r are clearly influenced by the bed slope for all the considered scenarios (Figure 14a) with a larger T_r for a steeper slope. Despite the larger maximum offshore migration rates (as shown Figure 13), the cumulative result is an increased T_r for steeper bed slopes as these high rates only occur in a relatively narrow cross-shore region. This confirms our idea that the morphodynamic feedback loop primarily governs T_r. Comparing the relative change in T_r compared to the averaged value for each series ($T_r/<T_r>$, Figure 14b), it can be seen that the sensitivity to the bed slope varies. The simulations with increased sediment size, wave angle and a reduced wave height result in a relatively reduced sensitivity to the bed slope, whereas an increased wave height shows an increased sensitivity.

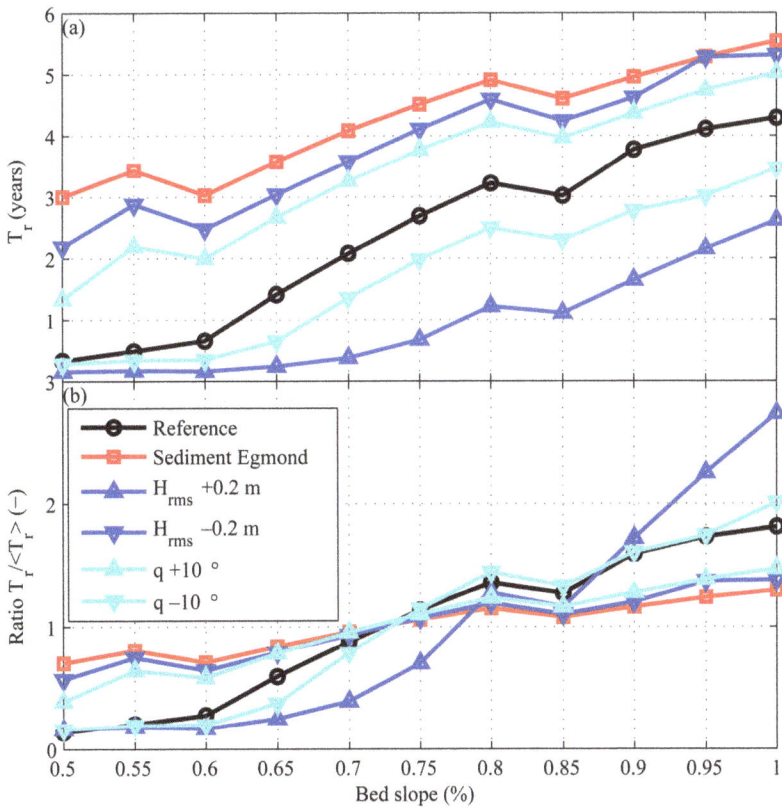

Figure 14. Absolute T_r (**a**) and the change in T_r relative to the T_r averaged over all considered bed slopes $T_r/<T_r>$ (**b**) as a function of the bed slope. The reference case is based on the Noordwijk environmental parameters.

The importance of the bed slope implies that h_{Xb} and the morphodynamic feedback loop primarily govern T_r. Despite more intense wave breaking and an initial enhanced offshore migration rate, the overall effect of a steeper profile is an increased T_r as it causes:

1) A relatively larger increase in h_{Xb} as a bar gradually migrates offshore which in turn causes fewer waves to break on the bar and consequently reduces the offshore bar migration.

2) Enhanced wave breaking results in relatively larger bars (e.g., see Figure 13b) that will also reduce the offshore migration (e.g. compare scenarios ENNN and NN in Table 5; see also [19]). Although a larger bar amplitude implies a somewhat smaller h_{Xb} at the same cross-shore location (and T_r), the increase in h_{Xb} as a bar migrates offshore dominates the T_r response.

3) An increased water depth where bar decay sets in due to more intense wave breaking. Combined with the more energetic wave climate this increases the bar zone width at Egmond by about 200 m compared to Noordwijk (as was both observed (Figure 2) and predicted (Figure 8)). Therefore, it takes longer for the bars to migrate across this region (e.g., a mean offshore migration rate of 40 m/year would lead to a five year increase in T_r).

6. Discussion

The present study has provided a physics-based exploration of the known worldwide differences in bar cycle duration, with a focus on the Dutch sites Noordwijk and Egmond. Although the model underestimated T_r by about 30% for Egmond, the factor 2 difference in T_r relative to Noordwijk is remarkable and provided us with significant confidence to use the model as an exploratory tool. By using identical model settings, the detailed and consistent model predictions allowed us to study the contributions of individual environmental parameters in great detail. Especially the role of the morphological feedback loop in which changes in depth also affect the waves, currents and sediment transport, which in turn influence the profile evolution, could be identified clearly. Due to the importance of the water depth at the bar crest (h_{Xb}), this feedback loop proved to be of major importance to explain the effect of the bed slope on T_r. The complex and highly non-linear interaction between the forcing and the inter-annual bar behavior can thus result in gradually diverging profile evolution at sites with seemingly very similar characteristics (e.g., profile evolution at either side of the pier at Duck or bar switch, see [5,23]. Our model results indicate that the inter-annual bar evolution should be regarded as forced behavior. Despite the non-linearities, the dissipation of wave energy within the nearshore system and the subsequent morphological response can be attributed to the forcing. In our opinion the indications of free (*i.e.*, non-forced) behaviur as identified in some studies (e.g., [42]) are due to the inability in data analysis studies to couple the observed non-linear response behavior to the (combined) state of a range of environmental parameters.

The identified dependences of T_r on wave climate, bar size/volume, bar zone width (and depth range) and sediment size are consistent with previous data-based studies of inter-site bar behavior (e.g., [3] and references therein). The importance of the bed slope on T_r has been suggested in earlier studies (e.g., [19,24] and our work unraveled the underlying physical processes. In contrast, Ruessink *et al.* [3] found that the bed slope did not appear to control inter-site differences in geometric and long-term temporal bar variability. We suspect that the varying influence of the environmental parameters on T_r for different bed slopes (Figure 14) and the limited amount of datasets/sites that could be considered in [3] are the primary reasons for this discrepancy.

7. Conclusions

Consistent with some earlier findings from field observations, our numerical model simulations illustrate that the bar cycle duration (T_r) is found to be positively correlated with sediment diameter and bar size, while T_r is negatively correlated with the wave forcing and profile slope. The simulations starting from composite profiles in which bar size, profile slope and sediment size were varied, clearly identified that the bed slope in the barred zone is the most important parameter that governs T_r. The sensitivity of T_r to this upper profile slope arises from the importance of the water depth above the bar crest (h_{Xb}) for sandbar response. As a bar migrates seaward, a steeper slope results in a relatively larger increase in h_{Xb}, which reduces wave breaking and subsequently causes a reduced offshore migration rate. Therefore, we conclude that the morphodynamic feedback loop is significantly more important than the initially larger offshore bar migration due to the more intense wave breaking in case of a steeper profile slope.

The application of the Egmond instead of the Noordwijk wave climate reduces T_r by a factor 3 to 4. However, the predicted T_r at Egmond is about two times larger, which is primarily originating from the difference in the upper profile slope and the larger sediment diameter at Egmond. These opposing effects further emphasize the importance of the upper bed slope and sediment diameter on T_r and illustrate that the net offshore bar migration is due to the highly non-linear two-way interaction between the wave forcing and the evolving profile morphology.

Acknowledgments: D.J.W. was supported by Deltares' Coastal Systems (CERM) strategic research program. G.R. was funded by the Dutch Technology Foundation STW, which is part of the Netherlands Organisation for Scientific Research (NOW), and which is partly funded by the Ministry of Economic Affairs (project number 12397).

Author Contributions: D.J.W. and G.R. developed the model and the modeling approach, while D.A.W. and E.C.D. performed most of the model simulations. D.J.W. wrote the paper. D.A.W., E.C.D. and G.R. all provided feedback during the writing.

Conflicts of Interest: The authors declare no conflict of interest.

References

1. Greenwood, B.; Davidson-Arnott, R.G.D. Sedimentation and equilibrium in wave-formed bars: A review and case study. *Can. J. Earth Sci.* **1979**, *16*, 312–332. [CrossRef]
2. Wijnberg, K.M.; Kroon, A. Barred beaches. *Geomorphology* **2002**, *40*, 103–120. [CrossRef]
3. Ruessink, B.G.; Wijnberg, K.M.; Holman, R.A.; Kuriyama, Y.; van Enckevort, I.M.J. Intersite comparison of interannual nearshore bar behavior. *J. Geophys. Res. Oceans* **2003**, *108*, 3249. [CrossRef]
4. Zenkovich, V.P. *Processes of Coastal Development*; Oliver and Boyd: White Plains, NY, USA, 1967.
5. Walstra, D.J.R.; Ruessink, B.G.; Reniers, A.J.H.M.; Ranasinghe, R. Process-based modeling of kilometer-scale alongshore sandbar variability. *Earth Surf. Process. Landf.* **2015**, *40*, 995–1005. [CrossRef]
6. Kroon, A. Three-dimensional morphological changes of nearshore bar system along the coast near Egmond aan Zee. *J. Coast. Res.* **1990**, *9*, 430–451.
7. Quartel, S.; Ruessink, B.G.; Kroon, A. Daily to seasonal cross-shore behaviour of quasi-persistent intertidal beach morphology. *Earth Surf. Process. Landf.* **2007**, *32*, 1293–1307. [CrossRef]
8. Walstra, D.J.R.; Brière, C.D.E.; Vonhögen-Peeters, L.M. Evaluating the PEM passive beach drainage system in a 4-year field experiment at Egmond (The Netherlands). *Coast. Eng.* **2014**, *93*, 1–14. [CrossRef]
9. Van Duin, M.J.P.; Wiersma, N.R.; Walstra, D.J.R.; van Rijn, L.C.; Stive, M.J.F. Nourishing the shoreface: Observations and hindcasting of the Egmond case, The Netherlands. *Coast. Eng.* **2004**, *51*, 813–837. [CrossRef]
10. Ojeda, E.; Ruessink, B.G.; Guillen, J. Morphodynamic response of a two-barred beach to a shoreface nourishment. *Coast. Eng.* **2008**, *55*, 1185–1196. [CrossRef]
11. Van der Spek, A.; Elias, E. The effects of nourishments on autonomous coastal behaviour. In Coastal Dynamics, Proceedings of the 7th International Conference on Coastal Dynamics; pp. 1753–1763.
12. Kroon, A. Sediment Transport and Morphodynamics of the Beach And Nearshore Zone Near Egmond, the Netherlands. Ph.D. Thesis, Utrecht University, Utrecht, The Netherlands, 1994.
13. Ruessink, B.G.; Kuriyama, Y.; Reniers, A.J.H.M.; Roelvink, J.A.; Walstra, D.J.R. Modeling cross-shore sandbar behavior on the timescale of weeks. *J. Geophys. Res. Earth Surf.* **2007**, *112*, 1–15.
14. Van Enckevort, I.M.J.; Ruessink, B.G. Video observations of nearshore bar behavior. Part 1: Alongshore uniform variability. *Cont. Shelf Res.* **2003**, *23*, 501–512. [CrossRef]
15. Ruggiero, P.; Walstra, D.J.R.; Gelfenbaum, G.; van Ormondt, M. Seasonal-scale nearshore morphological evolution: Field observations and numerical modeling. *Coast. Eng.* **2009**, *56*, 1153–1172. [CrossRef]
16. Dubarbier, B.; Castelle, B.; Marieu, V.; Ruessink, B.G. Process-based modeling of cross-shore sandbar behavior. *Coast. Eng.* **2015**, *95*, 35–50. [CrossRef]
17. Ruessink, B.G.; Kroon, A. The behaviour of a multiple bar system in the nearshore zone of Terschelling: 1965–1993. *Mar. Geol.* **1994**, *121*, 187–197. [CrossRef]

18. Wijnberg, K.M.; Terwindt, J.H.J. Extracting decadal morphological behavior from high-resolution, long-term bathymetric surveys along the Holland coast using eigen function analysis. *Mar. Geol.* **1995**, *126*, 301–330. [CrossRef]

19. Shand, R.D.; Bailey, D.G.; Shephard, M.J. An inter-site comparison of net offshore bar migration characteristics and environmental conditions. *J. Coast. Res.* **1999**, *15*, 750–765.

20. Kuriyama, Y. Medium-term bar behavior and associated sediment transport at Hasaki, Japan. *J. Geophys. Res. Oceans* **2002**, *107*, 3132. [CrossRef]

21. Walstra, D.J.R.; Reniers, A.J.H.M.; Ranasinghe, R.; Roelvink, J.A.; Ruessink, B.G. On bar growth and decay during interannual net offshore migration. *Coast. Eng.* **2012**, *60*, 190–200. [CrossRef]

22. Ruessink, B.G.; Terwindt, J.H.J. The behaviour of nearshore bars on the time scale of years: A conceptual model. *Mar. Geol.* **2000**, *163*, 289–302. [CrossRef]

23. Plant, N.G.; Holman, R.A.; Freilich, M.H.; Birkemeier, W.A. A simple model for inter-annual sandbar behavior. *J. Geophys. Res. Oceans* **1999**, *104*, 15755–15776. [CrossRef]

24. Wijnberg, K.M. Environmental controls on decadal morphologic behaviour of the Holland coast. *Mar. Geol.* **2002**, *189*, 227–247. [CrossRef]

25. Lippmann, T.C.; Holman, R.A.; Hathaway, K.K. Episodic, non-stationary behaviour of a double bar system at Duck, North Carolina, USA. *J. Coast. Res.* **1993**, *15*, 49–75.

26. Shand, R.D.; Bailey, D.G.; Shephard, M.J. Longshore realignment of shore-parallel sand-bars at Wanganui, New Zealand. *Mar. Geol.* **2001**, *179*, 147–161. [CrossRef]

27. Van Rijn, L.C.; Ribberink, J.S.; van der Werf, J.; Walstra, D.J.R. Coastal sediment dynamics: Recent advances and future research needs. *J. Hydraul. Res.* **2013**, *51*, 475–493. [CrossRef]

28. Van Rijn, L.C.; Ruessink, B.G.; Mulder, J.P.M. *Coast3D Egmond: The Behaviour of a Straight Sandy Coast on the Time Scale of Storms and Seasons*; Aqua Publications: Amsterdam, The Netherlands, 2002.

29. Hinton, C.; Nicholls, R.J. Spatial and temporal behaviour of depth of closure along the Holland coast. In Proceedings of the 26th International Coastal Engineering Conference, Copenhagen, Denmark, 22–26 June 1998; pp. 2913–2925.

30. Pape, L.; Plant, N.G.; Ruessink, B.G. On cross-shore sandbar behavior and equilibrium states. *J. Geophys. Res. Earth Surf.* **2010**, *115*. [CrossRef]

31. Battjes, J.A.; Janssen, J.P.F.M. Energy loss and set-up due to breaking of random waves. In Proceedings of the 16th International Conference on Coastal Engineering, New York, NY, USA, 27 August–3 September 1978; pp. 570–587.

32. Nairn, R.B.; Roelvink, J.A.; Southgate, H.N. Transition zone width and implications for modelling surfzone hydrodynamics. In Proceedings of the 22nd International Conference on Coastal Engineering, Delft, The Netherlands, 2–6 July 1990; pp. 68–91.

33. Roelvink, J.A.; Meijer, T.J.G.P.; Houwman, K.; Bakker, R.; Spanhoff, R. Field validation and application of a coastal profile model. In Proceedings of the Coastal Dynamics Conference, Gdansk, Poland, 4–8 September 1995.

34. Ruessink, B.G.; Walstra, D.J.R.; Southgate, H.N. Calibration and verification of a parametric wave model on barred beaches. *Coast. Eng.* **2003**, *48*, 139–149. [CrossRef]

35. Reniers, A.J.H.M.; Thornton, E.B.; Stanton, T.P.; Roelvink, J.A. Vertical flow structure during sandy duck: Observations and modeling. *Coast. Eng.* **2004**, *51*, 237–260. [CrossRef]

36. Roelvink, J.A.; Stive, M.J.F. Bar-generating cross-shore flow mechanisms on a beach. *J. Geophys. Res.* **1989**, *94*, 4785–4800. [CrossRef]

37. Ribberink, J. Bed-load transport for steady flows and unsteady oscillatory flows. *Coast. Eng.* **1998**, *34*, 52–82. [CrossRef]

38. Van Rijn, L.C. *Principles of Sediment Transport in Rivers, Estuaries and Coastal Seas*; Aqua Publications: Amsterdam, The Netherlands, 1993.

39. Dean, R.G. *Equilibrium Beach Profiles: U.S. Atlantic and Gulf Coasts*; Technical Report No. 12; Department of Civil Engineering, University of Delaware: Newark, DE, USA, 1977.

40. Horel, J.D. Complex principal component analysis: Theory and examples. *J. Clim. Appl. Meteorol.* **1984**, *23*, 1660–1673. [CrossRef]

41. Walstra, D.J.R.; Hoekstra, R.; Tonnon, P.K.; Ruessink, B.G. Input reduction for long-term morphodynamic simulations in wave-dominated coastal settings. *Coast. Eng.* **2013**, *77*, 57–70. [CrossRef]
42. De Vriend, H.J. On the predictability of coastal morphology. In Proceedings of the 3rd European Marine Science and Technology Conference, Lisbon, Portugal, 23–27 May 1998.

Journal of
*Marine Science
and Engineering*

MDPI

Article

Dynamics of Small-Scale Topographic Heterogeneity in European Sandy Salt Marshes

Kelly Elschot and Jan P. Bakker *

Conservation Ecology Group, University of Groningen. PO Box 11103, 9700CC Groningen, The Netherlands; kellyelschot1@gmail.com
* Correspondence: j.p.bakker@rug.nl; Tel.: +31-50-3636133

Academic Editor: Gerben Ruessink
Received: 7 December 2015; Accepted: 24 February 2016; Published: 3 March 2016

Abstract: Heterogeneity can boost biodiversity, as well as increase the resilience of an ecosystem to changing environmental conditions; therefore, it is important to understand how topographic heterogeneity in ecosystems is formed. Sandy tidal marshes have a repetitive pattern of higher elevated hummocks surrounded by lower elevated depressions, representing topographic heterogeneity at the scale of a few square meters. The aims of this study were to determine when this topographic heterogeneity forms, how it is structured, and whether it persists during marsh development. The soil topography of marshes consists of coarse-grained sediment formed before marsh vegetation development, with an overlaying fine-grained sediment layer formed after initial marsh development. To gain insight into the formation of topographic heterogeneity, we studied the underlying soil topography of four European sandy marshes, where topographic heterogeneity at a scale of a few square meters was present. The differences in elevation between hummocks and depressions can either be caused by heterogeneity in the coarse-grained sediment or by heterogeneity in the top layer containing the fine-grained sediment. Our results showed that the largest percentage of elevational differences between hummocks and depressions could be attributed to heterogeneity in the underlying coarse-grained substratum. Therefore, we conclude that the patterns in all four marshes were primarily formed before marsh development, before fine-grained sediment was deposited on top of the coarse-grained sediment. However, a smaller percentage of the elevational difference between hummocks and depressions can also be explained by the presence of thicker fine-grained sediment layers on top of hummocks compared with depressions. This implies that marsh accretion rates were higher on hummocks compared with depressions. However, this result was limited to very early stages of marsh development, as marsh accretion rates estimated on marshes ranging between 15- and 120-years-old showed that depressions actually accreted sediments at a significantly faster rate than hummocks. Eventually, the patterns of heterogeneity stabilized and we found similar marsh accretion rates on hummocks and in depressions in the 120-year-old marsh, which resulted in the persistency of these topographic patterns.

Keywords: accretion; coarse-grained sediment; depression; fine-grained sediment; hummock

---

1. Introduction

Heterogeneity can have large impacts on the functioning of ecosystems [1], boosting biodiversity [2,3], as well as increasing the resilience of an ecosystem and its associated species to changing environmental conditions [4,5]. When spatial heterogeneity is present within an ecosystem, it can result in more niches, which increases the number of plant species able to co-exist on a smaller spatial scale [2,4]. As different plant species will respond differently to rapid changes in environmental conditions, this will increase the resilience of an ecosystem to cope with these rapid changes [1]. Understanding the formation and persistence of spatial heterogeneity, therefore, is important now since climate change is threatening many ecosystems on a global scale and many species are at risk of becoming extinct [6]. This is especially true for coastal ecosystems, which are particularly vulnerable to the effects of climate change and resulting enhanced sea-level rise [7,8]. In this study, we focused on the formation, structure, and persistence of small-scale topographic heterogeneity in sandy marshes.

Salt-marsh development is initiated when pioneer vegetation establishes on a bare coarse-grained intertidal flat or sand bank, and fine-grained sediment (silt) deposition starts to accumulate [9]. As a consequence, the soil profile of sandy minerogenic marshes generally consists of coarse-grained sediment covered with a thin layer of fine-grained sediment [9]. Vegetation has been shown to increase the sediment deposition rate and reduce the erosion rate by stabilizing the soil [10–12]. Fine-grained sediments deposited by tides and local vegetation determine the later morphology of the marsh platform [13–15].

Sandy marshes in Europe have fine patterns at the scale of a few square meters in marsh morphology [16–19]. These patterns consist of a repetitive pattern of higher elevated, sandy hummocks covered with a thin fine-grained sediment layer surrounded by lower elevated depressions. The hummocks range between a few centimeters to a few meters in diameter, and elevational differences between hummocks and depressions can be up to 30 cm. The plant community of hummocks and depressions significantly differ in composition. In Britanny, hummock formation was shown to start with the perennial *Salicornia radicans*, followed by *Puccinellia maritima* and *Atriplex portulacoides*, whereas the annual *Salicornia herbacea* established in depressions. The hummocks appeared to be gradually coalesced over time, resulting eventually in more level ground [20]. In Wales, lateral extension of the pioneer species *Puccinellia maritima*, and later other species, filled the depressions between hummocks leading to hummock coalescence. Eventually, the general surface of the marsh became more even due to reduced sedimentation rates on the higher parts [21]. Both studies suggest these patterns are transitional. In the Netherlands, however, patterns of hummocks and depressions on the 100 year old salt marsh of Schiermonnikoog were observed (pers. observation). The origin and persistence of these patterns are still unclear, as is whether they emerge from similar processes in different marshes.

To gain more insight into the formation, structure, and persistence of these small-scale topographic patterns, we conducted a study on four different sandy marshes in Europe. We reasoned that topographic patterns can arise from heterogeneity of the underlying coarse-grained substratum and/or differences in the local fine-grained sediment layer. Two contrasting hypotheses were tested: 1) the patterns are formed in the pioneer stage before marshes develop, or 2) the patterns are formed after marsh development. To test these two hypotheses, we compared the vertical soil profile underlying heterogeneous marshes, with hummocks and depressions, to those underlying homogeneous marshes, with no clear hummocks and depressions. Additionally, we measured marsh accretion rates of hummocks and depressions at marshes between 15 and 120 years of age.

2. Methods

2.1. Study Sites

We included four sandy marshes located in Northwestern Europe (Figure 1) that featured hummocks and depressions: the Cefni Marsh (United Kingdom), Schiermonnikoog (The Netherlands), Terschelling (The Netherlands), and Skallingen (Denmark). Three different types of marshes were represented in that the Cefni marsh is located within the Cefni Bay, Schiermonnikoog, and Terschelling are back-barrier marshes located on islands, and Skallingen is on a peninsula. The soil profile on these sandy marshes consisted of a fine-grained, silty sediment layer on top of a coarse-grained, sandy sediment deposited before the marshes started to form [9]. The transition between these layers was very distinct [22] and allowed us to measure the thickness of the fine-grained sediment layer with great precision (up to a few millimeters).

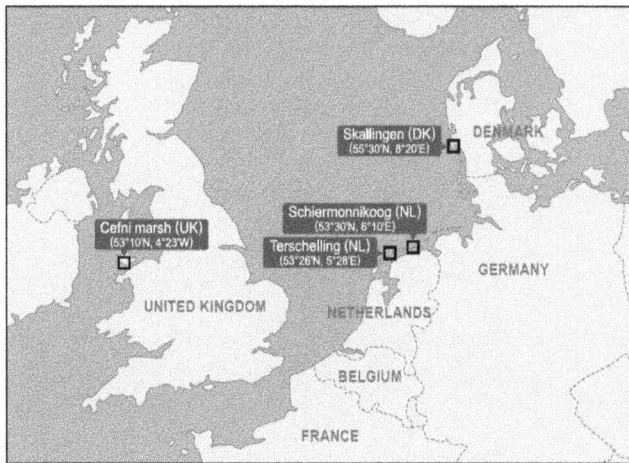

Figure 1. Field sites included in this study.

In Cefni Bay, salt-marsh development has only started since the 1960s [23]. Due to continuous expansion of the southern coast northward into the bay, a large pioneer zone is present in front of the Cefni marsh. In this pioneer zone, we studied higher elevated hummocks covered by *Puccinellia maritima* that were present on the otherwise bare intertidal flats. On Schiermonnikoog we studied the patterns along a natural chronosequence. A previous study used aerial photographs and topographic maps to identify marshes of different ages that were present adjacent to each other on Schiermonnikoog [9]. This chronosequence arose due to changing sea currents that have caused the island to grow eastwards, resulting in continuous new formation of dunes on the north side of the island and marshes behind them. Marsh age was determined from the first establishment of marsh vegetation identified from a time-series of maps and aerial photographs (for further detail see also [9]). We included marsh sites of approximately 15, 30, 45, 55, and 120 years of age in 2010. Marsh sites in this study were not grazed by livestock.

2.2. Patterns in Soil Morphology

To gain insight into the soil morphology and determine whether the underlying coarse-grained sediment differed between heterogeneous and homogeneous marshes (marsh without small-scale topographic heterogeneity), we sampled transects ranging from 70 m to 300 m in length. Both at the Cefni marsh (August 2011) and on Schiermonnikoog (May 2009), transects were measured starting on the marsh platform (underneath a dune on Schiermonnikoog and from the edge of a creek on the Cefni marsh) towards the intertidal flats. Along the marsh surface, we estimated the surface elevation and fine-grained sediment layer thickness every 0.5 m. We increased the number of measurements to every 0.25 m near transitions between hummocks and depressions to prevent missing any hummocks or depressions. The surface elevation was measured using an optical levelling instrument (Spectra Precision® Laser LL500 and Spectra Precision® Laser HR500 laser receiver by Trimble, Dayton, OH, USA) with an accuracy of ~5 mm. The fine-grained sediment layer thickness was measured using a small soil corer (diameter = 10 mm) with an accuracy of 5 mm. The corer was inserted vertically into the marsh platform and extracted, after which we could measure the thickness of the fine-grained sediment layer based on the soil profile exposed in the corner.

At the Cefni marsh, we compared two transects in the heterogeneous marsh (transect 1 and 2) with one transect in the homogeneous marsh (transect 3). All transects were located approximately 200 m apart from each other and covered both the marsh zone and part of the pioneer zone that is located in front of the marsh. On Schiermonnikoog, we compared one transect in the heterogeneous marsh with one transect in the homogeneous marsh. Both transects were measured on the 30 year-old-marsh, approximately 100 m apart from each other. Due to time constraints we could not measure similar transects on Skallingen or Terschelling.

2.3. Coarse-Grained vs. Fine-Grained Heterogeneity in Four Sandy Marshes

To study the generality of the topographic patterns (hummocks and depressions), we compared the soil topography of four European tidal marshes. We included the Cefni marsh (August 2011), Schiermonnikoog (May 2009), Terschelling (October 2010), and Skallingen (September 2009). We took pair-wise measurements of the marsh elevation on hummocks and in neighboring depressions with a distance of ~0.5 m between them. We decided upon a distance of ~0.5 m to be consistent in our methodology, and close enough to prevent site-specific differences within a paired sample. The sample sizes, tidal ranges, and dominant plant species are given in Table 1. We selected elevated hummocks, which ranged from a few centimeters up to a few meters in diameter. For each of these paired measurements, we estimated surface elevation according to Mean High Tide (MHT), measured fine-grained sediment layer thickness with a small corer (10 mm in diameter, similarly as mentioned in the previous section), and recorded the three most dominant plant species. The fine-grained sediment layer thickness was subtracted from the measured marsh elevation to determine the elevation of the underlying coarse-grained sediment. At the Cefni marsh, we selected hummocks and neighboring depressions alongside the three transects mentioned in the previous section, covering a large marsh surface area (~6 ha). The samples were taken in the pioneer zone, as well as the marsh zone. Vegetated hummocks located on bare intertidal flats, the flats that did not yet have a fine-grained sediment layer were referred to as in the pioneer zone. When vegetated hummocks were surrounded by vegetated depressions that had a fine-grained sediment layer, then we referred to them as in the marsh zone. On Schiermonnikoog, we included five marsh sites of different ages: 15, 30, 45, 55, and 120 year-old marshes, which prevents any age bias. All samples of the different marsh ages estimated on Schiermonnikoog were pooled together for further analyses. On Terschelling and Skallingen, we selected hummocks and neighboring depressions over a large marsh surface area of a few hectares.

Table 1. Characteristics and sampling effort at marsh sites.

	n	Tidal Range (m)	Most Dominant Plant Species	2nd Most Dominant Plant Species	3rd Most Dominant Plant Species
Hummocks					
Cefni marsh		4.7 *			
Pioneer zone	95		*Puccinellia maritima*	Bare soil	*Armeria maritima*
Marsh zone	60		Bare soil	*Armeria maritima*	*Festuca rubra*
Terschelling	40	2	*Festuca rubra*	*Puccinellia maritima*	
Skallingen	41	1.3	*Festuca rubra*	*Artiplex portulacoides*	
Schiermonnikoog		2.3			
15 year-old marsh	55		*Limonium vulgare*	*Festuca rubra*	*Atriplex portulacoides*
30 year-old marsh	55		*Festuca rubra*		
45 year-old marsh	62		*Festuca rubra*	*Artemisia maritima*	*Puccinellia maritima*
55 year-old marsh	38		*Festuca rubra*	*Artemisia maritima*	*Elytrigia atherica*
120 year-old marsh	66		*Festuca rubra*	*Puccinellia maritima*	*Artemisia maritima*
Depressions					
Cefni marsh		4.7 *			
Pioneer zone	95		Bare soil		
Marsh zone	60		Bare soil	*Puccinellia maritima*	*Plantago maritima*
Terschelling	40	2	*Limonium vulgare*	*Atriplex portulacoides*	*Aster tripolium*
Skallingen	41	1.3	*Atriplex poartulacoides*		
Schiermonnikoog		2.3			
15 year-old marsh	55		Bare soil	*Limonium vulgare*	*Atriplex portulacoides*
30 year-old marsh	55		*Limonium vulgare*	*Atriplex portulacoides*	Bare soil
45 year-old marsh	62		*Limonium vulgare*	*Salicornia europaea*	*Atriplex portulacoides*
55 year-old marsh	38		*Limonium vulgare*	*Atriplex portulacoides*	*Festuca rubra*
120 year-old marsh	66		*Atriplex poartulacoides*	*Festuca rubra*	*Salicornia europaea*

* The Cefni marsh was located inside Cefni Bay and the tidal range was measured outside the Bay. Dampening of the amplitude can be expected with increasing distance to the mouth of the Bay.

2.4. Marsh Accretion Rates during Marsh Development

To compare long-term marsh accretion rates between hummocks and depressions, sediment and erosion bars (SEBs, see also [24,25]) were placed along the natural chronosequence on Schiermonnikoog in 2001 at the 15, 30, 45, 55, and 120 year-old marshes. Each SEB consisted of two poles that were placed 2 m apart on the marsh platform, with one pole located on top of a hummock and one pole located within a depression. This set-up was duplicated three times per site. For stabilization, each pole was inserted at least 1.0 m into the underlying coarse-grained sediment. An aluminum bar with 17 holes 0.1 m apart along the entire length of the bar was placed on top of the two poles during measurement. We estimated the elevation of the marsh platform by inserting a small pin vertically through each hole until it touched the marsh platform and measured the length of the pin left above the aluminum bar. Between 2001 and 2011, we estimated marsh accretion rates yearly. Due to unrealistic accretion rates of ~10 cm found in the 2003 data, we removed all measurements taken in 2003 from further analyses.

2.5. Data Analysis

To analyze the SEB data, we visually assigned each individual measurement in the field to hummock, depression, or transition state, *i.e.*, located on the edge of a hummock. In the following analyses, we only included the measurements taken from hummocks and depressions, omitting data from transition states. An average annual marsh accretion rate was first calculated for each SEB individually by averaging over the 17 holes and over all the years measured. Ultimately, this resulted in three marsh accretion rates (cm· year^{-1}) per treatment (hummock or depression) for each marsh site. Thereafter, we tested for any significant effects between treatments using an ANOVA with marsh age and treatment (hummock or depression) as categorical predictors.

3. Results

When comparing the heterogeneous and homogeneous marshes in Cefni and Schiermonnikoog, all transects showed that the morphology of the marsh platform was similar to the elevational heterogeneity found in the underlying coarse-grained sediment (Figure 2). Transects measured in heterogeneous marsh sites with hummocks and depressions present had the same elevational heterogeneity in the underlying coarse-grained sediment (Figure 2a, first transect, and Figure 2b, first and second transect). Furthermore, in marsh sites that were relatively homogeneous in marsh morphology (Figure 2a, second transect, and Figure 2b, third transect), we found a similarly homogeneous elevation in the underlying coarse-grained sediment.

When comparing all four marshes, elevational differences between hummocks and depressions ranged from 6.9 cm on Terschelling to 12.5 cm at the Cefni marsh (Figure 3). The largest percentage of the elevational difference was caused by heterogeneity in the underlying coarse-grained sediment, ranging between 55% on Schiermonnikoog to 92% at the Cefni marsh. A smaller percentage of the elevational differences between hummocks and depressions could also explained by the fine-grained sediment layer (Figure 3). On all four marshes, thicker fine-grained sediment layers were found on top of hummocks compared to depressions. At one extreme, hummocks on Schiermonnikoog had a 3.4 cm thicker fine-grained sediment layer than depressions, whereas this difference was limited to only 1.0 cm at the Cefni marsh. At the Cefni marsh, up to 11.5 cm of the elevational difference was caused by the underlying coarse-grained sediment. Consistently at all four marshes, the hummocks consisted of higher elevated sand bodies and this original topography was conserved under a fine-grained sediment layer. The marsh accretion rates estimated along the natural chronosequence on Schiermonnikoog (Figure 4) differed significantly between hummocks and depressions, and changed with marsh age. Accretion rates decreased as marshes aged ($F_{4,24} = 7.56$, $p < 0.001$) and were significantly higher in depressions compared to hummocks ($F_{1,24} = 10.14$, $p < 0.01$). We found no significant interaction effect between marsh age and treatment (hummocks or depressions).

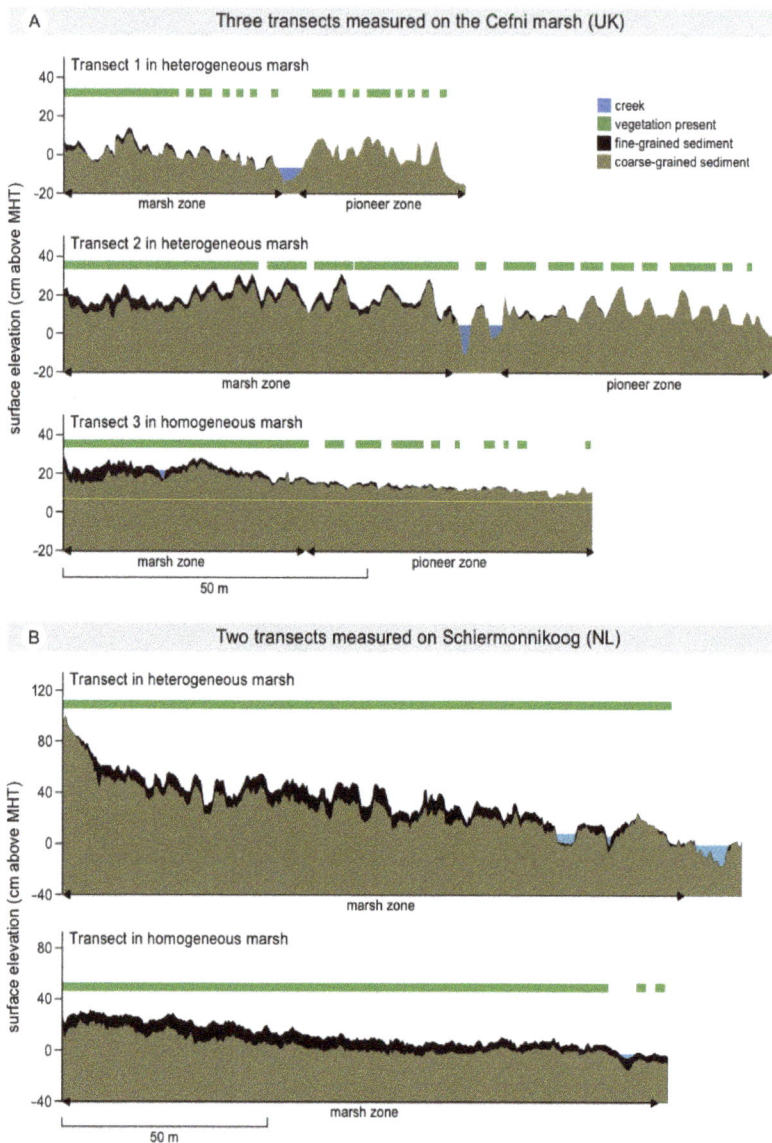

Figure 2. Transects on the Cefni marsh (**a**) and Schiermonnikoog (**b**). Light brown represents the coarse-grained sediment and dark brown represents the fine-grained sediment layer. When both hummocks and depressions were vegetated and had accumulated fine-grained sediment, we referred to it as the marsh zone. When vegetated hummocks were located on the bare intertidal flats, we referred to it as pioneer zone.

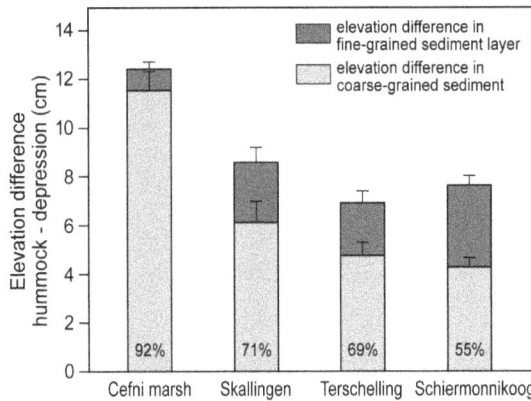

Figure 3. Elevational differences between hummocks and depressions estimated at four European sandy salt marshes. The elevational difference can arise from both heterogeneity of the underlying coarse-grained sediment and differences in thickness in the top fine-grained sediment layer. Percentages represent how much of the elevational difference, on average, were explained by the morphology of the coarse-grained sediment.

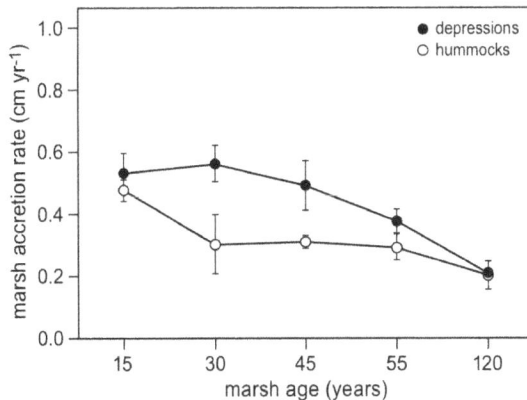

Figure 4. Marsh accretion rates (cm· year^{-1}) based on annual measurements between 2001 and 2011 on hummocks and depressions along the natural chronosequence of Schiermonnikoog.

4. Discussion

Our results support the first hypothesis that the patterns are formed on the intertidal flat before fine-grained sediment accumulated during marsh development. In all four salt marshes, more than 50% of the heterogeneity was explained by heterogeneity in the coarse-grained sediment (Figure 3). Furthermore, the transects measured at the Cefni marsh and on Schiermonnikoog also showed that the marsh platform followed the topography of the underlying coarse-grained substrate (Figure 2). Therefore, we conclude that the small-scale topographic heterogeneity was formed in the pioneer stage before the morphology of the intertidal flat was conserved under a layer of fine-grained sediment. We reject the second hypothesis that the patterns are formed after marsh development started. However, the topographic heterogeneity was enhanced during early marsh development, as we found a thicker fine-grained sediment layer on top of hummocks compared to the depressions in all four marshes (Figure 3). The marsh elevation determines for a large part whether salt-marsh plant species can successfully establish [26], and vegetation is known to increase the sediment deposition rate and reduce the erosion rate by stabilization of the soil [10–12]. Higher elevated hummocks covered by vegetation in the pioneer zone (see Figure 2, transects 1 and 2 in the pioneer zone in front of the marsh zone) will have a higher marsh accretion rate compared to the adjacent bare intertidal flat.

From the 15 year-old marsh up to the 120 year-old marsh, we found a significantly higher marsh accretion rate within the depressions (Figure 4). This higher accretion rate was mainly present in the intermediate marsh ages of 30 and 45 years old (Figure 4). This would result in dampening of the topographic heterogeneity over time. At the 55- and 120-year-old marshes, we actually found similar marsh accretion rates. All these results lead to the following conclusions: 1) heterogeneity in marsh topology was formed before the marsh developed; 2) during early marsh development, the vegetated hummocks accumulated more fine-grained sediments compared to the bare intertidal flat, which enhanced the elevational heterogeneity; 3) higher marsh accretion rates in the depressions caused the topographic heterogeneity to dampen; and 4) at mature marshes, similar marsh accretion rates between the hummocks and the depressions allowed for the patterns to persist in the marsh platform.

The persistence of small-scale topographic heterogeneity in marshes depends, for the large part, on the marsh accretion rate. The marshes included in this study were all sandy marshes with a thin fine-grained sediment layer on top of coarse-grained sediment and average marsh accretion rates of several mm· year^{-1} [23,27] depending on marsh elevation and marsh age [27]. Many mainland and estuarine marshes have very high accretion rates, up to 40 mm· year^{-1} [28–30]. In the literature, a few studies have addressed hummock formation in estuarine marshes [19,31] and marshes located within a bay [17,32]. In these studies, hummock formation occurred by active sediment trapping driven by *Puccinellia maritima* [17,32] or *Spartina anglica* [19,31]. This is in line with the results that we found that the hummocks in the pioneer zone were dominated by *Puccinellia maritima* (Table 1). According to a previous study [33], *Puccinellia maritima* and *Spartina anglica* will outcompete each other for space and light within the pioneer zone, and these authors concluded that *Puccinellia maritima* will prevail in more sandy marshes, whereas *Spartina anglica* will prevail in more clayish marshes. Therefore, hummock formation in the pioneer zone occurred in both sandy marshes with low accretion rates as well as in more clayish marshes with high accretion rates and, depending on soil type, either *Puccinellia maritima* or *Spartina anglica* caused hummocks to form in the pioneer zone. However, with increasing time, *Spartina anglica* tended to form large monospecific stands on the clayish marshes, whereas on the more sandy marshes, smaller scattered hummocks dominated by *Puccinellia maritima* remain presently [33].

This study contributes to general knowledge on the formation of marsh morphology and how topographic heterogeneity in marshes forms. Environmental heterogeneity within ecosystems can be important to boost biodiversity, which is one of the key objectives in conservation ecology [2,34]. The presence of this small-scale topographic heterogeneity in heterogeneous salt marshes likely increases biodiversity, compared to homogeneous salt marshes, and this could benefit not only primary (plant diversity) but also secondary diversity (e.g., herbivores). Additionally, heterogeneity is known to increase the resilience of ecosystems to changing environmental conditions [4,5]. Tidal marshes are very dynamic ecosystems, where the interplay between vegetation and sedimentation determines not only how the morphology of the marsh platform develops but also have a major impact on many important ecosystem functions, such as carbon sequestration [35,36], coastal protection [37], and the ability of marshes to cope with enhanced sea-level rise [8]. Understanding the feedbacks between vegetation and sedimentation, and their impact on marsh morphology, therefore, is key in the successful conservation of our coastal ecosystems.

Acknowledgments: We would like to acknowledge Harm Albers, Jan-Eise Wieringa, Elske Koppenaal and the Coastal Ecology Expedition for help with fieldwork, Dick Visser for help preparing the graphs and two anonymous reviewers for very useful comments on this manuscript. Esther Chang edited the manuscript. Many people have, throughout the years, contributed to the data collection of the SEB data on Schiermonnikoog: Dries Kuijper, Yzaak de Vries, Wolfgang Qual, Bregje Wesenbeeck, Alma de Groot and Elske Koppenaal, with special thanks to Dries Kuijper for installing the SEBs and Alma de Groot for organizing the database. This study

J. Mar. Sci. Eng. **2016**, *4*, 21

was funded by ZKO-NWO, project number 83908320, the Dutch Organization for Scientific Research as well as the Schure-Beijerinck-Popping Fund.

Author Contributions: Conceived and designed the study: K. Elschot and J.P. Bakker. Performed the study: K. Elschot. Collection of field data: K. Elschot. Analysed the data: K. Elschot. Wrote the paper: K. Elschot and J.P. Bakker.

Conflicts of Interest: The authors declare no conflict of interest.

References

1. Hooper, D.U.; Chapin, F.S.; Ewel, J.J.; Hector, A.; Inchausti, P.; Lavorel, S.; Lawton, J.H.; Lodge, D.M.; Loreau, S.; Naeem, S.; *et al.* Effects of biodiversity on ecosystem functioning: A consensus of current knowledge. *Ecol. Monogr.* **2005**, *75*, 3–35. [CrossRef]
2. Ricklefs, R.E. Environmental heterogeneity and plant species diversity: A hypothesis. *Am. Nat.* **1977**, *111*, 376–381. [CrossRef]
3. Costanza, J.K.; Moody, A.; Peet, R.K. Multi-scale environmental heterogeneity as a predictor of plant species richness. *Landsc. Ecol.* **2011**, *26*, 851–864. [CrossRef]
4. Snyder, R.E.; Chesson, P. How the spatial scales of dispersal, competition, and environmental heterogeneity interact to affect coexistence. *Am. Nat.* **2004**, *164*, 633–650. [CrossRef] [PubMed]
5. Godfree, R.; Lepschi, B.; Reside, A.; Bolgers, T.; Robertson, B.; Marshall, D.; Carnegie, M. Multiscale topoedaphic heterogeneity increases resilience and resistance of a dominant grassland species to extreme drought and climate change. *Glob. Change Biol.* **2011**, *17*, 943–958. [CrossRef]
6. Thomas, C.D.; Cameron, A.; Green, R.E.; Bakkenes, M.; Beaumont, L.J.; Collingham, Y.C.; Erasmus, B.F.N.; Ferreira de Siqueira, M.; Grainger, A.; Hannah, L.; *et al.* Extinction risk from climate change. *Nature* **2004**, *427*, 145–148. [CrossRef] [PubMed]
7. Fitzgerald, D.M.; Fenster, M.S.; Argow, B.A.; Buynevich, I.V. Coastal impacts due to sea-level rise. *Annu. Rev. Earth Planet. Sci.* **2008**, *36*, 601–647. [CrossRef]
8. Kirwan, M.L.; Guntenspergen, G.R.; D'Alpaos, A.; Morris, J.T.; Mudd, S.M.; Temmerman, S. Limits on the adaptability of coastal marshes to rising sea level. *Geophys. Res. Lett.* **2010**, *37*, L23401. [CrossRef]
9. Olff, H.; de Leeuw, J.; Bakker, J.P.; Platerink, R.J.; van Wijnen, H.J. Vegetation succession and herbivory in a salt marsh: Changes induced by sea level rise and silt deposition along an elevational gradient. *J. Ecol.* **1997**, *85*, 799–814. [CrossRef]
10. Peralta, G.; van Duren, L.A.; Morris, E.P.; Bouma, T.J. Consequences of shoot density and stiffness for ecosystem engineering by benthic macrophytes in flow dominated areas: A hydrodynamic flume study. *Mar. Ecol. Progr. Ser.* **2008**, *368*, 103–115. [CrossRef]
11. Mudd, S.M.; D'Alpaos, A.; Morris, J.T. How does vegetation affect sedimentation on tidal marshes? Investigating particle capture and hydrodynamic controls on biologically mediated sedimentation. *J. Geophys. Res.* **2010**, *115*, 1–14. [CrossRef]
12. Day, J.W.; Kemp, G.P.; Reed, D.J.; Cahoon, D.R.; Boumans, R.M.J.; Suhayda, J.M.; Gambrell, R. Vegetation death and rapid loss of surface elevation in two contrasting Mississippi delta salt marshes: The role of sedimentation, autocompaction and sea-level rise. *Ecol. Eng.* **2011**, *37*, 229–240. [CrossRef]
13. Temmerman, S.; Bouma, T.J.; van de Koppel, J.; van der Wal, D.; de Vries, M.B.; Herman, P.M.J. Vegetation causes channel erosion in a tidal landscape. *Geology* **2007**, *35*, 631–634. [CrossRef]
14. Fagherazzi, S.; Kirwan, M.L.; Mudd, S.M.; Guntenspergen, G.T.; Temmerman, S.; D'Alpaos, A.; van de Koppel, J.; Rybczyk, J.M.; Reyes, E.; Craft, C.; *et al.* Numerical models of salt marsh evolution: Ecological, geomorphic, and climatic factors. *Rev. Geophys.* **2012**, *50*, RG1002. [CrossRef]
15. Vandenbruwaene, W.; Bouma, T.J.; Meire, P.; Temmerman, S. Bio-geomorphic effects on tidal channel evolution: Impact of vegetation establishment and tidal prism change. *Earth Surf. Process. Landf.* **2013**, *38*, 122–132. [CrossRef]
16. Gray, A.J.; Bunce, R.G.H. The ecology of Morecambe Bay. VI. Soils and vegetation of the salt marshes: A multivariate approach. *J. Appl. Ecol.* **1972**, *9*, 221–234. [CrossRef]
17. Langlois, E.; Bonis, A.; Bouzillé, J.B. Sediment and plant dynamics in saltmarshes pioneer zone: Puccinellia maritima as a key species? *Estuar. Coast. Shelf Sci.* **2003**, *56*, 239–249. [CrossRef]

18. Stribling, J.; Cornwell, J.; Glahn, O. Microtopography in tidal marshes: Ecosystem enginering by vegetation? *Estuar. Coasts* **2007**, *30*, 1007–1015. [CrossRef]

19. Balke, T.; Klaassen, P.C.; Garbutt, A.; van der Wal, D.; Herman, P.M.J.; Bouma, T.J. Conditional outcome of ecosystem engineering: A case study on tussocks of the salt marsh pioneer Spartina anglica. *Geomorphology* **2012**, *153–154*, 232–238. [CrossRef]

20. Oliver, F.W. The Bouche d'Erqui in 1907. *New Phytol.* **1907**, *6*, 244–252.

21. Yapp, R.H.; John, D.; Jones, O.T. The salt marshes of the Dovey Estuary. Part II. The salt marshes. *J. Ecol.* **1917**, *5*, 65–103. [CrossRef]

22. De Groot, A.V.; Veeneklaas, R.M.; Kuijper, D.P.J.; Bakker, J.P. Spatial patterns in accretion on barrier-island salt marshes. *Geomorphology* **2011**, *134*, 280–296. [CrossRef]

23. Packham, J.R.; Liddle, M.J. The Cefni salt marsh, Anglesey, and its recent development. *Field Stud.* **1970**, *3*, 331–356.

24. Boumans, R.M.J.; Day, J.W. High precision measurements of sediment elevation in shallow coastal areas using a sedimentation-erosion table. *Estuaries* **1993**, *16*, 375–380. [CrossRef]

25. Nolte, S.; Koppenaal, E.C.; Esselink, P.; Dijkema, K.S.; Schuerch, M.; de Groot, A.V.; Bakker, J.P.; Temmerman, S. Measuring sedimentation in tidal marshes: A review on methods and their applicability in biogeomorphological studies. *J. Coast. Conserv.* **2013**, *17*, 301–325. [CrossRef]

26. Davy, A.J.; Brown, M.J.H.; Mossman, H.L.; Grant, A. Colonization of a newly developing salt marsh: Disentangling independent effects of elevation and redox potential on halophytes. *J. Ecol.* **2011**, *99*, 1350–1357. [CrossRef]

27. Van Wijnen, H.J.; Bakker, J.P. Nitrogen accumulation and plant species replacement in three salt marsh systems in the Wadden Sea. *J. Coast. Conserv.* **1997**, *3*, 19–26. [CrossRef]

28. Oenema, O.; Delaune, R.D. Accretion rates in salt marshes in the Eastern Scheldt, south-west Netherlands. *Estuar. Coast. Shelf Sci.* **1988**, *26*, 379–394. [CrossRef]

29. Dijkema, K.S.; Kers, A.S.; van Duin, W.E. Salt marshes: Applied long-term monitoring salt marshes. In *Wadden Sea Ecosystem No. 26*; Trilateral Monitoring and Assessment Group, Common Wadden Sea Secretariat: Wilhelmshaven, Germany, 2010; Volume 26, pp. 35–40.

30. Suchrow, S.; Pohlmann, N.; Stock, M.; Jensen, K. Long-term surface elevation changes in German North Sea salt marshes. *Estuar. Coast. Shelf Sci.* **2012**, *98*, 71–83. [CrossRef]

31. Van Wesenbeeck, B.K.; van de Koppel, J.; Herman, P.M.J.; Bouma, T.J. Does scale-dependent feedback explain spatial complexity in salt-marsh ecosystems? *Oikos* **2008**, *117*, 152–159. [CrossRef]

32. Langlois, E.; Bonis, A.; Bouzillé, J.B. The response of Puccinellia maritima to burial: A key to understanding its role in salt-marsh dynamics? *J. Veg. Sci.* **2001**, *12*, 289–297. [CrossRef]

33. Scholten, M.; Rozema, J. The competitive ability of Spartina anglica on Dutch salt marshes. In *Spartina Anglica, a Research Review*; Gray, A.J., Benham, P.E.M., Eds.; Institute of Terrestrial Ecology: London, UK, 1990; pp. 39–47.

34. Stein, A.; Gerstner, K.; Kreft, H. Environmental heterogeneity as a universal driver of species richness across taxa, biomes and spatial scales. *Ecol. Lett.* **2014**, *17*, 866–880. [CrossRef] [PubMed]

35. Mcleod, E.; Chmura, G.L.; Bouillon, S.; Salm, R.; Björk, M.; Duarte, C.M.; Lovelock, C.E.; Schlesinger, W.H.; Silliman, B.R. A blueprint for blue carbon: Toward an improved understanding of the role of vegetated coastal habitats in sequestering CO_2. *Front. Ecol. Environ.* **2011**, *9*, 552–560. [CrossRef]

36. Elschot, K.; Bakker, J.P.; Temmerman, S.; van de Koppel, J.; Bouma, T.J. Ecosystem engineering by large grazers enhances carbon stocks in a tidal salt marsh. *Mar. Ecol. Progr. Ser.* **2015**, *537*, 9–21. [CrossRef]

37. Gedan, K.B.; Kirwan, M.L.; Wolanski, E.; Barbier, E.B.; Silliman, B.R. The present and future role of coastal wetland vegetation in protecting shorelines: Answering recent challenges to the paradigm. *Clim. Change* **2011**, *106*, 7–29. [CrossRef]

Journal of
Marine Science and Engineering

MDPI

Article

Assessing Embayed Equilibrium State, Beach Rotation and Environmental Forcing Influences; Tenby Southwest Wales, UK

Tony Thomas [1,*], **Allan Williams** [1,2], **Nelson Rangel-Buitrago** [3], **Michael Phillips** [1] and **Giorgio Anfuso** [4]

[1] Coastal and Marine Research Group, University of Wales Trinity Saint David (Swansea), Mount Pleasant, Swansea, Wales SA1 6ED, UK; allan.williams@uwtsd.ac.uk (A.W.); mike.phillips@uwtsd.ac.uk (M.P.)
[2] Centro Interdisciplinar de Ciências Sociais, Universidade Nova de Lisboa – Faculdade de Ciências Sociais e Humanas (CICS.NOVA.FCSH.UNL), Lisboa 1069-061, Portugal
[3] Facultad de Ciencias Básicas, Programa de Física, Grupo de Geología, Geofísica y Procesos Litorales, Km 7 Antigua vía Puerto Colombia, Barranquilla, Atlántico 080020, Colombia; nelsonrangel@mail.uniatlantico.edu.co
[4] Ciencias de la Tierra, Universidad de Cadiz, Puerto Real 11510, Spain; giorgio.anfuso@uca.es
* Correspondence: tony.thomas@uwtsd.ac.uk; Tel.: +44-1792-481463

Academic Editor: Gerben Ruessink
Received: 9 November 2015; Accepted: 29 March 2016; Published: 8 April 2016

Abstract: The morphological change of a headland bay beach—Tenby, West Wales, UK—was analysed over a 73-year period (1941–2014). Geo-referenced aerial photographs were used to extract shoreline positions which were subsequently compared with wave models based on storm event data. From the 1941 baseline, results showed shoreline change rates reduced over time with regression models enabling a prediction of shoreline equilibrium *circa* 2061. Further temporal analyses showed southern and central sector erosion and northern accretion, while models identified long-term plan-form rotation, *i.e.*, a negative phase relationship between beach extremities and a change from negative to positive correlation within the more stable central sector. Models were then used in conjunction with an empirical 2nd order polynomial equation to predict the 2061 longshore equilibrium shoreline position under current environmental conditions. Results agreed with previous regional research which showed that dominant south and southwesterly wave regimes influence south to north longshore drift with counter drift generated by less dominant easterly regimes. The equilibrium shoreline was also used to underpin flood and inundation assessments, identifying areas at risk and strategies to increase resilience. UK shoreline management plans evaluate coastal vulnerability based upon temporal epochs of 20, 50 and 100 years. Therefore, this research evaluating datasets spanning 73 years has demonstrated the effectiveness of linear regression in integrating temporal and spatial consequences of sea level rise and storms. The developed models can be used to predict future shoreline positions aligned with shoreline management plan epochs and inform embayed beach shoreline assessments at local, regional and international scales, by identifying locations of vulnerability and enabling the development of management strategies to improve resilience under scenarios of sea level rise and climate change.

Keywords: morphological change; beach rotation; GIS platform; equilibrium assessments Environmental forcing influences

1. Introduction

In nature, many coastline sections are located in the lee of natural or artificial headlands that control beach evolution and feature curved shoreline geometry, best described as a zeta log spiral or parabolic curve. More than 50% of the world's coastlines are representative of this morphology [1]. Within this environment, a number of factors contribute and influence complex behavioral patterns that cause reshaping of both the beach profile and plan-form. These include underlying geology, sediment volume and composition, and external environmental conditions, such as incident wave characteristics *i.e.*, height, period, and particularly direction [2]. These determine induced sediment transport both in onshore/offshore and alongshore directions [3–5]. The nearshore bathymetry and the shelter induced by the beach headlands and local offshore islands further complicate beach behavior [6]. Additionally, morphological variability occurs at temporal scales that vary from a few seconds to several years [7]. Research often focuses on beaches in micro/mesoscale tidal environments and at regional scales, with multiple beaches studied at decadal timescales. Morphological responses of embayed beaches to storm and gale forcing have also been studied in the Northern Hemisphere (for example, [2,8–10]) and the Southern Hemisphere by, amongst others, [11] and [12]. However, few investigations involve varying spatial and temporal scales, particularly within macrotidal coastal environments, some notable exceptions being [13] and [14].

Unlike macrotidal beach work carried out in this research field, almost all embayed beach studies are carried out on beaches with microtidal or mesotidal ranges. Research on macrotidal embayed beaches is required to establish behavior under wide ranges of wave and tidal conditions [15]. Recent worldwide micro/mesotidal range studies focused on small groups of embayed beaches, with varying coastal aspects and geological constraints [15–26]. Apart from, for example, [11] and [25], few comparative studies detail single embayed beaches, notable macrotidal exceptions being [22] and Thomas *et al.*'s [27–29] work within the present study region. A typical characteristic of an embayed beach is the close correspondence between beach planform and refraction patterns associated with prevailing waves [1]. Consequently, a comparison of observed and predicted bay geometry can reveal the stability of embayed beaches, *i.e.*, the parabolic beach concept.

Beach rotation refers to periodic lateral sediment movement towards alternating ends of embayed beaches, causing shoreline realignment in response to shifts in incident wave direction [30]. Waves from one direction produce longshore sediment movement that accumulates against the downdrift headland resulting in erosion at the updrift. Waves from another direction can produce the reverse and the net result is an apparent rotation of the beach planform [15]. Rotational trends can be seasonal [30] or longer term related to climate variation [12,28]. Most research has been conducted on beaches with microtidal or mesotidal ranges, but the multi-decadal level changes in this paper are focused on the beach subaerial zone (based on the vegetation line). In this environment, seminal studies on sandy beaches have been made by [11,18–20,24,31] and on a gravel beach by [32].

This research assesses long-term shoreline evolution expressed through cross-shore migration, rotation and consecutive realignment, utilizing the vegetation line as a proxy shoreline change indicator (see for example [33]). Results were compared and contrasted with historic wind, and more recent storm forcing variables, to identify cause and effect. Identified long lasting changes in coastal processes led to development of temporal and spatial regression models describing the shoreline evolution. These established links and relationships have important consequences for embayed beach management strategies.

2. Physical Background

The Bristol Channel on the West coast of The United Kingdom (Figure 1A) separates Wales from Southwest England. There are a number of large embayments along the margins of the outer Bristol Channel (Figure 1B), Barnstaple, Bridgewater, Swansea and Carmarthen. Carmarthen Bay is a long sweeping embayment (30 km), described by [16] as displaying highly curved geometry (Figure 1C). Tenby Peninsula, on the western side of the bay (Figure 1C), is characterized mainly by rocky cliffs and small embayments that contain pocket beaches formed as a consequence of erosion of the softer mudstone rich Carboniferous Coal Measures and Millstone Grit [34].

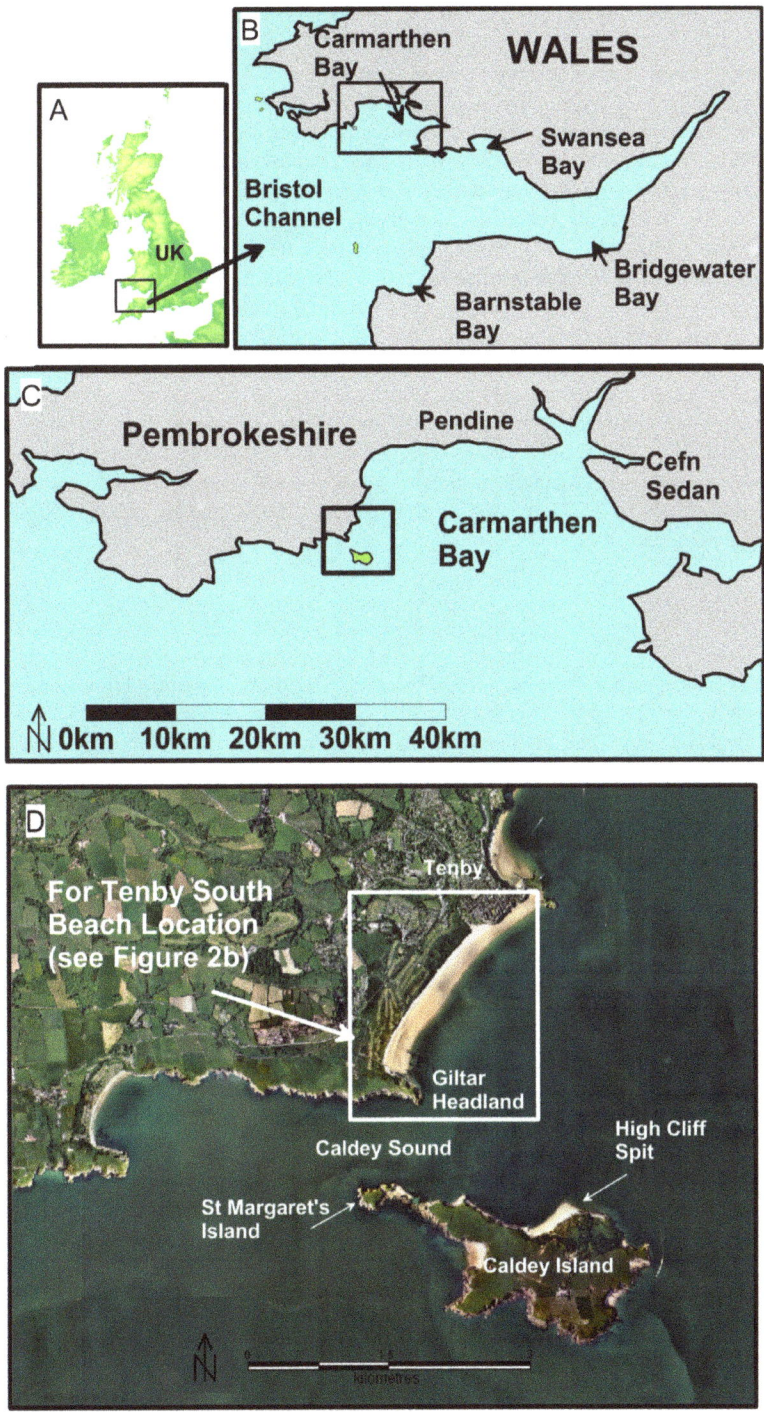

Figure 1. Locality of the study area, (**A**) United Kingdom; (**B**) Bristol Channel (**C**) Carmarthen Bay and (**D**) The study region.

The study area (51°39′36″ N; −4°42′36″ W) is located between two Carboniferous Limestone headlands, Giltar to the south and Tenby to the north [35], the distance between headlands being approximately 2 km. The embayment profile is shallow and concave, with a wide (*circa* 250 m) sandy intertidal zone. This gives way to a limestone shingle backshore overlain by a dune system (920×10^3 m^2; [27]), shingle is periodically exposed during storms and high spring tides, and extensive vegetation retards sediment movement from the dune field to the intertidal zone. The seaside town of Tenby to the north is a heavily urbanized coastal area, where tourist activity strongly supports the regional economy. To the south, the dunes, marshes and Giltar Headland promontory are ecologically important conservation areas. Semi-diurnal and macrotidal, the region has a mean spring tidal range of 7.5 m [28], with a Mean High Water Spring Tide level (MHWST) of 5 m Above Ordnance Datum (AOD). Incident offshore waves generally approach from the southwest with an average wave height of *circa* 1.2 m and associated mean periods of 5.2 s [27]. Storm waves of 7 m, with periods of 9.3 s, constitute less than 5% of the wave record. Longshore drift from south to north is influenced by heavily refracted southwesterly Atlantic swell waves which undergo diffraction as they encounter the south Pembrokeshire coast and offshore islands (Caldey and St Margaret's; Figure 1D). Between November 2013 and March 2014, a total of 32 storms (average $h_s > 3.4$ m) were recorded generating average waves that reached 4.7 ± 1.26 m with associated periods of 7.9 ± 1.00 s, and some waves reaching 9.3 m with periods of 12 s. These events caused widespread erosion and structural damage along the Pembrokeshire coastline.

3. Methods

3.1. Shoreline Change Modelling (1941–2014)

This paper builds upon Thomas *et al.*'s [27] centurial work, by utilizing additional recent data to assess morphological change between 1941 and 2014. Sixteen aerial photographs (1941–2014; Figure 2a), all geo-rectified in a Geographic Information System (Mapinfo®, Pitney Bowes Sofware Inc, New York, NY, USA) to the British grid reference system, were used to extract shoreline position. The figures vary, some incorporate a narrow aerial extent *versus* a wider aerial extent. Errors in aerial photographs can be of the order of 7.5 m–8.9 m caused by distortion and the digitizing process itself [36,37]. In this paper, the former was mitigated using 600 dpi images and the latter assessed for accuracy using Root Mean Square Error (RMSE) (see [38] for theoretical interpretations and [27,39,40] for practical applications). Survey control points (Figure 2b) established using RTK Network GPS 1200+ with an average of 200 readings taken at every control point ensured accuracy. Subsequently, RMSE was calculated using $\text{RMSE} = [(\sum(N_c - N_t)^2 + \sum(E_c - E_t)^2)/n]^{1/2}$, where; N_t and E_t are calculated co-ordinates from the photo transformation, N_c and E_c are control coordinates and n is the total number of data points. Table 1 shows respective source document scales and RMSE values. The average RMSE error for the aerial photographs was 1.76 m. The vegetation line was chosen as shoreline change indicator, as this could be easily identified on all aerial photographs and is valuable for investigating long term trends [33]. The corresponding extracted shoreline position was imported into the Regional Morphological Analysis Programme (RMAP; see [41]) a module within the Coastal Engineering Design and Analysis System (CEDAS), where inter-survey and cumulative shoreline changes were evaluated. Temporal change together with rotation analysis was achieved using the 1941 shoreline position as a proxy baseline. The shoreline positions measured from aerial photographs were extracted along 12 theoretical transects (T1–T12), spaced approximately 150 m apart (Figure 2c). Linear regression and correlation analysis within the dataset constituted of the 12 shoreline signals were used to characterize the shoreline planform evolution.

(a)

(b)

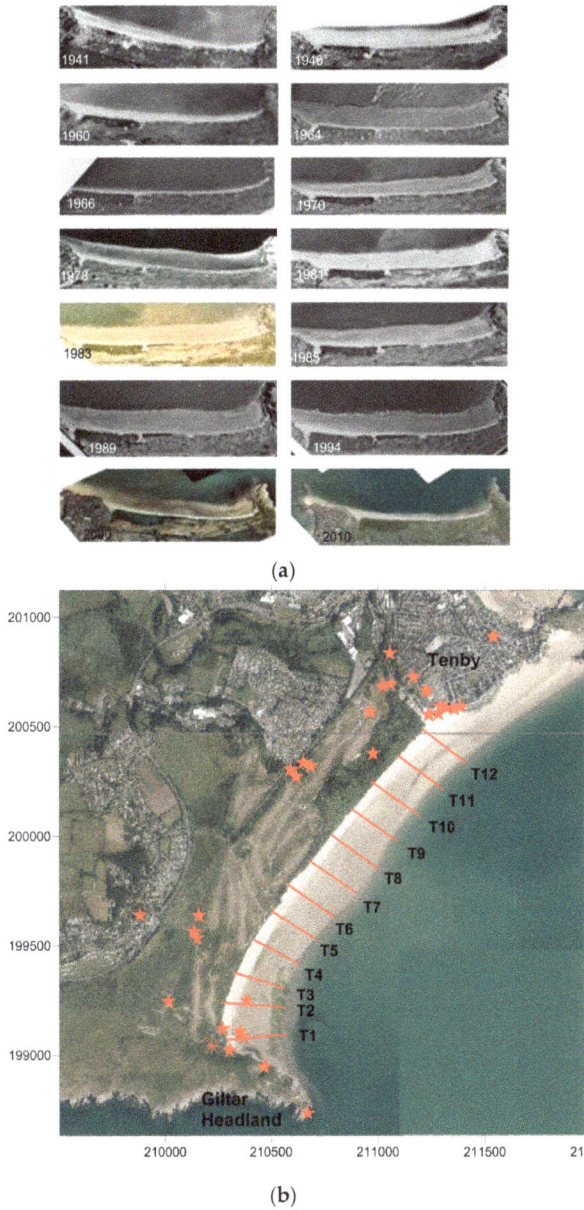

Figure 2. (**a**) aerial photographs depicting South Beach, Tenby prior to the geo-referencing process for the period 1941–2010 and utilized in this research and (**b**) 2014 aerial photograph showing the study area detailing the position of the permanent control points (red stars) used to aid geo-rectification and check RMSE results and transect locations from which morphological variables were computed.

Table 1. Aerial photographic source document scales and RMSE results after digitizing.

Year	Type	Source	RMSE	Scale
1941	Aerial Photograph	Welsh Assembly	1.5	1:10,000
1946	Aerial Photograph	Welsh Assembly	2.4	1:10,000
1960	Aerial Photograph	Welsh Assembly	1.9	1:10,000
1966	Aerial Photograph	Welsh Assembly	1.6	1:10,000
1970	Aerial Photograph	Welsh Assembly	1.1	1:10,000
1978	Aerial Photograph	Welsh Assembly	1.5	1:10,000
1981	Aerial Photograph	Welsh Assembly	1.2	1:10,000
1983	Aerial Photograph	Welsh Assembly	1.3	1:10,000
1985	Aerial Photograph	Welsh Assembly	2.2	1:10,000
1989	Aerial Photograph	Welsh Assembly	1.5	1:10,000
1994	Aerial Photograph	Welsh Assembly	0.9	Digitised 40 cm resolution
2000	Aerial Photograph	Getmapping	2.4	Digitised 40 cm resolution
2006	Aerial Photograph	Ordnance Survey	1.5	Digitised 40 cm resolution
2010	Aerial Photograph	Welsh Assembly	1.5	Digitised 40 cm resolution
2014	Aerial Photograph	Welsh Assembly	1.5	Digitised 40 cm resolution

3.2. Wind and Wave Data Characterization

Constant beach profile and plan-form reshaping is caused primarily by incident wave characteristics, *i.e.*, height, period, and particularly direction. These determine wave induced sediment transport both in onshore/offshore and alongshore directions [1–3]. Waves and their directional components are used as direct input into coastal engineering or coastal zone management calculations [30]. However, wind is the underlying cause of most sources of coastal flood and erosion risk but wind data is rarely used in these calculations [5]. Within the region of study, synthesized wind and wave time series from meteorological numerical models that were suitable to be inputted into wave prediction models have only recently been made available. However, wind speed and directional data from the early 1940s were available and obtained from the UK Meteorological Office, enabling direct comparisons to be made between shoreline change and these environmental forcing agents. Offshore wind speed and direction data was captured at approximately 3-h intervals at a point southeast of the study area (51°24′00″ N; −5°00′00′ W). Some of the early data was missing but, nevertheless, the dataset contained *circa* 147,000 independent values.

Based on calculated storm wave statistics from south-easterly, southerly and south-westerly directions, storm wave statistics were characterized using significant wave height, period and direction. Data was captured at 1 h intervals between 1998 and 2013 from a waverider buoy located southwest of the study location (NOAA station number 62303; 51°36′00″ N; 4°34′48″ W). Subsequently, wave energy, alongside storm frequency, power and class were computed using the methodologies of [40–47], with a detailed discussion and practical use given by [5] and, in all, a total of 267 storm events were identified during the assessed 15-year period. This approach differs from that adopted by [42] and [43] within the same study region as they used a minimum wave height of 1.5 m when characterizing waves that were capable of imposing morphological change at Tenby and nearby Pendine and Cefn Sedan, respectively (Figure 1C). The storm climate waves used in this research were characterized using a minimum wave height of 3.4 m (*i.e.*, Hs ⩾ 3.4m) because they represented rare events in Carmarthen Bay only occurring 8% of the 15-year period recorded.

Wave data were subsequently modeled using the Regional Coastal Processes WAVE propagation model (RCPWAVE), a module within the Nearshore Evolution MOdelling System (NEMOS), which forms an integral part of the Coastal Engineering Design & Analysis System (CEDAS 4.03, Veritech Enterprises, Arlington, MA, USA). RCPWAVE is a two-dimensional steady-state and modified form of the "mild slope" equation for monochromatic waves and simulates linear plane wave propagation over arbitrary bathymetry. Originally developed by [44] and documented by [45], the model considers shoaling, refractive and bottom-induced diffractive effects outside the surf zone where wave reflection and energy losses are considered negligible. The wave model grid offshore boundaries were generated using the GRId GENeration (GRIDGEN) module within NEMOS. The offshore model boundary was restricted to the 26 m isobath, as this was the optimum depth at which the Meteorological Office originally computed wave height, period and directional components. A rectangular computational grid of 10 m × 10 m square mesh that encompasses the coastal region was then used to predict wave conditions at Mean High Water Spring Tide (MHWST) wave conditions. The established wave direction vectors were subsequently used to assess sediment pathways fronting South Sands.

3.3. Equilibrium Model

To predict shoreline equilibrium, data were input into an empirical formulae based upon the Parabolic Bay Shape Equation (PBSE) developed by [46] and [47]. In their manual application, a vertical aerial photograph is used to obtain the main model variables (β and R_0; Figure 3) for input into the equation. $R_0/R_n = C_0 + C_1 (\beta/\theta) + C_2 (\beta/\theta)^2$. In the case of a single up-coast headland, the distance R_0, *i.e.*, the length of a control line drawn from the end of the headland to the nearest point on the down-coast shoreline where the shoreline is parallel with the predominant wave crest, is estimated directly from the aerial photograph. In this research, the distance from the headland to the downcoast control point (R_0) and wave approach angle (β) was estimated by predicting the zero migration date from a linear regression trend. The established date is then used to extrapolate both northern shoreline position and the region of rotation, once again based upon linear regression trends. The angle β (30°) was sub-tended between a line joining the predicted region of rotation and northern shoreline position and the control line R_0 (1357 m). The distance R_n, measured from the end of the up-coast headland defines the shoreline location at a varying angle θ. The coefficients C_o (0.045), C_1 (1.146) and C_2 (−1.94) were derived from the seminal workings of [46] and [47]. Finally, a GPS topographic survey was performed to establish ground levels along the predicted southern shoreline position in order to assess both flood vulnerability to imposed morphological change.

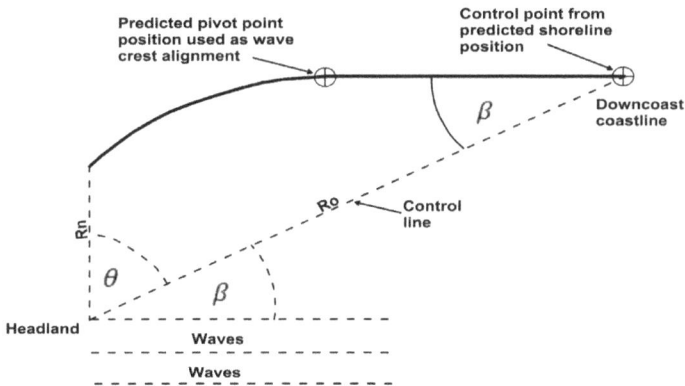

Figure 3. Definition sketch of the parabolic model used to assess equilibrium.

4. Results

The following qualitative and quantitative assessment of 2 km of shoreline using aerial photographic evidence provides an illustration of shoreline changes from 1941 to 2014.

4.1. Temporal Change (1941–2014)

Once the aerial photographs were geo-referenced, shoreline positions were extracted (Figure 4). The southern shoreline retreated consistently throughout the assessment. Two concrete groynes were constructed in the 1930s to protect a shooting range sited within the southern dune system, but their design caused downdrift erosion and arguably augmented recession rates. The shooting range was eventually relocated landward of the dune system and the groynes were eventually demolished in the early 1990s. All that remains of the original shooting range are partially collapsed butts (Figure 4 beach sector a). The shoreline now evolves naturally taking a typical embayed shape. Crucially, the construction of a gabion wall in the early 1980s has arguably prevented shoreline retreat (Figure 4 beach sector b). However, the structure was also outflanked and its presence caused downdrift erosion that exposed a large blow out to direct wave attack. The gabion wall was destroyed and the blow out collapsed during the 2013/2014 winter storms. Part of the gabion wall was reconstructed and the dune system around the blow out is already showing signs of recovery. The shoreline change trends within this sector changed from erosive to accretive resulting from additional sediment input from the 20 m high blow out. Apart from the 1960 shoreline that appeared to be eroded, the northern sector showed gradual accretion throughout the assessed period (Figure 4 beach sector c).

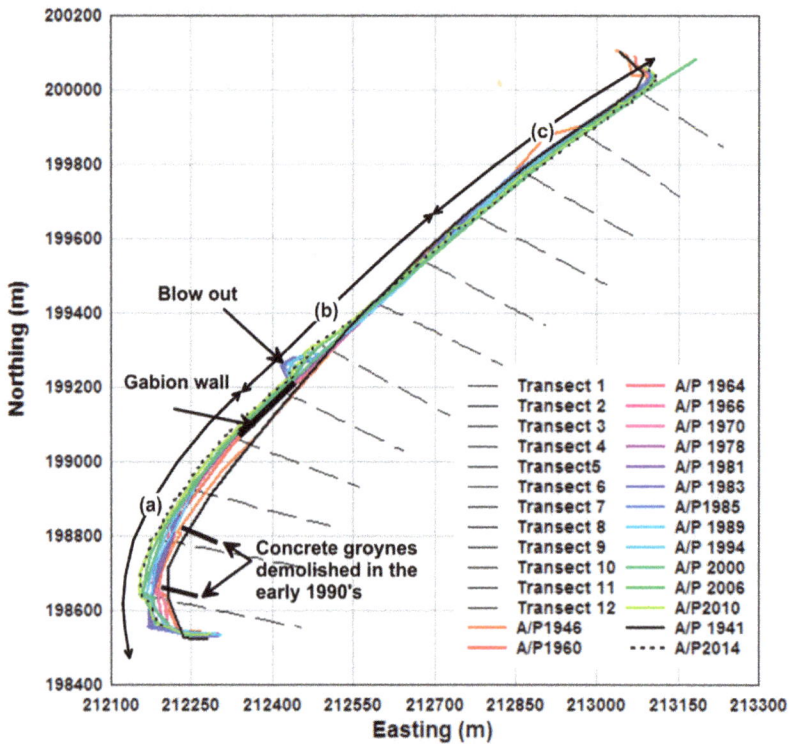

Figure 4. Shoreline positional change 1941–2014 produced from Table 4 represented graphically for (a) southern; (b) central and (c) northern beach sectors.

Table 2 was constructed by direct measurements of the shoreline position along each transect presented in Figure 2b compared to the 1941 baseline. Timeseries show a landward excursion of the southern shore with an average overall loss of 55 m (T1–T4; Figure 5a), in the central sector, the southernmost transects (T5–T6) eroded by *circa* 31m and the northernmost accreted by *circa* 10 m (T8; Figure 5b), the shoreline at T7 remained stable (−1 m). In contrast to the south, northern shores accreted albeit with more variation through time resulting in an average overall gain of 16 m (T9–T12; Figure 5c).

Table 2. Shoreline position by transect, in relation to the 1941 baseline (Note: negative values depict shoreline retreat, positive values depict shoreline advance and all values are in meters).

Timescale (Years)	Oscillation Point	T1	T2	T3	T4	T5	T6	T7	T8	T9	T10	T11	T12	Average T1-T4	Average T5-T8	Average T9-T12
1941		0	0	0	0	0	0	0	0	0	0	0	0	0	0	0
1946	630.5	-3	-9	-14	-14	2	2	22	24	-1	0	7	4	-10	13	3
1960	780.9	2	-19	-25	-27	-7	0	16	18	-1	4	-3	6	-17	7	1
1964	804.8	-10	-23	-36	-34	-12	3	21	24	4	1	1	14	-26	9	5
1966	796.5	-9	-22	-36	-36	-13	4	24	27	5	2	0	12	-26	11	5
1970	818.6	-19	-19	-36	-36	-16	3	26	26	5	4	2	13	-27	10	6
1978	841.4	-33	-20	-38	-37	-25	0	26	26	6	2	4	14	-32	7	6
1981	874.7	-37	-22	-40	-40	-28	-3	23	26	6	0	1	9	-35	5	4
1983	861.5	-37	-21	-39	-36	-27	-3	26	28	9	2	3	13	-33	6	7
1985	885.8	-36	-26	-40	-37	-25	-5	23	24	5	-2	-2	6	-34	4	2
1989	856.7	-40	-28	-40	-37	-26	-4	28	32	12	7	7	16	-36	7	10
1994	873.2	-45	-32	-42	-39	-26	-15	27	31	13	4	7	16	-39	4	10
2000	911.7	-53	-39	-41	-38	-23	-20	26	27	23	16	9	17	-43	2	16
2006	864	-53	-43	-40	-39	-28	-13	19	35	21	12	17	18	-43	3	17
2010	978.8	-54	-48	-45	-42	31	-25	-1	10	18	10	13	14	-47	4	14
2014	978.8	-55	-53	-49	-49	-31	-32	-1	10	21	8	18	17	-52	-14	16

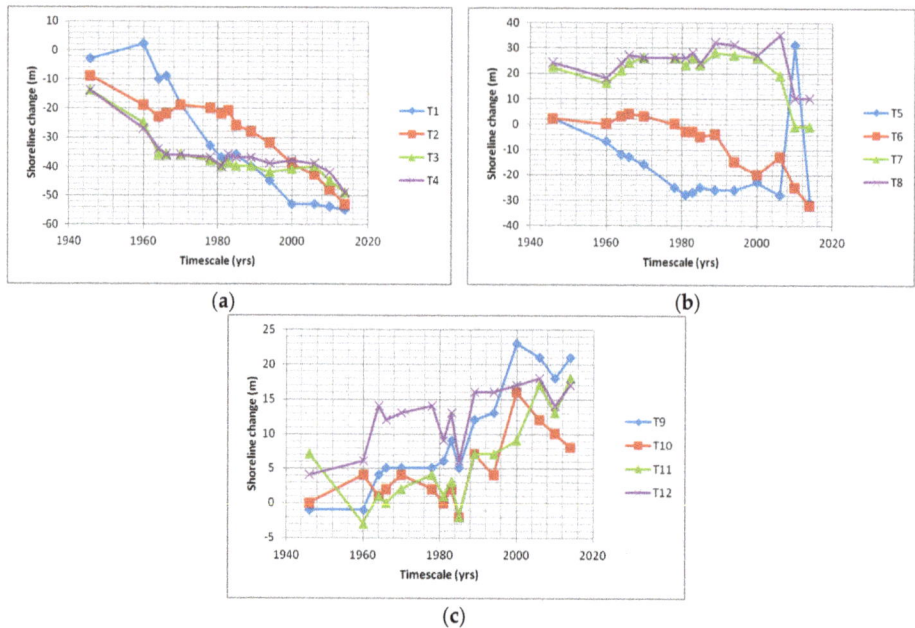

Figure 5. Transect change timeseries from the 1941 baseline, for (a) southern; (b) central and (c) northern beach sectors.

An assessment of temporal changes showed high correlation indicating a consistent trend of southern shoreline erosion given by the regression equation $y = -0.56x + 1071$, while the regression model coefficient of determination ($R^2 = 97\%$) showed that a significant percentage of spatial variation was explained by a constant migration rate ($p < 0.01$; Figure 6a). With lower correlation, northern shores accreted and the regression equation $y = 0.24x - 461$ ($R^2 = 73\%$) showed that a high percentage of spatial variation could be explained by a constant migration rate ($p < 0.01$; Figure 6b). Similar to the southern sector the central sector shoreline also eroded ($y = -0.23x + 466$) and even though the R^2 value explained just over half of the spatial variation through time ($R^2 = 58\%$), results were still statistically significant at the 95% confidence level (Figure 6c).

RMAP was utilized to compute the cumulative change rates shown in Table 3. The programme compares respective shoreline positions against a predetermined landward baseline at 10 m intervals along the beach frontage.

A regression model constructed using Table 3 cumulative data showed that a statistically high positive correlation and marked relationship existed (Figure 6d). The regression model demonstrated that between 1941 and 2014, a linear trend explained over half of the overall shoreline rates of change. ($R^2 = 55\%$; $p < 0.01$) and suggested that there was a reduction in shoreline retreat through time. Observed rates of change between 1941 and 1946 are substantial in relation to all other values. Therefore, a regression model was constructed with this value removed, once again highlighting with a very high positive correlation that a linear trend could explain a significantly high percentage in overall shoreline rates of change ($R^2 = 76\%$; $p < 0.01$). This indicates continued shoreline retreat, decreasing in severity over time. The results were heavily influenced by the initial shoreline response to the construction of the groynes in the southern sector.

The 2014 vegetation shoreline indicator, delineated by a solid red line, was superimposed upon the 1941 aerial photograph (Figure 7a) and highlights the significant erosive trend in the southern beach sector, central stability and northern advance throughout the assessment period. Overall shoreline rates of change between 1941 and 2014 (Figure 7b) confirmed previous trends showing that southern shores retreated at a maximum rate of 1.18 m/year, contrasting against a maximum northerly advance of 0.4 m/year. The rotation point was observed near the beach centre at *circa* 850 m alongshore from Giltar Headland. Overall, the frontage of South Beach showed a recession trend (*circa* 18 m; Table 3) throughout the 73-year period.

Table 3. The shoreline change rate record. (Note: negative values depict shoreline retreat, positive values depict shoreline advance and all values are in meters).

Timescale		Time Span from 1941 Baseline	Inter-survey Change Rate m· year^{-1}	Cumulative Change Rate m· Year^{-1}
From	To			
1941	1946	5	−1.23	−1.23
1946	1960	19	−0.12	−0.44
1960	1964	23	−0.50	−0.47
1964	1966	25	0.04	−0.41
1966	1970	29	−0.16	−0.38
1970	1978	37	0.05	−0.29
1978	1981	40	−0.95	−0.33
1981	1983	42	−0.28	−0.33
1983	1985	44	−0.73	−0.35
1985	1989	48	0.90	−0.2
1989	1994	53	−0.85	−0.26
1994	2000	59	0.32	−0.2
2000	2006	65	−0.22	−0.21
2006	2010	69	−0.50	−0.26
2010	2014	73	0.12	−0.24

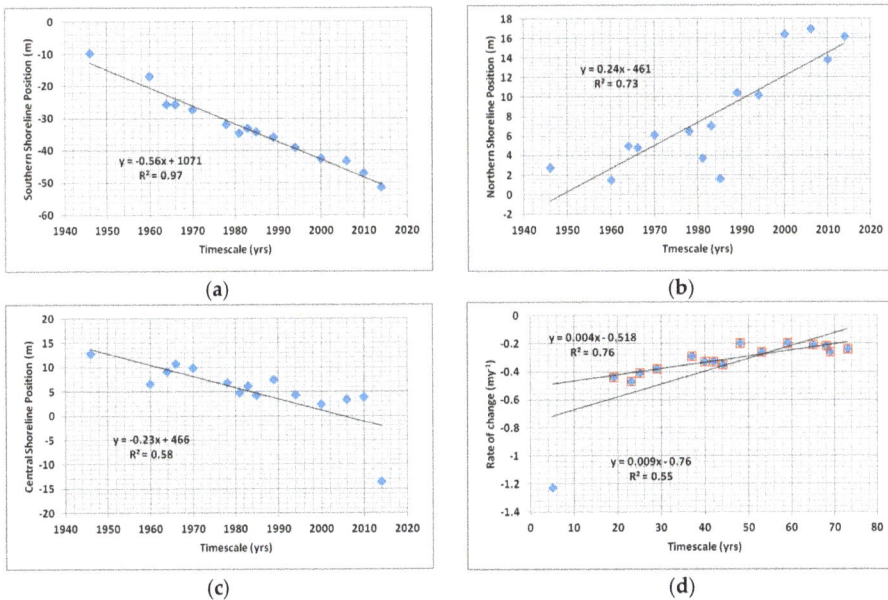

Figure 6. Temporal shoreline positional change between 1941 and 2014, (**a**) southern beach extremity (average T1–T4); (**b**) northern beach extremity (average T8–T12); (**c**) central beach region (average T5–T7) and (**d**) cumulative shoreline rates of change.

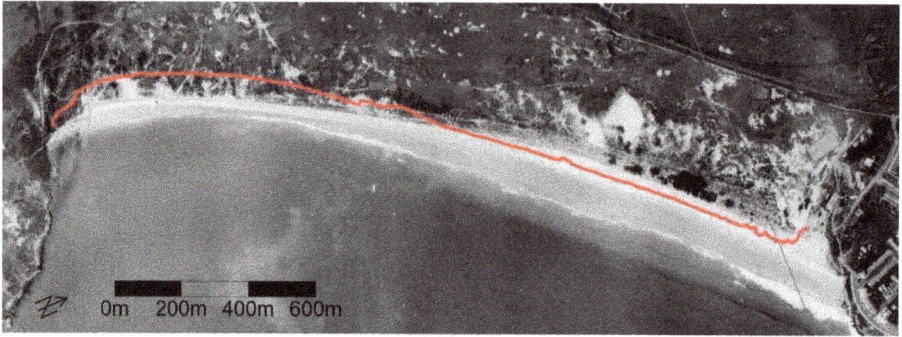

(a)

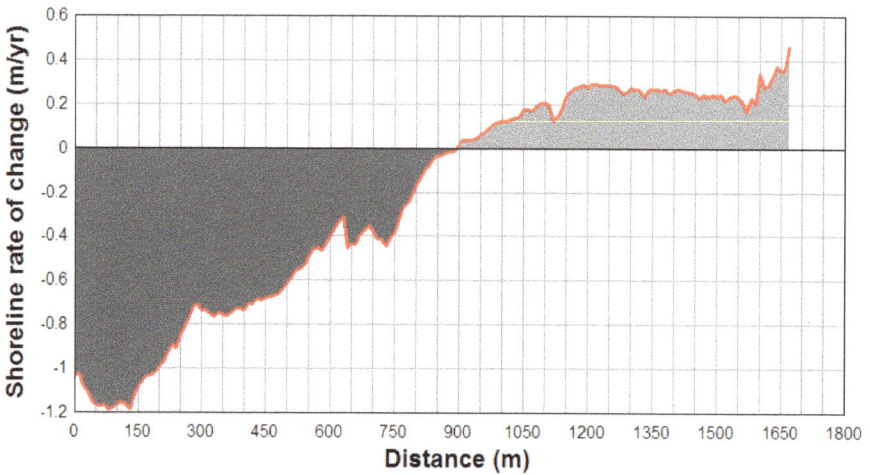

(b)

Figure 7. (a) Shoreline position of 2014 (red line) superimposed upon the 1941 aerial photograph and (b) a graphical representation of shoreline change between 1941 and 2014. Note:—light grey = accretion and dark grey = erosion.

4.2. Assessment of Beach Rotation (1941–2014)

A positive relationship existed between southern and central sectors given by the regression equation $y = 0.387x + 18.26$ and coefficient of determination (R^2) that explained 56% of data variation ($p < 0.01$; Figure 8a). Results indicated that when changes occur in the southern sector, they also occur in the central sector. In contrast, a negative phase relationship existed between northern and central sector cross-shore positions $y = 0.587x + 9.94$ ($R^2 = 33\%$, $p < 0.05$; Figure 8b), indicating that when changes take place within the northern sector the opposite would be true in the central sector. However, it is the statistically high negative phase relationship that existed between southern and northern beach extremities, given by the regression equation $y = 0.444x - 6.35$ explaining 69% of data variation ($p < 0.01$; Figure 8c), that is of most interest, as this indicates that beach rotation exists. A statistically high relationship also existed between the steady migration northward of the observed point of rotation and southern shoreline changes, given by the regression equation $y = 0.203x + 140$ that explained 74% of data variation ($p < 0.01$; Figure 8d).

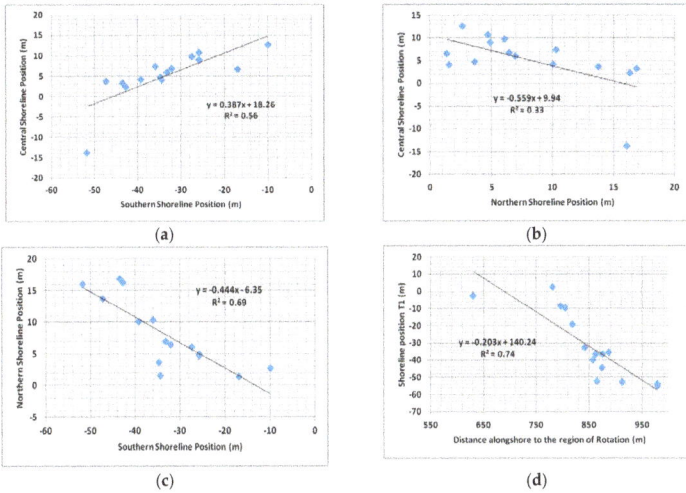

Figure 8. Spatial change 1941–2014 between, (**a**) north and south beach extremities; (**b**) central region and southern beach extremity; (**c**) central region and northern beach extremity and (**d**) Transect 1 shoreline position and the region of beach rotation.

Table 4 produced from Table 2 (Columns 3–14) shows a Pearson correlation matrix constructed to compare the temporal variation along each profile from the 1941 baseline. Positive high correlations signify that substantial relationships existed between southern profiles (T1–T4), indicating that when changes occur at one profile location they also occur on adjacent profiles ($p < 0.01$). A similar scenario existed within northern profiles (T8–T12) where positive correlations, varied between moderate and high. With the exception of the correlation between T8, all results were significant at 95% or 99% confidence ($p < 0.05$–$p < 0.01$; Table 3). The central profiles (T5–T8) showed statistically insignificant positive/negative correlations that varied from negligible to moderate ($p > 0.05$).With the exception of positive/negative high correlations between T6 and the southern/northern profiles, insignificant correlation existed between remaining central profiles and both southern and northern profile locations.

Table 4. Pearson correlation coefficients set at zero timelag compare longshore inter-survey shoreline positions by transect 1941–2014. Note: **bold**, $p < 0.05$, ***bold italic**, $p < 0.01$* and grey area highlights the negative relationships involved in the rotation process.

T2	*0.84*										
T3	*0.82*	*0.77*									
T4	*0.76*	*0.76*	*0.98*								
T5	0.28	0.04	0.32	0.33							
T6	*0.84*	*0.94*	*0.66*	*0.64*	0.02						
T7	0.22	*0.56*	0.26	0.34	−0.49	0.56					
T8	−0.11	0.17	0.04	0.12	−**0.56**	0.25	*0.86*				
T9	*−0.90*	*−0.92*	*−0.73*	*−0.68*	−0.19	*−0.88*	−0.27	0.10			
T10	*−0.66*	*−0.76*	**−0.44**	**−0.41**	−0.06	*−0.73*	−0.13	0.20	*0.86*		
T11	*−0.74*	*−0.82*	**−0.45**	**−0.42**	−0.07	*−0.82*	−0.38	0.05	*0.85*	*0.78*	
T12	*−0.70*	*−0.72*	*−0.74*	*−0.70*	−0.30	*−0.59*	−0.07	0.25	*0.83*	*0.71*	*0.69*
	T1	T2	T3	T4	T5	T6	T7	T8	T9	T10	T11

However, it is the statistically high and very high negative correlations that are of most interest as they signified marked and very dependable inverse relationships between north and south sectors *i.e.*, beach rotation (T1–T4 and T8–T12) statistically significant at the 95% or 99% confidence ($p < 0.05$–$p < 0.01$) confirming earlier regression model results (Figure 6). Negative correlation was also observed between the south and central sector. Coincidently, this also concurs with results shown by both [11] and [19] in studies of long-term rotational trends at Narabeen Beach, Australia. The fulcrum is observed at the change of correlation signs within the central region, and finds agreement with the work of [16] along the Brazilian coastline and [27] in the present region of study.

4.3. Shoreline Position Forecast

The equilibrium bay shape equation developed by [46,47] was used to estimate the expected shoreline position corresponding to a zero migration rate. The zero migration date (Z_{mr}) was extrapolated from the linear trend obtained in Figure 6d and given by Equation (1). The date was then inputted into the linear trends obtained for southern (S_{sp}) and northern (N_{sp}) shorelines (Figure 6a,b respectively) and 2061 shoreline positions computed (Equations (2) and (3)). The extrapolated southern shoreline position was inputted into the linear trend obtained for the region of rotation (Figure 8d) and the region of rotation (C_{rr}) computed (Equation (4)). To estimate the predominant wave direction a perpendicular line was drawn from the predicted northern shoreline position (*i.e.*, downcoast control point) to the predicted point of rotation. Figure 9b shows the southern shoreline sector, highlighting the 2061 shoreline position extrapolated from the linear trend obtained from Figure 6a (black cross within a circle) and the predicted bay shape using the 2nd order parabolic curve (red line) alongside the 2014 shoreline position (blue line). Results show the efficiency of using both linear trends and empircal bay shape equations.

$$Z_{mr} = [(0.5086/0.0044) + 1946] = 2061 \tag{1}$$

$$S_{sp} = -0.56\,(2061) + 1071 = -83.16\,\text{m (from 1946 baseline)} \tag{2}$$

$$N_{sp} = 0.24\,(2016) - 461 = 33.64\,\text{m (from 1946 baseline)} \tag{3}$$

$$C_{rr} = [-(-83.16) + 140.24)]/-0.203 = 1100.498\,\text{m (from T1)} \tag{4}$$

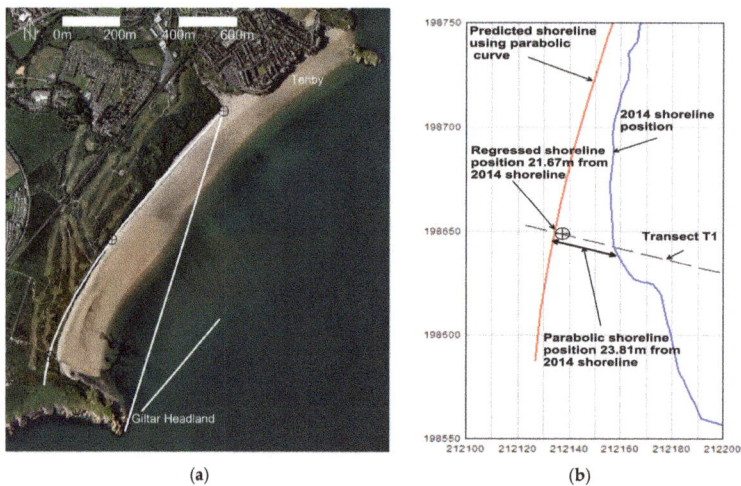

(a)   (b)

Figure 9. Equilibrium bay shape assessment using, (**a**) a parabolic curve fitted to the predicted downcoast and pivot point control points and superimposed upon the 2014 aerial photograph and (**b**) a comparison between the regressed and parabolic prediction of the equilibrium southern shoreline position in 2061.

4.4. Wave Models

In this region, extreme storm waves (>3.4 m) make up 6% of the record and expose the coast to waves and associated periods that can reach 7 m and 9.3 s, respectively. However, these storms are rare with on average of two occurrences a year. The majority of storm waves (*circa* 90% of the record) range between 4.7 m and 6.5 m with associated periods of 8 s to 8.6 s. Offshore Island location, bathymetry, Bristol Channel fetch limitation and the orientation of the shoreline narrows the range of wave directions experienced at frontage of South Beach. This was shown in the extreme wave results of both [42] and [43] who highlighted similar regional patterns of wave directional change irrespective of wave height and period variation (*i.e.*, 1:1 month, 1:1 year and 1:10 year assimilations). This present paper used extreme storm events from the same assessed directions (southwest, south and southeast); modelled vectors were once again similar irrespective of the event severity. Therefore, Figure 10 only shows simulation of the most extreme storm waves encountered in each of the assessed directions. Southwesterly waves are heavily diffracted around St Margaret's Island before entering Caldey Sound and impact Giltar Headland at an acute angle, where further diffraction takes place as waves enter South Sands littoral (Figure 10a). Wave energy is focused at an obtuse angle to the beach (Figure 10b), suggesting a northward sediment pathway (Tenby). Under southerly conditions generated waves are heavily diffracted around both Caldey and St Margaret's Islands before entering South Bay. The two wave trains meet and form a shadow zone along the trace of High Cliff spit (Figure 10c). It is reasonable to deduce that through wave energy loss; entrained sediments derived from Caldey Sound would be deposited explaining both continued sand spit growth and sediment loss in Caldey Sound reported by [29]. Further refraction once again refocuses waves at an angle along the frontage of South Beach. Therefore, it is also reasonable to deduce that southerly waves would be the cause of south toward north longshore drift (Figure 10d). In contrast, southeasterly waves diffract around the easternmost point of Caldey and on entering South Bay, become parallel to the island frontage. This has the effect of reducing wave impacts generated within Caldey Sound from St Margaret's Island (Figure 10e). These waves approach South Beach at a slight southward angle (toward Giltar Headland), and under these conditions, longshore sediment pathways would emanate (albeit weakly) from Tenby towards Giltar Headland (Figure 10f).

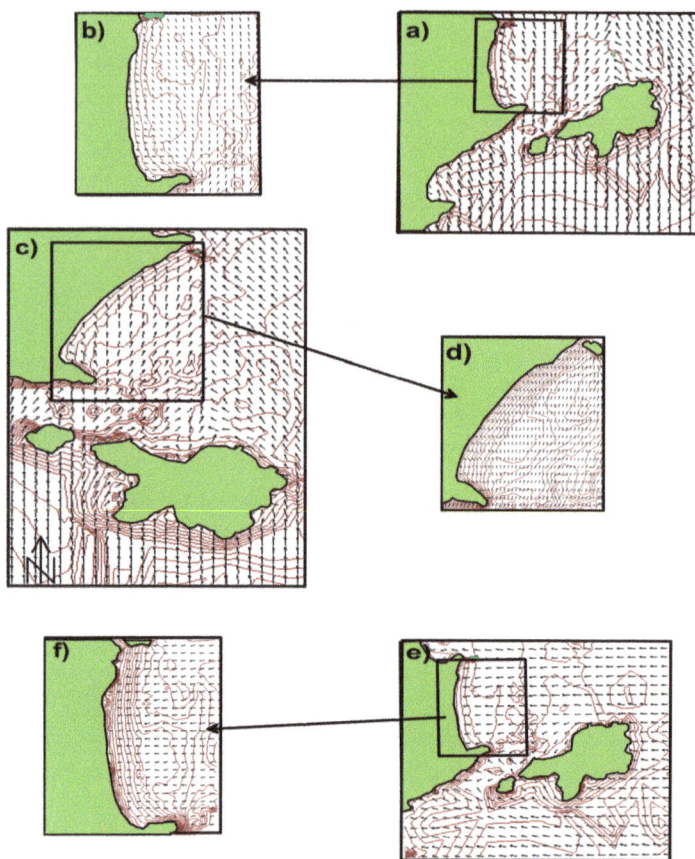

Figure 10. Illustrations of modelled wave vectors based upon the highest astronomical tidal range for, (a) southwesterly offshore waves; (b) southwesterly nearshore waves; (c) southerly offshore waves; (d) southerly nearshore waves; (e) southeasterly offshore waves and (f) southeasterly nearshore waves.

4.5. Shoreline Changes in Relation to Wind Conditions

Within the region of study, synthesized wind and wave timeseries from meteorological numerical models suitable to be used in qualitative and quantitative assessments have only been available since 1986. However, comprehensive sets of wind speed and directional data were available from the early 1940s and used to assess shoreline behavior against these imposed environmental forcing agents over an historic timeframe.

The largest landward excursion of the shoreline (Table 3) took place between 1941 and 1946 (-1.23 m/year) and occurred when both direction ($0 = 206 \pm 16.9°$, Figure 11a) and wind speed ($0 = 7 \pm 0.97$ m/s; Figure 11b) are below average, suggesting that winds predominate from south toward east, with a wind speed reduction as a consequence of the limiting Bristol Channel fetch. A reduction in shoreline retreat rates was observed between 1946 and 1960 (-0.12m/year), as winds predominated from a southeasterly direction (negative) in the early 1950s shifted to a southwesterly one towards the middle of the decade returning to a mean direction of around 210° at the end. The highest winds during this timeframe also occurred during the early 1950s followed by a period of below average wind speeds. Increasing wind speeds, as wind direction fluctuates above and below the average value, corresponded to a landward shoreline excursion (-0.5 m/year) between 1960 and 1964. Observed changes 1964–1966 and 1970–1978 indicated shoreline advances albeit small, which reversed the more general trend (0.04 m/year and 0.05 m/year, respectively); wind speed and direction fluctuated between negative and positive values. Between 1966 and 1970, a shoreline retreat of −0.16 m/year occurred during a period that is dominated by increasing wind speed and winds from a southwesterly direction.

The system returned to the more normal trends of shoreline retreat 1978–1981, 1981–1983 and 1983–1985 (−0.95 m/year, −0.28 m/year and −0.73 m/year, respectively). Once again, southwesterly winds and rising wind speeds dominated this period. Apart from high speed in 1984, wind speed trend is near and below the average value, and the direction was predominantly from the southeast to south during two periods of shoreline advance of 0.9 m/year and 0.32 m year (1985–1989 and 1994–2000, respectively). However, similar trends of wind speed and direction to previous values were observed between 1989 and 1994 that resulted in a shoreline retreat of −0.85 m/year. The shoreline is observed to have retreated between 2000 and 2010 (−0.5 m/year), when winds were lower than average and wind direction fluctuated between southwest and southeast. Extreme storms were recorded between late 2013 and early 2014 that caused erosion along the southern sector, *i.e.*, a landward excursion of the vegetation line. However, overall the frontage showed a slight increase (0.12 m/year) as the northern sector accreted. Regional wave data covering the period up until the end of 2013 showed that relatively weak wind speeds predominated, suggesting that the bulk of the erosion took place during the January/February 2014 storms.

Table 2 data was transformed to characterize inter-survey changes by beach sector (south/central/north) and compared to wind direction (Figure 11c) and wind speed (Figure 11d). The overall erosion trend highlighted in Table 2 between 1941 and 1946 was restricted to the southern sector as both central and northern sectors accreted under south/southeast (below average) and less energetic wind regimes. When southerly wind directions were encountered under a variable wind speed, there was reduction in the erosive trend within the southern sector and losses in both northern and central sectors (1946–1960). Under less energetic wind speed and directions emanating from south toward southwest, there is variable erosive/accretive behavior within all beach sectors (1960–1978). Southwesterly wind directions and variable below average wind speeds result in southern erosion, contrasted against northern accretion with central sectors varying between erosion and accretion throughout (1978–present).

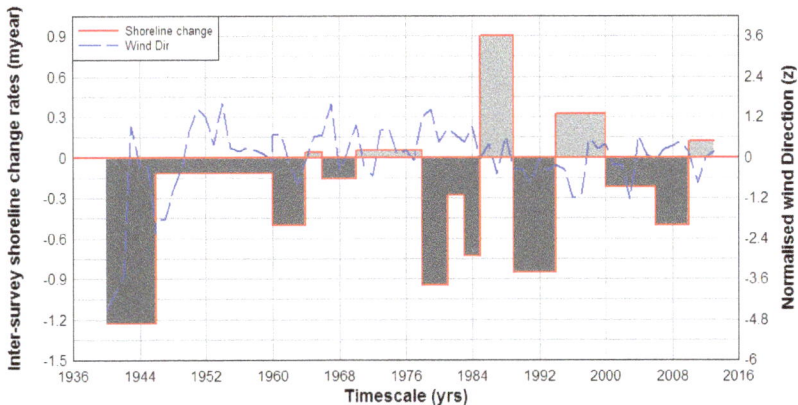

(a)

Figure 11. *Cont.*

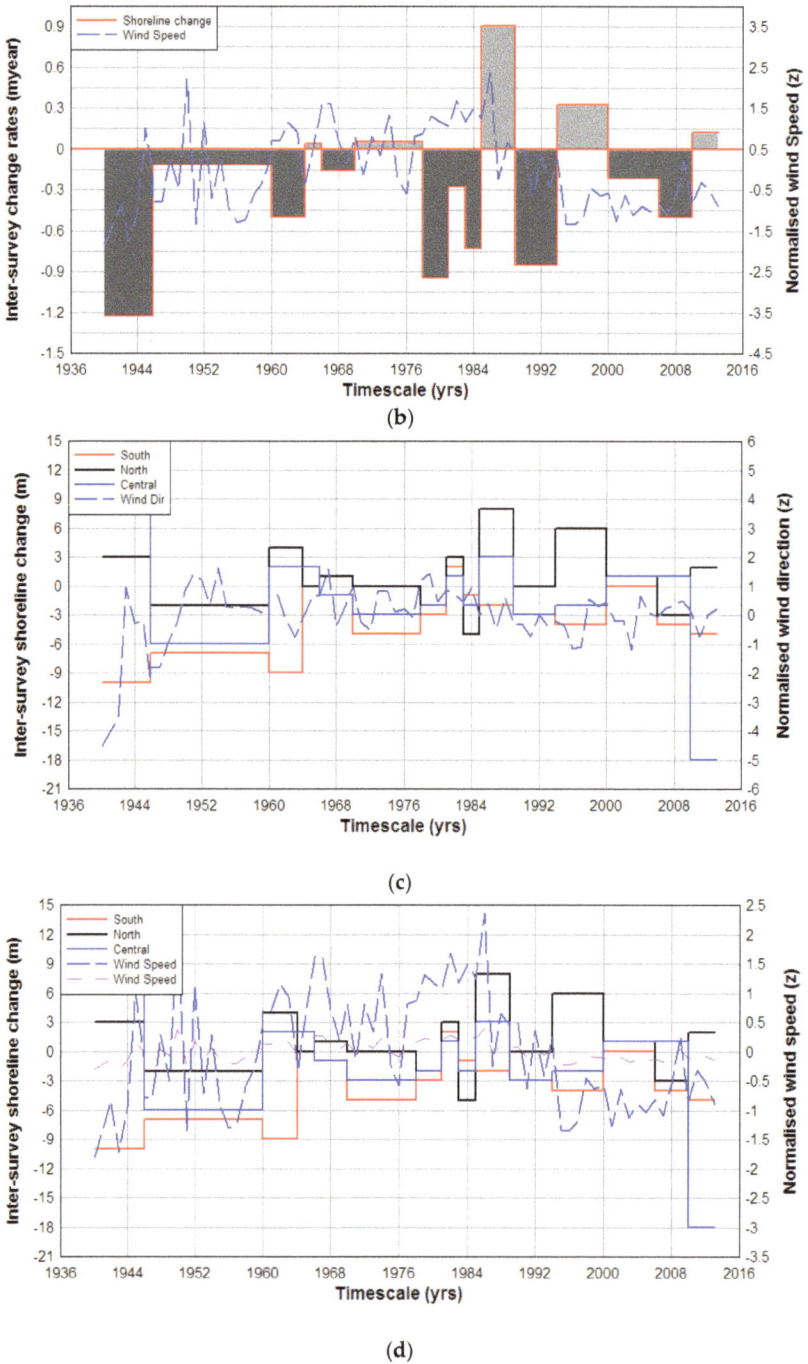

Figure 11. Comparison of Shoreline change rates (Table 2) and normalized (**a**) wind direction in degrees clockwise from true north; (**b**) wind speed and comparisons of average inter-survey shoreline changes by beach sector, with normalized (**c**) wind direction and (**d**) wind speed for the period 1941–2014. Note: dark grey and negative shoreline change values = erosion and light grey and positive shoreline change values = accretion.

5. Discussion

In times of accelerated sea level rise and increasing demands on beaches to provide defense against flood and coastal erosion, coastal practitioners need robust and "hands on" approaches that simplify beach management. This article describes a simple methodology particularly useful to embayed beach coastal management. Qualitatively, inter-survey change rates varied throughout the assessment period showing a mainly erosive trend. In the southern sector, two concrete groynes and gabion walling exacerbated erosive trends. The groynes were eventually demolished and the southern sector has evolved into a classic embayed beach shape, while the gabion walling was destroyed during the winter storms of 2013/14. Quantitatively, cumulative results showed erosive trends that reduced over time suggesting that the bay is slowly reaching equilibrium. The statistically significant ($R^2 = 76\%$; $p < 0.001$) regression models that were constructed to assess temporal trends enabled shoreline equilibrium to be predicted (2061) but the result should be treated with caution as the linear trend used for the prediction was heavily influenced by an accelerated retreat rate between 1941 and 1946. However, this does not diminish the importance of this research that proves the principle that the developed models can be used to predict shoreline position at any given temporal epoch. For example, large-scale assessments of the UK coastline use standardized epoch timescales of 20, 50 and 100 years and these are indoctrinated in all shoreline management plans [5,32]. When the present shoreline was compared with the 1941 aerial photograph, the south shoreline eroded (max = 1.1 m/year) and the northern shore advanced (max = 0.4 m/year), while the central sector remained stable. An assessment of temporal recession/accretion rates at specific locations alongshore, highlighted statistically significant southern and central erosive trends ($R^2 = 79\%$ and $R^2 = 58\%$; $p < 0.01$), northern accretive trends ($R^2 = 73\%$; $p < 0.01$).

Regression models were once again constructed to assess beach rotation and when the central sector was compared with the south a positive phase relationship existed indicating that when changes take place in the south, similar changes take place in the central region ($R^2 = 56\%$; $p < 0.05$). Conversely, when the central sector was compared with the north, a negative phase relationship was found suggesting that when changes occur in the central sector, the opposite would be true for the north ($R^2 = 33\%$; $p < 0.05$). What was of most interest was the negative phase relationship that existed between the southern and northern shores suggesting that when changes take place in one sector, the opposite would be true in the other sector (*i.e.*, beach rotation). Rotation phenomena relies upon the beach rotating about a central pivot point and the regression model representing spatial trends in southerly shoreline position, and the point of rotation quantitatively verified a temporal trend of northward migration ($R^2 = 74\%$; $p < 0.01$). This rotation point was recognized using correlation coefficients set at zero timelag, a point of rotation was centrally placed, at which a negative phase relationship changed to positive, confirmed by increasing variability in the regression model of the central beach sector (Figure 6c). These temporal models represent shoreline indicator variation (*i.e.*, vegetation line) and were used as a simple tool in the prediction of shoreline position at the expected time of equilibrium (*i.e.*, 2061). Analysis using these equations suggests that southern shorelines will retreat by *circa* 72 m; northern shorelines will advance by *circa* 25 m by 2061 (from the 1946 shoreline). The parabolic bay shape equation used to predict the equilibrium shoreline position compared favorably with extrapolated linear trend results.

Paucity of environmental data (wave height, period and direction) and the temporal spacing of the aerial photographic evidence make assessment of shoreline change influences difficult. However, qualitative assessments between inter-survey overall and sectored shoreline changes, wind speed and direction data do highlight that southeasterly regimes tended to be accretive and south/southwest erosive. Storm wave model results were based upon the south, southwest predominant and southeast subdominant directions. Wave models showed similar vector alignment irrespective of wave height. Results suggested that, in addition to erosion caused by south/southwesterly winds, the dominant south/southwesterly waves produce south toward north longshore drift resulting in erosion in the southern sector and accretion in the north. On the contrary, in association with southeasterly winds related to accretive trends, sub-dominant southeasterly waves produce a counter drift, albeit weak, that is reversing the general evolution trends. Additional short term assessments of wind and wave effects on the shoreline evolution are required to confirm the qualitative assessment of the present study.

Results were used in conjunction with topographic surveys of the hinterland to produce a flood map (Figure 12a) based upon the predicted southern shoreline position (2061) and the inland 5 m contour line (*i.e.*, the highest spring tide level). The map identified potential flood inlet points confirmed by a topographic survey along the equilibrium shoreline; one near the headland itself would allow water to access the hinterland along a low lying track, two at 100 m and 300 m alongshore along the line of a newly formed footway access to the beach and near an eroding blowout (500 m alongshore) that collapsed on the seaward side and is open to wave attack on most spring tides. The hinterland behind the dune system is already located below MHWST. A length of dune extending from the headland to the northern end of the gabion basket wall would be at risk along its entire length (*circa* 400 m). The area of flood potential extends from the dune field to the railway line that was constructed in an elevated position in relation to the MHWST level and would act as a barrier to further ingress. It has to be remembered that the area would only flood during the spring tidal range or during storm surge conditions. This could also be mitigated against by dune stabilization and infilling along the two potential flood routes.

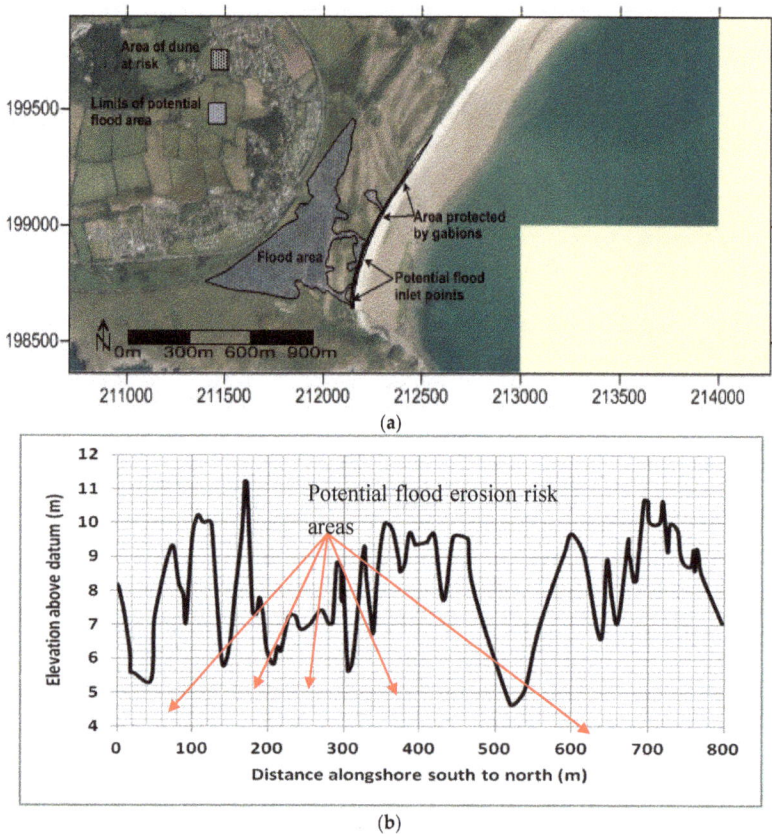

Figure 12. (**a**) 2014 Aerial photograph illustrating flood limits based upon the estimated shoreline position and (**b**) a graphical representation of the existing ground profile along the predicted shoreline position.

The developed regression models, graphical representations and flood maps are relatively simple tools that can be understood by all stakeholders, and, therefore, are an important addition to the management of coastal areas and should be repeated on a wider scale, especially in areas of risk. This work has global implications and can help in the development of embayed beach management strategies designed to improve resilience under various scenarios of sea level rise and climate change.

6. Conclusions

The morphology change of a headland bay beach—Tenby, West Wales, UK—has been analyzed between 1941 and 2014. Geo-referenced aerial photographic evidence was used to extract shoreline positions based upon the vegetation line and subsequently compared with environmental forcing. Shoreline change rates were shown to reduce over time suggesting the bay was reaching equilibrium. Extrapolation of the regression model results predicted that the shoreline would reach this equilibrium state in 2061. Temporal trend analysis also showed that the southern and central sectors were erosive and the northern sector accreted during the assessed timeframe. Further analysis identified long-term plan-form rotation confirmed by negative phase relationships between beach extremities and variations from negative to positive correlation within the more stable central sector. The point of rotation was identified to be migrating northward as southern shores eroded. Constructed models were used in conjunction with an empirical 2nd order polynomial equation to identify the equilibrium shoreline position in 2061 provided current environmental conditions prevail. In line with previous regional research, wave models and important environmental forcing agents indicate that dominant south and southwesterly regimes influence south to north longshore drift and a counter drift engendered by less dominant easterly regimes. Derived predictive linear trends from regression models produced a simple method of assessment of flood risk at Tenby South Sands and should be repeated on a wider scale, especially where coastal areas are deemed to be at risk, so that suitable coastal management policies can be developed.

Acknowledgments: The authors would like to thank Welsh Assembly Government Aerial Photographs Unit, Welsh Government, Crown Offices, Cathays Park, Cardiff, Wales, CF10 3NQ, for aerial photographs used in the research.

Author Contributions: Tony Thomas, and Nelson Rangel-Buitrago conceived and designed the experiments; Tony Thomas and Nelson Rangel-Buitrago performed the experiments; Tony Thomas, Nelson Rangel-Buitrago, Allan Williams Michael Phillips and Giorgio Anfuso analyzed the data; Tony Thomas, Nelson Rangel-Buitrago and Giorgio Anfuso contributed data/materials and analysis tools; Tony Thomas, Allan Williams and Mike Phillips wrote the paper.

Conflicts of Interest: The authors declare no conflict of interest.

References

1. Short, A.D.; Masselink, G. Embayed and Structurally Controlled Beaches. In *Handbook of Beach and Shoreface Morphodynamics*; Short, A.D., Ed.; John Wiley and Sons Ltd: Chichester, UK, 1999; pp. 230–250.

2. O'Connor, M.O.; Cooper, J.A.G.; Jackson, D.W.T. Morphological Behaviour of Headland-Embayment and Inlet-Associated Beaches, Northwest Ireland. *J. Coast. Res.* **2007**, *59*, 626–630.

3. Brown, E.; Colling, A.; Park, D.; Phillips, J.; Rothery, D.; Wright, J. *Waves, Tides and Shallow-Water Processes*; Butterworth Heinemann: Oxford, UK, 1999; p. 227.

4. Woodroffe, C.D. *Coasts, Form, Processes and Evolution*; Cambridge University Press: Cambridge, UK, 2002; p. 623.

5. Rogers, J.; Hamer, B.; Brampton, A.; Challinor, S.; Glennerster, M.; Brenton, P.; Bradbury, A. *Beach Management Manual*, 2nd ed.; CIRIA: London, UK, 2010; p. 915.

6. Benedet, L.; Klein, A.H.F.; Hsu, J.R.C. Practical insights and applicability of empirical bay shape equations. In Proceedings of the 29th International Conference Coastal Engineering, Lisbon, Portugal, 19–24 September 2004; pp. 2181–2193.

7. Carter, R.W.G. *Coastal Environments, an Introduction to the Physical, Ecological and Cultural Systems of Coastlines*; Academic Press: London, UK, 1988; p. 617.

8. Cooper, J.A.G.; Jackson, D.W.T. Geomorphological and dynamic constraints on mesoscale coastal response to storms, Western Ireland. In Proceedings of the 6th International Symposium on Coastal Engineering and Science of Coastal Sediment Processes, New Orleans, LA, USA, 13 May 2007; pp. 3015–3024.
9. Cooper, J.A.G. Temperate Coasts. In *Applied Sedimentology*; Perry, C.M., Taylor, K., Eds.; Blackwell Publishing: Oxford, UK, 2007; pp. 263–301.
10. Archetti, R. Quantifying the evolution of a beach protected by low crested structures using video monitoring. *J. Coast. Res.* **2009**, *25*, 884–899. [CrossRef]
11. Short, A.D.; Trembanis, A.C. Decadal Scale Patterns of Beach Oscillation and Rotation: Narabeen Beach, Australia. Time Series PCA and Wavelet Analysis. *J. Coast. Res.* **2004**, *20*, 523–532. [CrossRef]
12. Harley, M.D.; Turner, I.L.; Short, A.D.; Ranasinghe, R. A reevaluation of coastal embayment rotation: The dominance of cross-shore *versus* alongshore sediment transport processes, Collaroy-Narrabeen Beach southeast Australia. *J. Geophys. Res.* **2011**, *116*. [CrossRef]
13. Stone, G.W.; Orford, J.D. Storms and their significance in coastal morpho-sedimentary dynamics. *Mar. Geol.* **2004**, *210*, 1–5. [CrossRef]
14. Maspataud, A.; Ruz, M.H.; Harquette, A. Spatial variability in post-storm beach recovery along a macrotidal barred beach, southern North Sea. *J. Coast. Res.* **2009**, *56*, 88–92.
15. Loureiro, C.; Ferreira, O.; Cooper, J.A.G. Contrasting Morphological Behaviour at Embayed beaches in Southern Portugal. *J. Coast. Res.* **2009**, *56*, 83–87.
16. Klein, A.H.F.; Benedet Filho, L.; Schumacher, D.H. Short-term Beach Rotation Processes in Distinct Headland Bay Systems. *J. Coast. Res.* **2002**, *18*, 442–458.
17. Dolique, F.; Anthony, E.J. Short-Term Profile Changes of Sandy Pocket Beaches Affected by Amazon-Derived Mud, Cayenne, French Guiana. *J. Coast. Res.* **2005**, *21*, 1195–1202. [CrossRef]
18. Anthony, E.J.; Dolique, F. The influence of Amazon-derived mud banks on the morphology of sandy headland-bound beach in Cayenne, French Guiana: A short to long-term perspective. *Mar. Geol.* **2004**, *208*, 249–264. [CrossRef]
19. Ranasinghe, R.; McLoughlan, R.; Short, A.; Symonds, G. The Southern Oscillation Index, Wave Climate and Beach Rotation. *Mar. Geol.* **2004**, *204*, 273–287. [CrossRef]
20. Anthony, E.J.; Dolique, F. The influence of Amazon-derived mud banks on the morphology of sandy headland-bound beach in Cayenne, French Guiana: a short to long-term perspective. *Mar. Geol.* **2004**, *208*, 249–264. [CrossRef]
21. Sedrati, M.; Anthony, E.J. A Brief Overview of Plan-Shape Disequilibrium in Embayed Beaches: Tangier Bay (Morocco) 2007. *Mediterranee*, 108. 2007. Available online: http://mediteranee.revues.org/index190.html?file=1 (accessed on 21 September 2009).
22. Dehouck, A.; Dupuis, H.; Senechal, N. Pocket beach hydrodynamics: The example of four macrotidal beaches, Brittany, France. *Mar. Geol.* **2000**, *266*, 1–17. [CrossRef]
23. Jacob, J.; Gama, C.; Salgado, R.; Lui, J.T.; Silva, A. Shadowing effects on Beach Morphodynamics during Storm Events on Troia-Sines Embayed Coast, Southwest Portugal. *J. Coast. Res.* **2009**, *56*, 72–77.
24. Ojeda, E.; Guillen, J. Shoreline Dynamics and beach rotation of artificial embayed beaches. *Mar. Geol.* **2008**, *253*, 51–62. [CrossRef]
25. Norcross, Z.M.; Fletcher, C.H.H.; Merrifield, M. Annual and Iterannual changes on a reef-fringed pocket beach: Kailua Bay, Hawaii. *Mar. Geol.* **2002**, *190*, 553–580. [CrossRef]
26. Archetti, R.; Romagnoll, C. Analysis of the effects of different storm events on shoreline dynamic on an artificially embayed beach. *Earth Surf. Process. Landf.* **2011**, *36*, 1449–1463. [CrossRef]
27. Thomas, T.; Phillips, M.R.; Williams, A.T. Mesoscale evolution of a headland bay: Beach rotation Process. *Geomorphology* **2010**, *123*, 129–141. [CrossRef]
28. Thomas, T.; Phillips, M.R.; Williams, A.T.; Jenkins, R.E. Short-term beach rotation, wave climate and the North Atlantic Oscillation (NAO). *Prog. Phys. Geogr.* **2011**, *35*, 333–352. [CrossRef]
29. Thomas, T.; Phillips, M.R.; Williams, A.T.; Jenkins, R.E. Medium timescale beach rotation: Gale climate and offshore island influences. *Geomorphology* **2011**, *135*, 97–107. [CrossRef]
30. Turki, R.; Medina, R.; Gonzalez, M.; Coco, G. Natural variability of shoreline position: Observations at three pocket beaches. *Mar. Geol.* **2013**, *338*, 76–89. [CrossRef]

31. Short, A.D.; Trembanis, A.C.; Turner, I.L. Beach Oscillation, Rotation and the southern Oscillation. Coastal Engineering. In Proceedings of the 27th International Conference on Coastal Engineering, Sydney, Australia, 16–21 July 2000.

32. Ruiz de Alegria-Arzaburu, A.; Masselink, G. Storm response and beach rotation on a gravel beach, Slapton Sands, UK. *Mar. Geol.* **2010**, *278*, 77–99. [CrossRef]

33. Morton, R.A. Accurate shoreline mapping; past, present and future. In Proceedings of the Coastal Sediments 91, Seattle, DC, USA, 25–27 June 1991; pp. 997–1010.

34. Halcrow. Lavernock Point to St Ann's Head SMP2. 2010. Available online: http://www.southwalescoast.org/contents.asp?id=55#SMP2MainDocument (accessed on 20 June 2011).

35. Hillier, R.D.; Williams, B.P.J. The alluvial OLD Red Sandstone: Fluvial Basins. In *The Geology of England and Wales*; Brenchley, P.J., Rawson, P.F., Eds.; The Geological Society: London, UK, 2006; pp. 155–172.

36. Crowell, M.; Leatherman, S.P.; Buckley, M.K. Historical Shoreline Change: Error Analysis and Mapping Accuracy. *J. Coast. Res.* **1991**, *7*, 839–852.

37. Douglas, B.C.; Crowell, M. Long-term Shoreline position Predictions and Error Propagation. *J. Coast. Res.* **2000**, *16*, 145–152.

38. Maune, D.F. *Digital Elevation Model Technologies and Applications: The DEM Users Manual*, 2nd ed.; American Society for Photogrammetry and Remote Sensing: Bethesda, MD, USA, 2007; p. 655.

39. Thomas, T.; Phillips, M.R.; Williams, A.T.; Jenkins, R.E. A multi-century record of linked nearshore and coastal change. *Earth Surf. Process. Landf.* **2011**, *36*, 995–1006. [CrossRef]

40. Thomas, T.; Lynch, S.K.; Phillips, M.R.; Williams, A.T. Long-term evolution of a sand spit, physical forcing and links to coastal flooding. *Appl. Geogr.* **2014**, *53*, 187–201. [CrossRef]

41. Morang, A.; Batten, B.K.; Connell, K.J.; Tanner, W.; Larson, M.; Kraus, N.C. *Regional Morphology Analysis Package (RMAP), Version 3: Users Guide and Tutorial 2009*; Coastal and Hydraulics Engineering Technical Note ERDC/CHL CHETN-XIV-9; U.S. Army Research and Development Center: Vicksburg, MS, USA, 2009. Available online: http://acwc.sdp.sirsi.net/client/en_US/search/asset/1011302;jsessionid=F8BAB79430EB26551A26C896F880E523.enterprise-15000 (accessed on 6 June 2010).

42. Gibbard, B. *Tenby South Beach Erosion: Review of Wind/Wave Data, Beach Monitoring and Modelling*; Royal Haskoning: Peterborough, UK, 2005; p. 39.

43. Thomas, T.; Phillips, M.R.; Williams, A.T.; Jenkins, R.E. Links between environmental forcing, offshore islands and a macro-tidal Headland-Bound Bay Beach. *Earth Surf. Process. Landf.* **2014**, *39*, 143–155. [CrossRef]

44. Ebersole, B.A. Refraction, Diffraction Model for Linear Waves. *J. Water Port Coast. Ocean Eng.* **1985**, *3*, 939–953. [CrossRef]

45. Ebersole, B.A.; Cialone, M.A.; Prater, M.D. *Regional Coastal Processes Numerical Modelling System, Report 1, RCPWAVE-A linear Wave Propagation Model for Engineering Use*; Technical Report CERC-86-4; Coastal Engineering Research Centre, US Army Engineer Waterways Experiment Station: Vicksburg, MS, USA, 1986.

46. Hsu, J.R.C.; Evans, C. Parabolic Bay Shapes and Applications. *Proc. Inst. Civ. Eng.* **1989**, *87*, 557–570. [CrossRef]

47. Hsu, J.R.C.; Silvester, R.; Xia, Y.M. Generalities on Static Equilibrium Bays. *J. Coast. Eng.* **1989**, *12*, 353–369. [CrossRef]

Journal of
*Marine Science
and Engineering*

MDPI

Article

Observation of Whole Flushing Process of a River Sand Bar by a Flood Using X-Band Radar

Satoshi Takewaka

Division of Engineering Mechanics and Energy, University of Tsukuba, Tsukuba, Ibaraki 305-8573, Japan; takewaka@kz.tsukuba.ac.jp; Tel.: +81-29-853-5361; Fax: +81-29-853-5207

Academic Editor: Gerben Ruessink
Received: 25 December 2015; Accepted: 18 April 2016; Published: 4 May 2016

Abstract: Morphological changes during a flood event in July 2010 were observed with X-band marine radar at the mouth of Tenryu River, Shizuoka, Japan. Radar images were collected hourly for more than 72 h from the beginning of the flood and processed into time-averaged images. Changes in the morphology of the area were interpreted from the time-averaged images, revealing that the isolated river dune was washed away by the flood, the width of the river mouth increased gradually, and the river mouth terrace expanded radially. Furthermore, image analysis of the radar images was applied to estimate the migration speed of the brightness pattern, which is assumed to be a proxy of bottom undulation of the river bed. The migration was observed to be faster when the water level gradient between the river channel and sea increased.

Keywords: flushing of river sand bar; remote sensing; X-band radar

1. Introduction

Morphological data are essential to evaluating and understanding the long- and short-term behavior of a sandy river. Traditional *in situ* surveying, such as leveling and echo sounding, provides precise position data at measured points. It is, however, costly, time-consuming, and difficult to collect data during floods, and, therefore, provides only infrequent and low-density measurements. Data collection is limited to fair weather conditions and daytime periods, which makes it difficult to wholly track a sudden event like a flood. In this context, an X-band radar can be used as an alternative to remotely observe the behavior of river morphology. An X-band radar is an imaging radar that is capable of tracking the movements of wave crests over an area spanning several kilometers and has become popular in coastal studies in the last decades [1–3]. X-band radar provides distortionless images of a broad area at intervals of 2–3 s. The intensity of a pixel in the radar image corresponds to the relative amount of backscatter signal of the emitted radar beam reflected from the flood plain, such as vegetation, water surface, *etc.*, and hence it is usable during the night and under slightly rainy and high-wave conditions. However, one defect of the radar system is the difficulty it has in detecting color information: it is difficult to infer the condition of the water surface, existence of suspended materials, vegetation, *etc.*, which can be discriminated by the interpretation of visible images.

Recently, Holman and Haller discussed the various aspects of the different disciplines of nearshore remote sensing in a review article [4]. They compared merits and characteristics of active and passive remote sensors (cameras, radars, lidars, *etc.*) and platforms (fixed, flying, floating, and orbiting), and concluded that for nearshore oceanographic applications, fixed optical cameras and X-band radars are the most frequently used and best developed. This study demonstrates the potential of land-based X-band radar observation in a drastic developing flood event at a river mouth in Japan.

The River Tenryu flows to the Enshu Coast (lat/long: N-34.6472/E-137.7933). The coast suffers from severe erosion and an enhancement of sediment supply from the catchment is planned by authorities in order to mitigate the erosion. The river basin area is 5090 km^2, and the length of the river is 213 km. Observation by a land-based X-band imaging radar helps us understand the morphology and hydrodynamics by capturing spatial distributions and temporal variation of water lines of the river channel and coast lines, and wave propagation in the shallow area [5,6].

In this work, morphological changes during a flood event are described from the radar images collected continuously for more than 72 h from the beginning of the flood, which demonstrates the potential of X-band radar in tracking geomorphological processes during an extreme event. Morphological variations of the area were interpreted from the time-averaged radar images. Brightness patterns in the time-averaged radar images migrated in the down-flow direction during the flood, and image analysis of the pixel intensities was applied to estimate their migration speeds. The migration was faster when the water level gradient between the river channel and the sea level increased, which implies that the time-averaged images captured the migration of the bottom features of the river bed.

Pianica *et al.* [7] recently reported a video-based observation of bedform deformation of ebb delta over 23 days in the United States. They tracked the migration of bedform features, which was mainly induced by the tidal action under fair conditions and estimated their speed. This study also discusses migration of bedform features, but induced by a river flood, and shows another aspect of morphological dynamics at a river mouth.

2. Study Area and X-Band Radar Observation

The radar employed in this study is a conventional incoherent marine X-band radar for commercial use (JMA-3925-9 Japan Radio Co. Ltd. Tokyo, Japan, 3 cm wavelength, transmitting power 25 kw, HH-polarization, radar pulse length 0.08 µs), which is usually installed on fishing boats or ships. The 2.8 m antenna rotates with a period of approximately 2.6 s and transmits with a beamwidth of 0.8° in the horizontal and 25° in the vertical. The radar is installed on the roof of a sewage plant located on the right side bank close to the mouth of the river as shown in a satellite image in Figure 1a. The measurements started in June 2007 and continued until December 2015.

The echo signals are sampled along the radial direction and then converted to a rectangular image of 1024 pixels in the horizontal and vertical. Each pixel corresponds to a square of length 5.43 m, which is smaller than the theoretical spatial resolution of 7.5 m of the radar system determined from the pulse length of the emitted beam [3]. A total of 512 radar images captured every 2 s are processed into time-averaged radar images, an average over 17 min, as shown in Figure 1b. Time-averaged radar images are processed hourly, which enables interpretations of water lines of the river channel, dune locations, shore positions, breaker zones, river plume front formed by the river discharge, *etc.* [5,6]. Qualitative comparison with satellite data shows that the highest echo signals come from solid surfaces such as dunes and the floodplain, *etc.*, which are depicted with bright pixels, and water areas where depths are shallow also return relatively high signals from rippling on the water surface. Smooth sea and river water surfaces return only a small amount of the emitted beam, so they are dark in the time-averaged radar image. The figure also shows the coordinate system used in the study: *x*-coordinate for east-west extent and *y*-coordinate for north-south. The details of radar data processing are described in previous work [3].

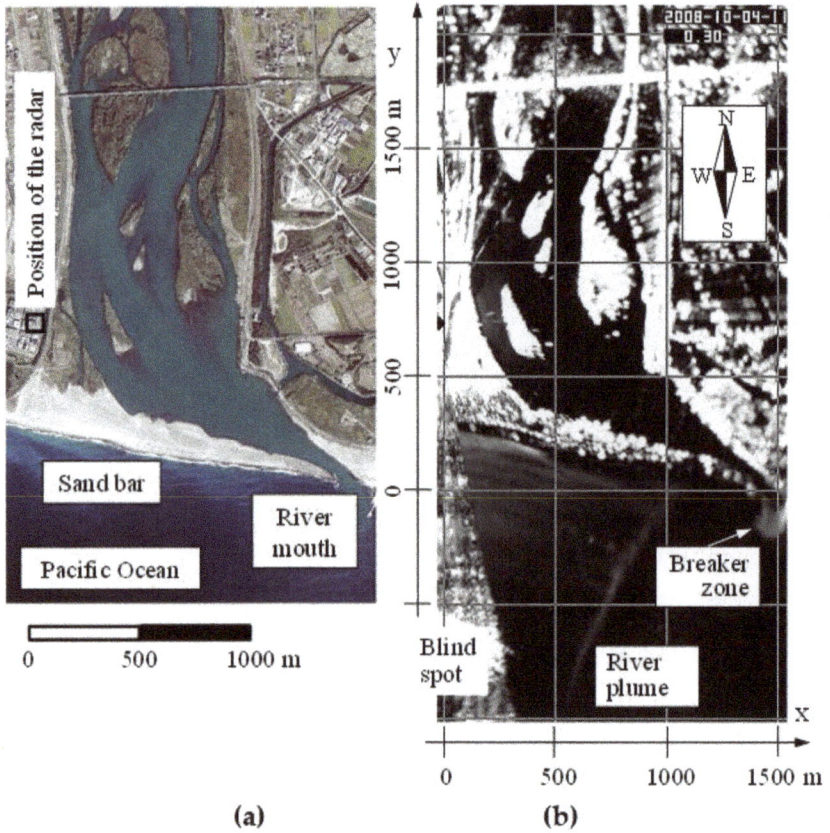

Figure 1. (a) Satellite visible image of the study area (True color composite display); and (b) time-averaged radar image. Satellite image: Ikonos-2. Acquisition date/time: 4 October 2008, 10:54 JST. Time-averaged radar image: average of images collected from 11:00 JST to 11:17 JST on 4 October 2008. Tide level of Omaezaki: 0.30 m (T.P.).

A Baiu front was active from 10 to 16 July 2010 along the Japanese main island Honshu and delivered heavy rain in the catchment of the River Tenryu. Figure 2 shows the variations of river water levels, tide level, flow rate and rainfall. River water levels in Tokyo Peil (T.P.) were measured at the Kakezuka station, 4 km from the river mouth in the tidal range, and the Nakanomachi station, 9 km from the river mouth, and the river flow rate was estimated at the Kashima station, 25 km from the river mouth. The water level record is missing from 16 h (JST) on 16 July 2010 at the Kakezuka station. The ocean sea level was measured at the Omaezaki tide station, 40 km to the east of the river mouth. The largest delay of the tidal propagation from the tide station to the river mouth was estimated approximately 5 min from the records of neighboring tide stations. The rainfall was measured at the Iwata weather observatory, 9 km to the north east of the river mouth.

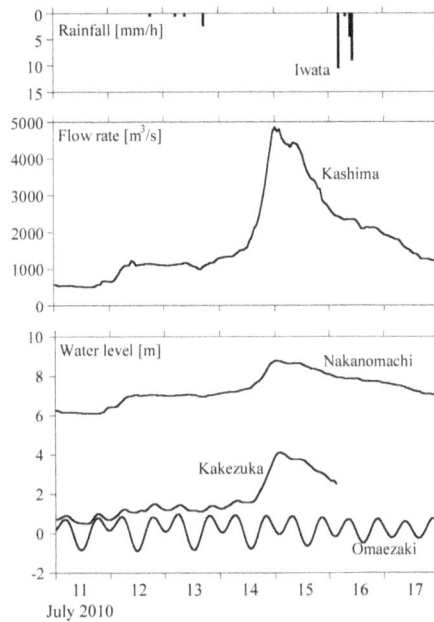

Figure 2. Variations of rainfall (upper panel), estimated flow rate (middle panel), and water levels in the river channel (Nakanomachi and Kakezuka) and sea (Omaezaki, bottom panel), observed during the flood.

According to the Hamamatsu Office of River and National Highway, the maximum amount of water released from the Funagira reservoir, located 5 km upstream from the Kashima station, was 4400 m^3/s at 22 h on 14 July 2010. Right after this time, the highest water levels were recorded at the Nakanomachi and Kakezuka stations. The return period of a flow rate of 4000 m^3/s for the River Tenryu is estimated to be approximately 30 years by the Japanese River Bureau. The maximum deviation of the observed tide level from the astronomic tide, which is estimated by the Japan Meteorological Agency, was approximately 0.15 m during this period. There were short-term rainfalls up to 10 mm/h on 18 July 2010, which did not contribute to the flood flow. Significant wave heights measured by a nearby wave station (approximately 5 km from the river mouth) were almost below 1 m during the period of interest. The author regards the effect of wind waves as negligible compared to the result of the flood flow. Detailed analyses on radar images for the days from 14 to 16 July during the flood will be described in the following sections.

3. Morphological Changes during the Flood

3.1. Interpretation of Time Averaged Radar Images

Figure 3a shows the time-averaged radar images for pre- and post- flooding, and Figure 3b for even hours during the flood when the river discharge exceeded 3000 m^3/s for 30 h. The lower portion of each image shows the water level variations measured at the Nakanomachi station in the river and the Omaezaki tide level station. The white vertical line in this lower portion denotes the temporal position in the radar acquisition. As shown in Legends 1 and 2, dunes in the river are named 'Dunes 1, 2, and 3' in sequence from the river mouth, dunes extending from east to west at the river mouth are labeled 'Sand bar', and a fan-shaped deposition at the river mouth is designated 'River mouth terrace'.

An animation processed from the time-averaged radar image has been uploaded to YouTube [8] and the results of image interpretation at different stages during the flood are summarized in Table 1.

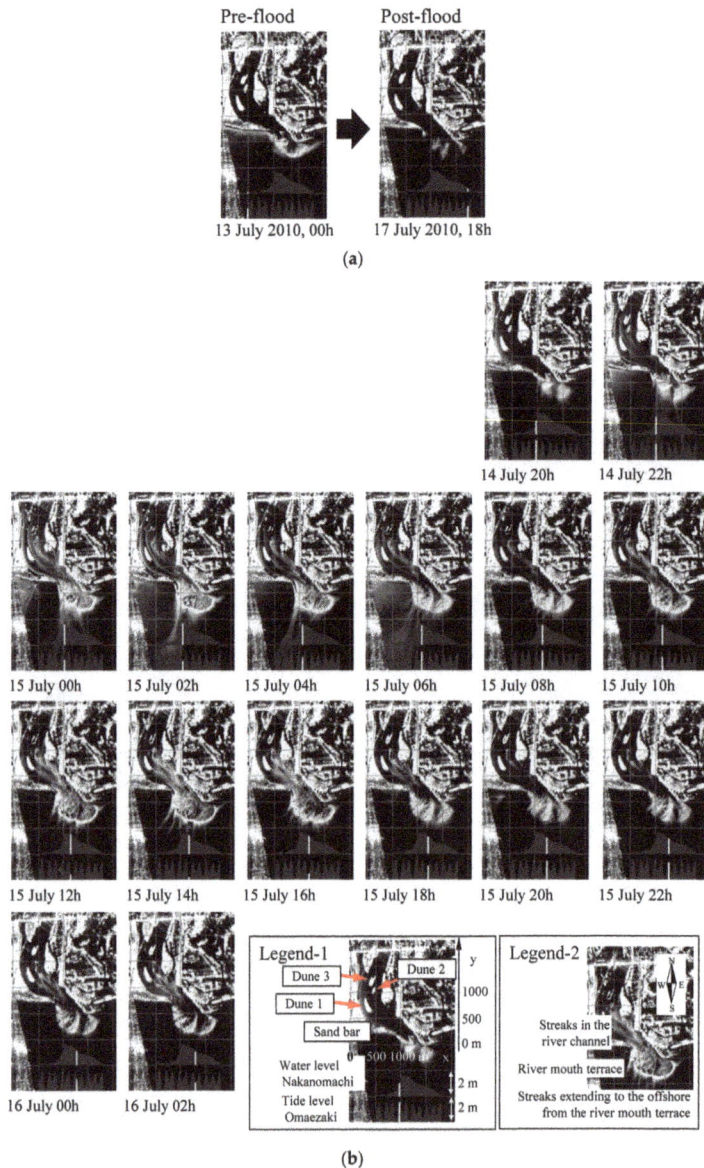

Figure 3. (**a**) Time-averaged images during the flood. Pre-flood: 13 July 2010, 00 h JST. Post-flood: 17 July 2010, 23 h JST. The lower portion of each image shows the water level variations measured at the Nakanomachi station in the river and the Omaezaki tide level station. The white vertical line in this lower portion denotes the temporal position in the radar acquisition; (**b**) Time-averaged images during the flood. From 14 July 2010, 20 h to 16 July 2010, 02 h.

Table 1. Morphological changes during the flood.

Time	Flood	Dunes 1, 2, and 3	Sand Bar and River Mouth Terrace
00 h 13 July 2010 (Figure 3a)	• Pre-flood.	• Dunes 1, 2, and 3 are clearly observable.	• Tip of the sand bar is located at $x = 900$ m. Width of the river mouth is approximately 190 m.
20 h–22 h 14 July (1st row of Figure 3b)	• Water level is rising.		• Ring-shaped river mouth terrace starts to expand offshore.
00 h–10 h 15 July (2nd row of Figure 3b)	• Water level increases rapidly and reaches the maximum during the flood.	• Dunes 2 and 3 are submerged for approximately ten hours and may have been eroded actively.	• Sand bar is eroded and the tip shifted to $x = 800$ m. Width of the river mouth increases to approximately 280 m. • Radial development of the river mouth terrace proceeds offshore from the river mouth.
12 h–16h 15 July (3rd row of Figure 3b)	• Water level of the river is decreasing, but the tide level is close to low tide and there is still a significant gradient in water level, which should have increased the sediment load in the river.	• Dune 3 is observable when the water level decreases, whereas Dune 2 does not appear, suggesting the lower portion of the dune is eroded by the flow.	• Position of the tip of the sand bar remains stagnant. • The fan-shaped river mouth terrace becomes clearly visible when the tide level descends. At the edge of the river mouth terrace, several streaks are extending radially to the offshore, which may be trace flow.
18 h 15 July–02 h 16 July (3rd and 4th row of Figure 3b)	• Final stage of the flood. Water level is decreasing. Morphological change becomes small.	• Upper portion of Dune 3 appears above the water. Similar to sand dune 1, the geometry is unchanged from the pre-flood state. Relative elevation of Dune 1 was high, and it was never submerged during the flood.	
18 h 17 July 2010 (Figure 3a)	• Post-flood.	• Dune 2 is washed away.	

Comparing pre- and post-flood time-averaged images shown in Figure 3a, the major changes due to the flood are that dune 2 was totally washed away and the width of the river mouth was widened to approximately 100 m. The configuration of dunes 1 and 3 and the main portion of the sand bar remain almost the same, suggesting that the river channel is not easily erodible even by a 30 year-return-period flood going through the area.

3.2. Erosion of the Sand Bar during the Flood

The waterline, boundary between the land and water surface, can be identified from the time-averaged radar images by manual digitization [3]. Based on empirical experiences, the distribution of pixel intensity shows an abrupt change at the waterline, higher brightness for the sand covered surface and darker for the water body. Using this empirical rule, the waterlines along the sand bar at different times have been identified from the time-averaged radar images by manual operation.

Figure 4 shows variation in the east-west position of the tip of the sand bar, with the water level difference between the river water level at the Kakezuka station and the tide level at the Omaezaki station. The variation in the water level difference may be regarded as a proxy of flow velocity change during the flood. The river mouth widened during the flood, as described in Table 1, and this process can also be tracked from the display: when the difference in the water level between the river and sea starts to increase on 14 July 2010, the tip position begins to shift westwards, or erosion starts with a speed of approximately 25 m/h. After the water level difference reaches a maximum on 15 July 2010, the shifting and erosion of the tip of the sand bar has stopped. After this period, variation in the main channel becomes prominent, which will be described in the next section.

3.3. Migration Speeds of Radar Light-Dark Brightness Patterns along the Channel

Careful inspection of the sequence of time-averaged radar images and animation introduced above reveals that streaky light-dark patterns migrated downstream in the river channel, especially during the periods when the river water level was descending. As mentioned previously, these light-dark patterns are the result of strong backscattering from the water surface roughness, which becomes larger when the near surface flow is turbulent. The water flow may become turbulent in shallow areas, where active interaction between the water surface and river bottom occurs. Thus, we surmise that migrations of light-dark patterns are proxies of river bed variation during the flood. Unfortunately, exact determination of the origin of this light-dark pattern is difficult since no on-site measurement was done during the flood. To assess this speculation, migration speeds of the patterns were estimated by processing a time-stack image.

Figure 5 shows a time-averaged image on the left and a time stack image on the right. To obtain the time stack image, which has spatial extent in the vertical and temporal extent in the lateral, pixel intensities along the oblique grey bands shown in the time-averaged image, whose width in the lateral or east-west direction is 15 pixels or 81 m, were extracted and stacked in the vertical at specific times. This extraction was repeated for the following time-averaged images and stacking positions shifted in the lateral. In the lower right below the time stack image, the river water level variation measured at Nakanomachi and the tide level at Omaezaki are also shown.

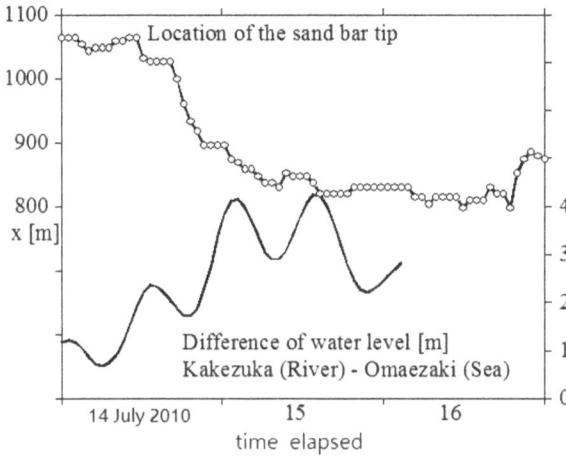

Figure 4. Variation of the east-west location of the tip of the sand bar, and the water level difference between the river and the sea. x denotes east-west coordinate shown in Figure 1b.

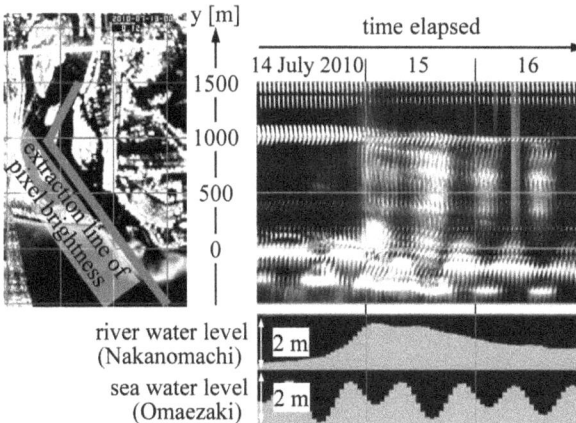

Figure 5. Migration of streaky light-dark patterns downstream in the river channel: Time-averaged radar image (left) and time stack image (right). y denotes north-south coordinate shown in Figure 1b.

In the time stack, oblique propagations towards the lower right can be seen, which indicate that light-dark patterns in the time-averaged image traveled in the down-flow along the extraction line. This becomes prominent for 15 and 16 July 2010 when the flood was decaying, especially during low tide. The local gradient of the oblique features seems to change with the water level variation.

A PIV (Particle Image Velocimetry)-like method was applied to detect local gradients of the oblique features in the time stack image [9]. First, a 33 m or 6 pixel-long spatial template extending in the vertical in the times stack image was established at time t_0. Next, the most similar pattern was found at time $t_0 + 1$, by calculating the correlation coefficient, to estimate the hourly migration distance of the local light-dark pattern. Figure 6 shows the variation of migration speeds averaged over 500 m in the north-south direction and the difference between the river water level at the Kakezuka station and the tide level at the Omaezaki station. Estimation at some times and locations, mostly when the tide level was high and rainy periods, are missing since the patterns were unclear to establish a correlation. We should notice that the velocity estimate here is a one-dimensional assessment: the estimated velocity is the migration speed of the light-dark pattern along the extraction line shown in Figure 5. In other words, an assumption has been made that the light-dark pattern moves mainly along the extraction line.

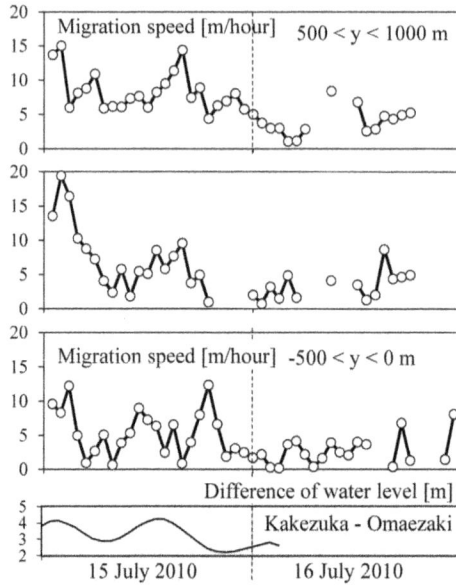

Figure 6. Variation of migration speeds of light-dark patterns in the time-averaged radar image and the water level difference between the river and the sea.

Estimated migration speeds of the light-dark pattern were of the order of 10 m/h (~0.003 m/s) and varied in accordance with the water level gradient, which supports our speculation, or the proxy hypothesis, that light-dark patterns can be regarded as proxies of river bed variation during the flood.

Gaeuman and Jacobson [10] made measurements on the change of bottom profiles in the Missouri River at a portion where the river width is several hundred meters. During brief water rises with flow rates of 5464 m^3/s, 4303 m^3/s, 4048 m^3/s, and 2831 m^3/s, they found migration speeds of the bed undulations were in the range of 1 to 6 m/h, which is of the same order as the results shown above. This further supports our proxy hypothesis.

4. Concluding Remarks

A flood event at the Tenryu River mouth was observed with X-band radar. Hourly time-averaged radar images over 72 h were processed to understand the morphological changes due to the flood. Image interpretation qualitatively described the processes of morphological transitions. The variation of tip location of the sand bar read from the time-averaged radar images shows that the erosion at the sand bar tip occurred mostly before the river water level reached the maximum. PIV-like image analyses were applied to estimate the down-flow migration speed of possible morphological features, which varied in accordance with the water level gradient between the river and the sea.

The estimated migration speeds from the radar may be used in formulae that relate sediment loads to migration rates of river bed undulations to estimate river sediment loads. For example, Gaeuman and Jacobson [10] tried to estimate the bed-load transport using the down-stream propagation speed obtained from the field measurement and correlated with flow velocity. A similar approach may be attempted using the present data; however, this is beyond our scope at this time.

The results of the field observation demonstrate that an X-band radar is a powerful imaging device to track a flood event in detail and continuously during day- and nighttime periods. Although it cannot detect color information and therefore is hard to infer the condition of the water surface, existence of suspended materials, vegetation, *etc.*, the morphological variation can be assessed robustly throughout an extreme event.

Acknowledgments: The author is grateful for the support provided by Takahashi Ryo, a former graduate student, and Sato Shinji and Tajima Yoshimitsu, the University of Tokyo, for their help with the field observations. This research is supported by the Special Coordination Funds for promoting Science and Technology of Ministry of Education, Culture, Sports, Science and Technology and the River Fund, The Foundation of River & Watershed Environment Management.

Conflicts of Interest: The author declares no conflict of interest.

References

1. Bell, P.S. Shallow water bathymetry derived from an analysis of X-band marine radar images of waves. *Coast. Eng.* **1999**, *37*, 513–527. [CrossRef]
2. Ruessink, B.G.; Bell, P.S.; Van Enckevort, I.M.J.; Aarninkhof, S.G.J. Nearshore bar crest location quantified from time-averaged X-band radar images. *Coast. Eng.* **2002**, *45*, 19–32. [CrossRef]
3. Takewaka, S. Measurements of shoreline positions and intertidal foreshore slopes with X-band marine radar system. *Coast. Eng. J.* **2005**, *47*, 91–107. [CrossRef]
4. Holman, R.; Haller, M.C. Remote sensing of the nearshore. *Ann. Rev. Mar. Sci.* **2013**, *5*, 95–113. [CrossRef] [PubMed]
5. Takewaka, S.; Takahasi, Y.; Tajima, Y.; Sato, S. Observation of morphology and flow motion at the river mouth of Tenryu with X-band radar. In *Proceedings of Coastal Dynamics 2009*; Mizuguchi, M., Sato, S., Eds.; World Scientific: Singapore, 2009.
6. Takewaka, S. Visibility of river plume fronts with an X-band radar. *J. Sens.* **2016**. [CrossRef]
7. Pianca, C.; Holman, R.; Sieglea, E. Mobility of meso-scale morphology on a microtidal ebb delta measured using remote sensing. *Mar. Geol.* **2014**, *357*, 334–343. [CrossRef]
8. Flood observed at the mouth of Tenryu River. Available online: http://www.youtube.com/watch?v=msu-7cEHH_I (accessed on 25 December 2015).
9. Elsayed, M.G.; Takewaka, S. Longshore migration of shoreline mega-cusps observed with X-band radar. *Coast. Eng. J.* **2008**, *50*, 247–276.
10. Gaeuman, D.; Jacobson, R.B. Field assessment of alternative bed-load transport estimators. *J. Hydraul. Eng.* **2007**, *133*, 1318–1319. [CrossRef]

Journal of
*Marine Science
and Engineering*

MDPI

Article

A Conceptual Model for Spatial Grain Size Variability on the Surface of and within Beaches

Edith Gallagher [1,*], Heidi Wadman [2], Jesse McNinch [2], Ad Reniers [3] and Melike Koktas [3]

[1] Department of Biology, Franklin and Marshall College, Lancaster, PA 17603, USA
[2] Field Research Facility, Coastal Hydraulics Laboratory, USACE, Duck, NC 27949, USA;
     frf.wadman@gmail.com (H.W.); frf.mcninch@gmail.com (J.M.)
[3] Hydraulic Engineering, Delft University of Technology, 2628 CD Delft, The Netherlands;
     a.j.h.m.reniers@gmail.com (A.R.); koktas.melike@gmail.com (M.K.)
*   Correspondence: edith.gallagher@fandm.edu; Tel.: +1-717-291-4055

Academic Editor: Gerben Ruessink
Received: 1 February 2016; Accepted: 22 April 2016; Published: 25 May 2016

Abstract: Grain size on the surface of natural beaches has been observed to vary spatially and temporally with morphology and wave energy. The stratigraphy of the beach at Duck, North Carolina, USA was examined using 36 vibracores (~1–1.5 m long) collected along a cross-shore beach profile. Cores show that beach sediments are finer (~0.3 mm) and more uniform high up on the beach. Lower on the beach, with more swash and wave action, the sand is reworked, segregated by size, and deposited in layers and patches. At the deepest measurement sites in the swash (~−1.4 to −1.6 m NAVD88), which are constantly being reworked by the energetic shore break, there is a thick layer (60–80 cm) of very coarse sediment (~2 mm). Examination of two large trenches showed that continuous layers of coarse and fine sands comprise beach stratigraphy. Thicker coarse layers in the trenches (above mean sea level) are likely owing to storm erosion and storm surge elevating the shore break and swash, which act to sort the sediment. Those layers are buried as water level retreats, accretion occurs and the beach recovers from the storm. Thinner coarse layers likely represent similar processes acting on smaller temporal scales.

Keywords: grain size; stratigraphy; morphology; morphodynamics; storms; sediment; beach; swash; shore break; depth of disturbance

1. Introduction

Grain size varies on a natural beach and observations of different sand sizes and their spatial and temporal distributions are well documented (e.g., [1–11]). Recently, Gallagher *et al.* [10] found that the spatial distribution of sand grain size on the surface of the beach at Truc Vert, France was correlated with morphology. Specifically, coarse sediment was found around rip channels, where flows were strong, and the finest sediment was found high on the intertidal beach where wind was the most common transporting mechanism. These spatial patterns were also observed to change with changing morphology. For example, as a rip channel and shoal system migrated, sediment was constantly reworked and redistributed by the spatially variable surf zone energy field. To examine grain size variations at a higher temporal resolution, an experiment was carried out along two cross-shore profiles in Monterey, CA [11] where the beach was relatively steep (slope 1:7.5) and had a vigorous shore break (where surf zone waves crash dramatically and finally, driving the swash run-up, and where there is often a morphological step). During that experiment, grain size was observed to be largest in that energetic shore break. In addition, the coarse patch moved up and down the beach with the shore break as a function of tidal fluctuations [11], similar to observations by Ivamy and Kench [6]. Attempts to collect stratigraphic information during the Monterey experiment failed [11]. The limited observational studies connecting dynamic morphology change to stratigraphy (e.g., [8,9,12]) indicate that wave and swash interactions with morphology have been observed to induce changes in sediment stratigraphy over a range of temporal and spatial scales, from wave-by-wave and cm-scale changes [8], to tidal and 10 s of cm-scale changes [5,9], to storm and seasonal and m-scale changes [12]. However, most experiments acknowledge that stratigraphy sampling should have been deeper [5,9], or more time or higher frequency measurements would have been useful [8], or they didn't have direct hydrodynamic measurements [12].

Despite the basic understanding that sediment across a beach is often highly variable in space and time, a mean grain size and spatial uniformity is often assumed in morphology modeling studies to simplify sediment transport calculations (e.g., [13–21]). In fact, Soulsby *et al.* [21] identified grain size as having the largest uncertainty of any sediment transport model input parameter. What remains unclear is the severity of the penalty, with respect to morphology modeling skill and, in nature, how sensitive beach and nearshore morphodynamics are to variations in sediment characteristics both on the surface and in the shallow stratigraphy. Gallagher *et al.* [22] modeled cross shore bar migration and found that the model had higher skill when cross shore-varying (surface) grain size was used instead of a single mean size for the whole beach profile. Ruessink *et al.* [23] modeled a coarse foreshore as immobile, because the relatively fine mean size used in the model predicted unrealistic, excessive erosion in that energetic region. Preliminary work by Gallagher *et al.* [24] showed that by including a coarse-grained, less-mobile patch in the swash, the dynamics of the offshore sand bar were altered. That study suggests that local grain size and its patchiness may impact the dynamics of the whole beach profile. If true, one would expect that variable layers, which are distributed throughout the shallow stratigraphy and the result of earlier time periods with differing wave energy and swash location, will impact overall profile dynamics as well.

Reniers *et al.* [11] used a multi-grain size module in Xbeach (an open source nearshore morpho- and hydrodynamics model; [18]) to include spatially and temporally varying grain size in beach profile predictions. They modeled the coarse patch that was observed in Monterey and found that the high levels of turbulence and strong (but intermittent) currents in the shore break and swash were effective at winnowing and moving finer sediment both up and down the beach, leaving coarse sediment in place at the base of the swash. Reniers *et al.* [11] used a uniform distribution (equal amounts of the different sizes) in each layer to commence each model run because information about stratigraphic layers at Monterey was not available. In that study, modeled flows would encounter well-mixed sediment and redistribute it, successfully recreating observed surface sediment spatial variations.

In the present study, stratigraphy in the beach at Duck, NC was measured to document the presence and extent of grain size spatial variation within the bed on a natural beach and to begin to understand how the waves and swash create and then interact with layers of different sediment sizes. Specifically, spatial and temporal variations in the beach elevation and the shallow stratigraphy were examined over a course of a week, before and after a storm event as well as over a tidal cycle, using a suite of sediment cores and beach trenches. These stratigraphic observations were examined in the context of changing beach topography and water levels to understand the role nearshore hydrodynamics may play in creating and/or destroying beach stratigraphy. A conceptual stratigraphic model (including shore break sorting, wave/infragravity-scale bed level fluctuations and shore break translation) is proposed to describe the relationship between swash hydrodynamics at the shore break and the resulting sediment stratigraphy across the beach and foreshore.

2. Materials and Methods

2.1. Experiment

In March–April, 2014, a field experiment was conducted to test an enhanced coring technique, developed to overcome difficulties of collecting sediment cores in the swash, and to sample the stratigraphy across the beach. The experiment was conducted at the U.S. Army Corps of Engineers

Coastal and Hydraulic Laboratory's Field Research Facility (FRF) in Duck, NC. Cores were collected at 8 locations along a cross-shore profile, which spanned from the bottom of the dunes into the deep swash (Figures 1a,c and 2a). The profile was sampled on five occasions: before and after a storm on 25 March and 27 March, after a smaller wave event on 28 March, and at high and low tide on 1 April (Figure 1b). After the coring was finished (2–3 April), two large trenches were dug in the beach to examine and photograph stratigraphy and to put the core observations in context (Figure 1a). From 27 March to 2 April, a simple experiment was conducted to examine the depth of disturbance of the surface sediment (Figure 1a). RTK-GPS (Real Time Kinematic Global Positioning System) elevation surveys of the experiment area were conducted daily both with a backpack and the CRAB (Coastal Research Amphibious Buggy; www.frf.usace.army.mil) (Figure 1a,c). In addition, over 27–28 March, a lidar was mounted on the dune to measure waves, swash run-up, and beach elevation changes along a cross-shore profile line adjacent to the core cross-shore profile. In addition to these experiment-specific measurements, current and wave data were being collected by the FRF in 2 m, 4 m, and 8 m water depths.

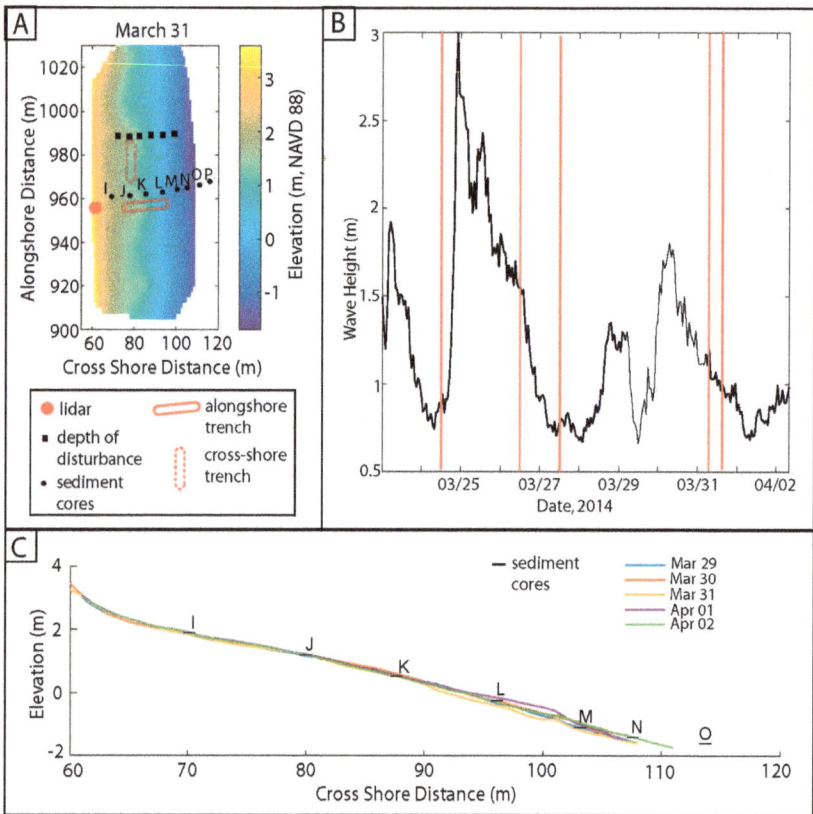

Figure 1. (**A**) Experiment layout is shown on a map of beach elevation, with the depth of disturbance experiment shown as small squares and the core locations as small circles. The red lines show the approximate locations of the cross-shore and alongshore trenches (solid and dashed, respectively), while the solid red circle indicates the position of the dune-mounted lidar; (**B**) Wave height measured in 11 m water depth during the experiment with the time of the individual coring efforts indicated with vertical red lines; (**C**) Elevation profiles across the beach along with the core locations shown as horizontal black lines.

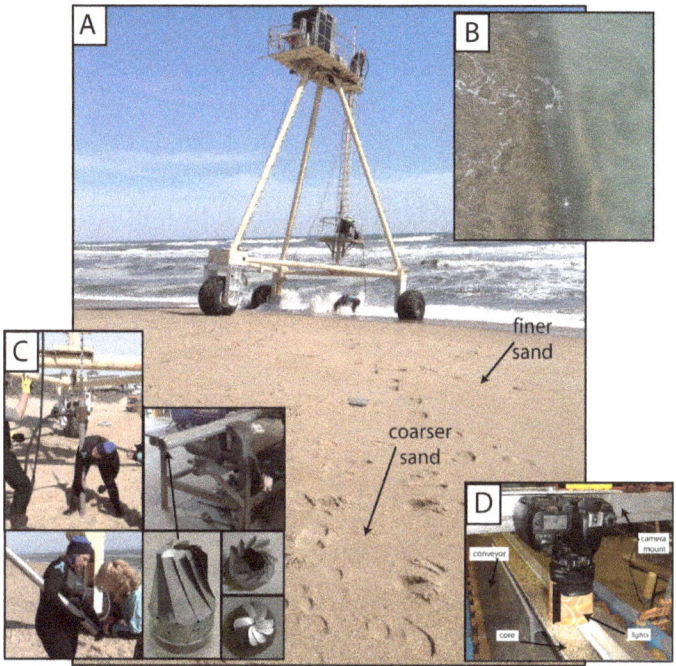

Figure 2. Various images from the experiment including: (**A**) sampling from the DI in the swash; (**B**) close-up of the coarse-gravel step at the base of the swash zone; (**C**) coring and core catchers; and (**D**) DIS camera mounted above the conveyor belt with a core beneath it.

2.2. Cores

Previous efforts to collect cores in the swash met with mixed results. Because the sediment in this region at Duck (and in Monterey) is coarse and highly fluidized by the constant motion of the shore break and swash (Figure 2b), so retaining sediment in core tubes after collection proved to be difficult. To address this, aluminum core barrels were fitted with custom-designed, reusable core catchers, which were riveted into the base of each barrel (Figure 2c). Crafted from a single sheet of stainless steel, the catchers are fitted with a cutting edge on the base, an internal rim to brace the barrel, and taper to a series of thin, laser-cut "fingers". The fingers unfold and press along the internal wall of the barrel by the force of sediment entering the barrel during coring. Then, they subsequently fold closed by the weight of the sediment in the core barrel, effectively forming a solid plug once core extraction begins (Figure 2c). In addition, prior to extraction, the fully penetrated core was filled to overflowing with seawater to create suction and sealed with a standard 3-inch plumbers plug. Once extracted, a plastic cap was taped over both ends to further seal the core. The cores were kept upright until they were processed.

In addition to collecting surveys, the stable, amphibious CRAB was used as the sampling platform for vibracore collection (Figure 2a). A Briggs and Stratton 5 Hp engine with a modified concrete shaker (to vibrate the cores for penetration) was mounted on the CRAB ~3 m off of the ground (Figure 2), allowing the engine to stay dry throughout the coring process, even when collecting cores in the surf. The operation still required personnel to be in the water to guide the core barrel into the sand, therefore core collection was constrained to ~1 m water depth (Figure 2a,c). Using these special adaptations, a total of 36 cores were collected on four different days over about a week. Cores were not collected at exactly the same locations on successive days, but instead were spaced 1–2 m from the previous spot, alongshore and on the same elevation. This was done in part because the coring process disturbs the stratigraphy, and in part because reoccupying the exact same positions is nearly impossible. Accordingly, cross-shore and alongshore trenches were dug to examine small-scale spatial variability in stratigraphy, and are described more in Section 2.4, below.

2.3. Core Processing

Once returned to the FRF, the core barrel was cut just above the sediment line to allow the excess water to flow out. The top of each barrel was stuffed with paper towels to minimize sediment movement, and then the cores were temporarily placed horizontally in order to remove the core catchers for future use. Once removed, the base of each barrel was also stuffed with paper towels to minimize sediment movement and slow water loss, then capped and taped. If not opened immediately, the cores were stored upright in a refrigerator.

For processing, each core was laid horizontally in a steel trough and cut in half with a skill saw by cutting just the metal barrel on each side. Care was taken not to allow the blade to penetrate into the sediment. Cores were split open, photographed and described qualitatively. One half of the core was sampled and bagged for later analysis. Samples were collected every 10 cm as well as at significant changes in lithology. The other half of the core was placed on a conveyor belt and photographed using a digital imaging system for estimating grain size (Figure 2d).

2.4. Digital Imaging System

A fixed frame above the conveyor belt described in Section 2.3, above, was used as the reference/photo point (Figure 2d) for the digital imaging system (DIS). The core was moved using the conveyor belt and images were taken every cm moving down the core. Each image is approximately 2×2 cm (or ~2000 × ~2000 pixels), allowing an approximately 50% overlap between successive down-core images.

For analysis, each image is split into horizontal (cross-core) sub-images or plaquettes that are 200×2000 pixels or 0.2×2 cm. The plaquettes are used to estimate mean grain size for that small flat layer. There is 75% overlap between each plaquette, giving 34 overlapping plaquettes and resulting in a running average of grain size moving down the image. Ultimately, each image provides a profile from top to bottom, with 34 values of mean grain size. Because the different images overlap by approximately 50%, multiple estimates of grain size are provided in different depth bins. In addition, images were taken side-by-side along the open core face, further increasing the number of estimates in each depth bin. DIS data have been found to be noisy, but averaging, overlap, and multiple photos have been shown to significantly improve DIS mean grain size estimates [10]. For this study, final smoothing was done by averaging all estimates in 0.5 cm depth ranges. Grain size was estimated from images using the autocorrelation technique developed by [25], together with recently developed grain size image calibration curves for Duck, NC.

2.5. Trenches

To put the core observations in context, two large trenches were dug in the dry beach using a backhoe. One trench was dug perpendicular the shoreline (trench 1) and the other was dug parallel to the shoreline (trench 2). The total length of trench 1 was about 15 m and it cut across the side of the horn of a beach cusp (see Figure 1a for approximate locations). At its seaward end, trench 1 was about 1 m deep. At its landward end, it was ~1.5 m deep. The seaward-most extent of the trench was limited by the water table, which caused slumping within ~10 m of the swash. The landward extent of the trench was constrained by the dune. The alongshore trench was dug just above the high tide line on the berm and was dug across the cusp trough and to the crest of the next beach cusp. Trench 2 was ~30 m long, ~2 m deep, and reached down to the water table. Once dug, the stratigraphy was photographed every ~1 m along each trench, and general features were described and drawn.

2.6. Changes in Beach Elevation Change and the Active Layer

A simple test was conducted to examine thickness of the active layer of sediment in the swash together with changes in beach surface elevation. Metal washers were placed over metal rebar rods, which had been driven into the sand on a cross-shore profile approximately 20 m north of the coring profile (Figure 1a). The distance from the top of the rod to the sand level, and from the sand level to the depth of the washer, was measured on a daily basis. Because the stainless steel washer is denser than the sediment, it will settle to the bottom of the moving layer as each swash crosses the beach, lifting, moving, and re-depositing the active sediment. This simple test gives a measurement of the deepest depth of sediment motion (*i.e.*, the active layer or depth of disturbance) of the bed over the time period of measurement (here: 1 day). This method is similar to previous efforts to quantify active layer thickness (e.g., [6]).

To examine changes in surface elevation at finer scales than that permitted by the washer technique (and without the potential for scour) a Riegel terrestrial lidar scanner (LMS-z390i, 1550 nm laser with a 0.3 mrad beamwidth) was used to measure surface elevation of the beach at 2 Hz for 30 min starting at the top of each hour. Specific details regarding lidar operation and limitations of the specific system are detailed in [26]. To estimate beach elevation, data were transformed from the scanner coordinate system (angle and range) to rectified Cartesian coordinates (local coordinates for the horizontal and NAVD88 coordinates for the vertical) using a transformation matrix determined from scans of GPS-surveyed reflectors. It should be noted that because the lidar measurements were made ~35 m south of the depth of disturbance experiment, detailed comparisons between the methods are not appropriate. Accordingly, only general trends will be compared.

3. Results

3.1. Core Observation

Figure 3 shows the results from the DIS grain size estimation, giving profiles of mean grain size with depth in the cores at the different times and locations. The wide range of grain sizes observed varies not only spatially and temporally between core sites but also within individual cores. This illustrates the inadequacy of using a single grain size to characterize the beach.

The length of the cores, and the thickness and position of observed layers, may be somewhat altered due to compaction created during the coring process. As the core tube is vibrated and driven into the ground, the sediment in the tube is compressed. In addition, when the core is removed from the ground, water can flow out of the barrel, potentially disturbing the sediment layers within the barrel. The numbers shown in red in Figure 3 represent the difference between the ground surface along the edge of the core barrel (the actual depth of penetration; marked on the barrel prior to the barrel being extracted from the ground) and the top of the sediment inside of the core barrel. Compaction is common during coring and the amount of compaction in each core measured during this experiment is consistent with compaction measured previously during other sandy beach coring efforts using similar methodology (e.g., [27–29]). Because compaction happens nonlinearly in the core, stretching the core data generally is not done.

Unfortunately, during the coring process, vibration of loose sediment may jostle, rework, and ultimately damage some of the finer layers ($\leqslant$1 cm). Evidence for the destruction of finer layers comes from the comparison of the cores (both visual inspection and DIS measurements) and observations from the trenches, where cm-scale stratigraphic layers were visually observed. Accordingly, visual and DIS data were used together to help define four distinct sedimentary units, as preserved in the cores: (1) Homogenous Sand (HS), a fine-grained sand (<0.5 mm grain size) often with minor coarse sediment, either as individual grains or as thin (<5 mm) laminae, as well as occasional heavy mineral laminations; (2) Poorly Sorted coarse Sand (PSS), a medium- to very coarse (~0.5–1.3 mm grain size) sand with abundant granules and pebbles but no discernible layering, and that often coarsens with depth down-core; (3) Laminated Sand (LS), a unit comprised of discrete layers (on the order of 1 cm) of both HS and PSS; and (4) Coarse Gravel (CG), moderately well to well-sorted, well-rounded coarse gravel (1–2 mm grain size), with minor med-coarse sand and varying amounts of fragmented shell (see Figure 4 for examples of the sedimentary units). Although discrete laminae of PSS and HS of 1–2 cm or finer were observed in the sediment cores, for the purposes of the core descriptions (Figure 5), only layers of PSS or HS that were at least 5 cm in thickness were considered a discrete stratigraphic unit. Finer layering is defined as LS.

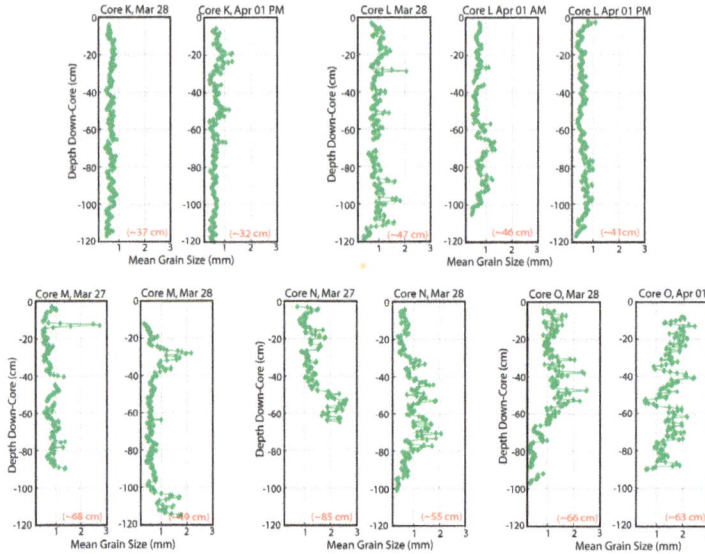

Figure 3. Examples of grain size data measured with the DIS, as preserved in sediment cores, at different locations during the experiment. Red numbers indicate measured compaction. See text for full explanation and discussion.

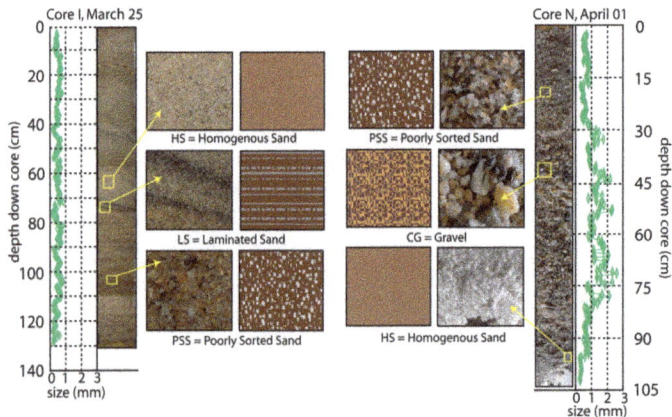

Figure 4. Two different cores are shown to illustrate the four units (represented by the different brown fill patterns) used in describing the stratigraphy. Detailed photographs show enlarged sub-sections, and the DIS profiles are shown in green. These general units are chosen owing to their prevalence as well as their dynamical significance.

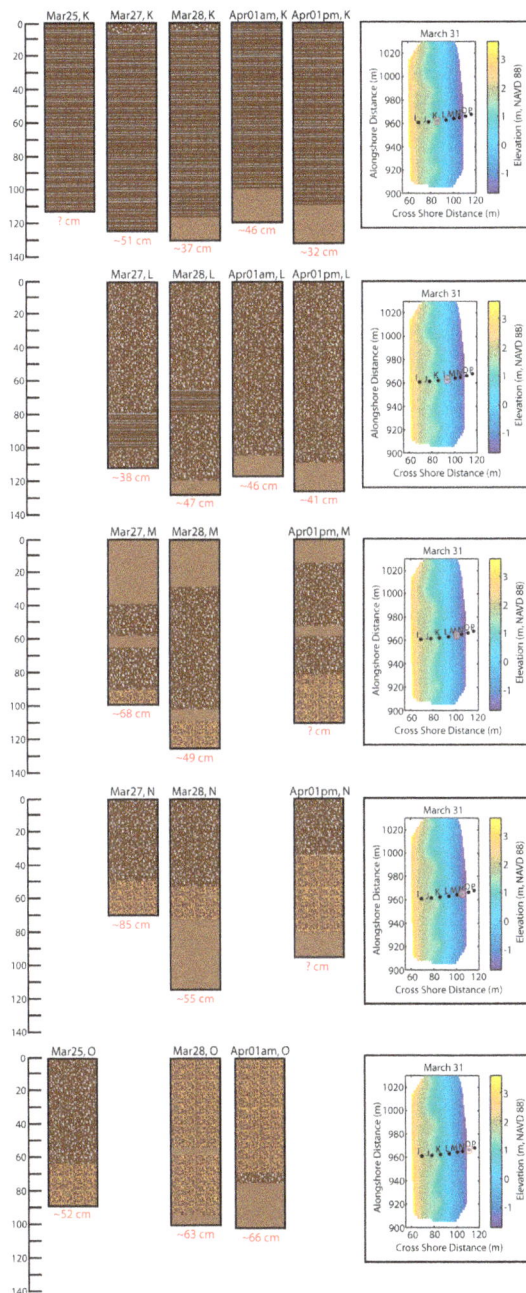

Figure 5. Cores described using their sedimentary units at five cross-shore locations (top to bottom: maps on the right show specific location), and for the five different sampling days/times (left to right). See Figure 4 for definition of the sedimentary units. Scale on the left is depth in the core in cm and red numbers indicate measured compaction.

A selection of cores is shown in Figures 3 and 5 that illustrates the major observations from the cores. Higher on the beach, grain size is dominated by finer grains, generally <0.5 mm, and is more uniform with depth. Location K is shown as representative of these more uniform observations at locations I, J and K (Figures 3 and 5). At all three locations, LS is the dominant unit. At K, a thin layer (<10 cm) of PSS is seen on the surface after the large wave event (26 March, Figure 1b) deposited a small amount of coarser material high on the beach. The HS near the bottom of some of the cores at location K is consistent with the observation of finer and relatively uniform sediment high on the beach because HS is a component of LS and was likely deposited during an extended quiescent period. It also is possible that HS is below LS at many or all locations and times, but was not reached with some of the cores.

With increasing distance seaward from the dune, the sediment becomes coarser and more varied, both spatially and temporally. The coarsest material (~2 mm) dominates at the deepest coring location shown (location O) and this layer of coarse material is relatively thick, at least 60–80 cm. Farther offshore, the sediment is once again fine-grained ([30], [11] and from our personal observations, but not sampled). At location N, landward of O but still frequently in the swash zone and shore break (Figure 1c), the coarse layer is also visible, but lies below a PSS layer. This is likely owing to location N being in a highly variable location, sometimes in the lower swash and sometimes directly in the intense shore break.

These data also suggest that over the course of the experiment, the top 20–40 cm of the beach surface was highly variable in time (see the discussion of the active layer in Section 3.3). This is well illustrated by the cores at locations M, N, and O where the top layer of the cores is quite different from day to day (Figure 3). For example, at location N, on 27 March, the grain size in upper 30 cm of the core ranged from 1–2 mm, whereas on 28 March, the upper 30 cm was finer, ~0.7 mm (Figure 3).

The more variable, intermediate elevations on the cross-shore profile are well illustrated by the cores at locations L and M (Figures 3 and 5). Location L is dominated by poorly sorted, downward coarsening sediment. A minor LS unit is observed mid-core on 27 March and 28 March, but not on 1 April, and this could be due to horizontal spatial variability of the layer. The L core collected on 27 March does not show the HS unit at the base (Figure 5, found on 28 March and 1 April), but the layer may have existed below the base of the collected core. The cores at location M have fine material (HS) on the surface, poorly sorted sediment (PSS) over most of their depth, and coarse gravel (CG) near the bottom. These PSS layers, together with the vertical variability at M, represent the reworked and highly changeable sediment observed at this location, which can be in the upper, mid or lower swash or in the shore break, depending on tides, waves and storms.

3.2. Trench Observations

Layering of many scales was observed in the trenches (Figure 6). Thick (~2–10 cm) gravel/sand layers were visible (CG or PSS) and are generally attributed to elevated sea level and swash action, likely owing to storm events (e.g., the thick, continuous coarse layer below the "7", "8", "9" and "10" in the bottom panel in Figure 6). Thinner laminae (1–2 cm) were also visible but were less distinct and not as well sorted. For example, immediately below the "10" in the top, center photo of Figure 6, but above the thick coarse layer of PSS, are four thin (⩽1 cm) layers of finer PSS and HS sediments. The larger, coarser sediment layers are outlined in Figure 7 to emphasize their extent and shape. Overall, these layers form "lenses" of coarse sediment with a cross-shore extent of almost the length of the trench (~10–15 m), and a thickness of roughly 10 cm (Figure 6). These lenses are thought to be owing to shore break and swash acting higher on the beach, and are described in more detail in Section 4.

Figure 6. Images of the cross-shore trench that was dug near the end of the experiment. The bottom image is a compilation of multiple images, showing the whole trench. The other images show details (like layer thickness and extent) at various locations in the trench. Note the numbers scratched in the sand wall (and outlined in black) of the trench approximately every meter for reference.

Figure 7. This image is the same as the bottom panel in Figure 6 but with specific coarse layers highlighted to illustrate their size, shape, and extent.

3.3. Depth of Disturbance Observations

In the shore break and swash, each wave suspends, moves and deposits sediment, thus the processes acting to sort and redistribute different grain sizes are constantly at work. Observations in the literature (e.g., [6,8,9,31–33]) indicate that the beach face in the swash is alternatingly eroding and accreting over relatively short time scales. Figure 8 shows about two days of data from the dune-mounted lidar (26–28 March). The time series shows the elevation every hour (the minimum elevation is taken each hour to represent the time window) at three different core locations. The beach elevation is seen to fluctuate by 20–40 cm over the course of a few hours. Indeed, variations in elevation in the swash of up to 15 cm over the course of about 5 min are visible in the lidar data ([31,32] and in the present data set, not shown).

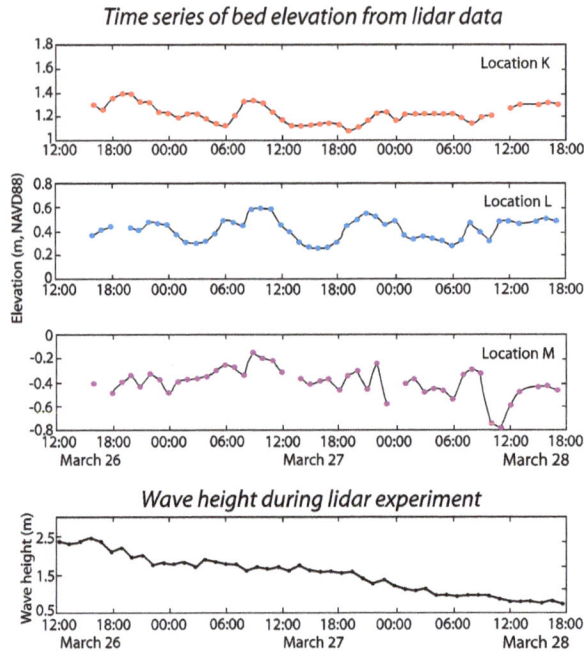

Figure 8. Two days of time series of bed elevation from the lidar at locations K, L, and M. Bottom panel shows wave height during the same period.

Figure 9 shows the results of the daily depth of disturbance measurement from the rebar array, and this simple experiment also provides insight into the movement of the bed. The blue line shows the daily elevation of the beach surface, which is under-sampled when compared with Figure 8, but still indicates that variations in surface elevation in the lower, active swash are at least 40 cm, even when the waves are small (e.g., 27–29 March, locations 4 and 6; Figures 1 and 9). In the region of the upper swash, those variations are smaller, but still reach ~20 cm at location K (Figures 8 and 9). These data (as well as the traditional cross shore profile surveys in Figure 1c) indicate that the top layer of the beach is highly active and being eroded and re-deposited over relatively short time periods. In addition, the timing and amplitude of the observed elevation change does not seem to be directly related to offshore wave height (Figure 9).

The green, red, and black lines in Figure 9 represent the maximum depth to which the washers moved. Vertical lines indicate when the washer was removed and placed back on the beach surface, while dashed lines indicate that the washer was not found at the next sampling. Lost washers were buried too deeply to retrieve via digging along the rebar, though it should be noted that, in the swash, digging deeper than about 30 cm is difficult because water and sand refill the hole as fast as one can dig. These results indicate that at least the top 10–30 cm of the sand bed in the swash is active, being suspended and re-deposited regularly: on at least a daily, and likely hourly, basis.

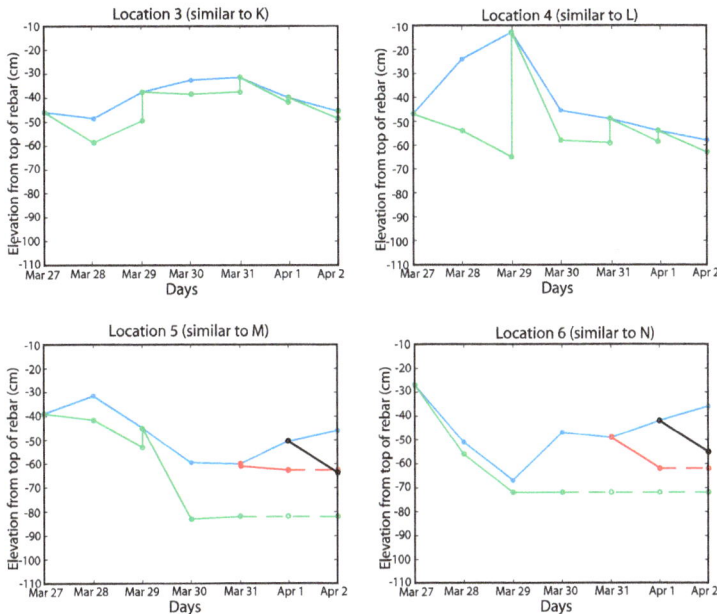

Figure 9. Depth of disturbance measurements from four different locations on the beach (the lowest four circles in Figure 1). The blue lines show the daily vertical position of the beach surface with respect to the top of the rebar. The green lines (and red and black) show the depth below the sand surface to which the washer was buried. Dashed lines represent times when the washer could no longer be found, and color changes (*i.e.*, red and black) indicate when a new washer was placed on the same rebar. Vertical lines indicate when the washer was removed and replaced on the surface. It should be noted that the elevations are not relative to NAVD88 but instead are relative to the top of the rebar.

4. Discussion

The shore break, where waves break at the base of the foreshore before running up the beach and where there is often a morphological step (Figure 2b), creates intense, turbulent velocities that can suspend finer sediment, allowing it to be transported landward and/or seaward, while the coarser fraction, which is less mobile, remains in place [5–7,9,11,20]. This process of sorting with each wave in the shore break is the fundamental process in the conceptual stratigraphic model that is presented here. It can be thought of as the sorting engine and it is illustrated in Figure 10a. The depth of disturbance experiment and the lidar reveal topographic variations of at least 10–40 cm owing to waves reworking the surface in the swash. This erosion/accretion happens over a range of timescales from minutes to hours and reflects the forcing of individual waves varying with the wave-groups and infragravity motions and the response of the beach, slope, and grain size [8,31–36]. One effect of this vertical change in beach elevation is to thicken the sorted layer and this illustrated in Figure 10b. Astronomical tides and storm surge translate the shore break across the beach profile (Figure 10c), horizontally extending the region where the shore break sorting engine is working [5,11]. Sediment in this region is systematically sorted by these processes, with finer material being moved away, leaving a lens of coarse sediment (Figure 10a–c). This coarse-grained, well-sorted sediment layer, usually represented by the CG unit (Figures 4 and 5), is thickest where the shore break resides for the longest period of time. Where it was observed at locations N and O (Figures 3 and 5), it can reach thicknesses in excess of 50 cm.

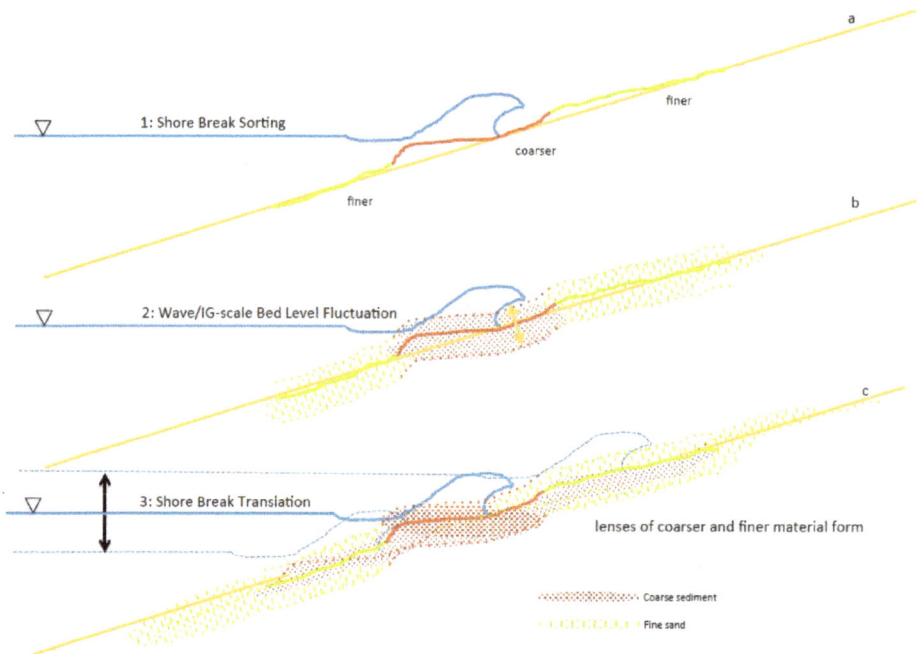

Figure 10. Mechanisms at work in sorting, separating and depositing sediment. (**a**) Shore Break Sorting: the turbulence in the shore break acts to suspend sediment; coarser material stays in place while finer material is moved on- and off-shore. (**b**) Wave/IG-scale Bed Level Fluctuations: because the natural beach elevation varies on minute to hourly scales (Figure 9), the sorted sediment layer becomes thicker. (**c**) Shore Break Translation: as the water level changes owing to tides, surge and wave set-up, the shore break is translated up and down the beach, extending the layers of sorted sediment.

The PSS unit can be thought of as being a poorly sorted precursor to the CG unit. When the sorting engine has not had time to fully sort the sediment or at the distal ends of the shore break sorting region, PSS will result.

Away from the shore break, at the upper edge of the swash, where the water decelerates (or accelerates) as it runs up (or down) the foreshore, sediment also is sorted [8]. As the flows decelerate (on their way up the beach), sediment falls from suspension, coarser sands first, then medium sands and the finest grains are deposited farthest from the shore break, falling out of suspension last. High on the beach, these finer-grained, depositional layers are exemplified by the LS unit (Figure 4). The layers are thin and may be deposited (or eroded) with each wave [8]. Core observations reveal that these layers may vary in the vertical from coarse sand to fine sand because of differences in transport potential from wave to wave, over infragravity periods and position in the swash. Indeed, these two processes ((1) vertical stirring, winnowing and sorting in the shore break and (2) horizontal sorting higher in the swash) are the same mechanically, but the first happens over many waves, reworking and transporting the grains as exemplified in Figure 10. The second happens with every wave, and the separation is immediate and slightly less distinct.

During a storm, large waves often remove sand from the beach, cutting the surface down to a lower elevation, which results in an onshore translation of the water line (and a redistribution of sediment). Additionally, elevated water levels owing to storm surge and wave set-up, translate the mechanisms outlined in Figure 10 farther onshore (Figure 11a). This is the primary mechanism for creating the coarse lenses (Figure 7) observed farther onshore than the swash position would occur under the normal tidal translation (Figure 10c). As water level fluctuates during a storm event, from either falling tides or waning surge, the coarse layer may extend across the beach (either CG or PSS, depending on the time frame). In addition, the mechanisms creating thin, graded layers (LS) may be at work at the highest extent of the swash, also higher on the beach than usual, and finer sediment may be moved seaward and deposited on top of coarser sediment, where the swash usually resides (see, for example, location M on 28 March at 100 cm depth, Figures 3 and 5).

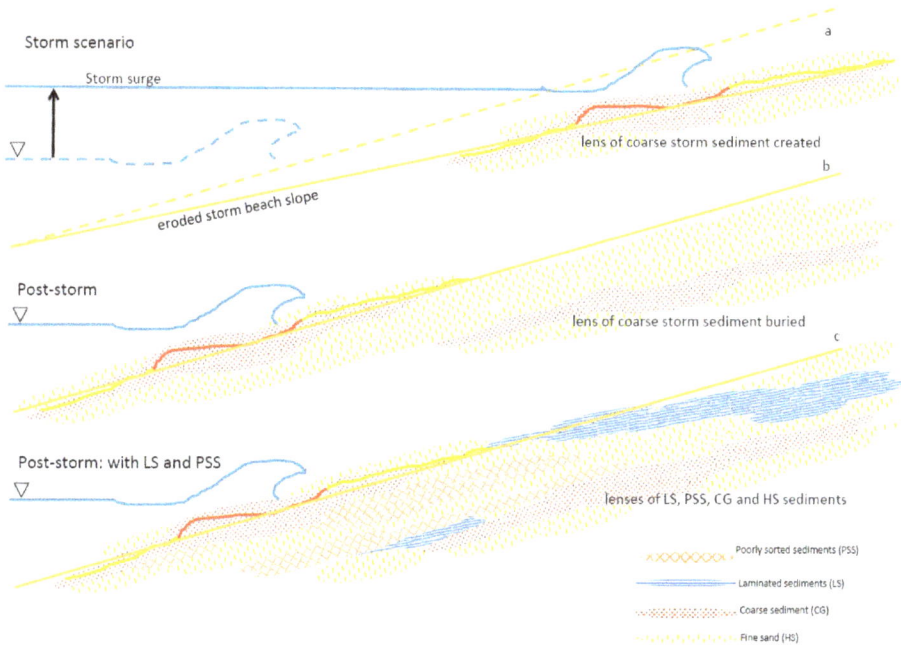

Figure 11. Illustration of storm deposition mechanism. (**a**) When water level is elevated and the beach has eroded, the shore break is translated onshore and the three processes (Figure 10) act in that location to create a lens of coarse sediment. (**b**) When the storm recedes and the beach recovers, that coarse lens is preserved. (**c**) Rather than only two sand units (fine and coarse), including the four sedimentary units (Figure 4) is a natural extension of the conceptual model and gives realistic pictures of beach stratigraphy.

While deposition is important for any observation of stratigraphy, it is commonly assumed that erosion during storm conditions frequently results in the destruction of storm layers in the foreshore and lower beach. Recently, however, researchers have observed beach recovery (accretion) beginning during the height of the storm, as wave heights begin to wane [34–36]. These observations suggest that accretion may be occurring even while storm waves and surge are working high on the beach, thus allowing the accumulation and preservation of the coarser storm layer as described in Figure 11a,b.

During more quiescent conditions, data indicate that the top ~10–40 cm or so of the beach at Duck is reworked regularly. A number of recent observations conducted on similar sandy beaches indicate that those beaches eroded and accreted significantly over the course of a few waves to an hour, with elevations fluctuating up and down in the swash (Figures 8 and 9 and studies by [31–33]). If there is no net change in beach elevation, these small waves will repeatedly accrete and erode sediment within the active layer. During times of net accretion, such as a summer beach accreting under the influence of small waves, preservation of these thin layers will occur, resulting in LS stratigraphy.

The different stratigraphic units (Figure 4) were defined in order to characterize not only the observed grain sizes but also the temporal and spatial variability of the responsible depositional processes. We suggest that these processes reflect a continuum of forcing processes, which can be represented by these characteristic sedimentary layers. In Figure 11c, the conceptual model is qualitatively redrawn using the four units and possible cross-shore and vertical positions as a function of sorting hydrodynamics. PSS is a poorly sorted unit (Figure 4) and likely reflects sorting and depositional processes that occur between conditions responsible for the quiescent LS and the energetic CG formation. CG is well-mixed owing to strong turbulence in the shore break. LS tends to form at the upper reaches of the swash where more gentle velocities allow grain size sorting on an intra-wave basis. HS layers result from suspension and transport away of well-sorted, finer grain sizes from the most turbulent regions of the shore break, and Aeolian processes likely also play a role in sediment sorting and accretion. Thus, accretion during a post-storm recovery might result in the accumulation of a coarser LS layer or a PSS layer in the mid-swash, CG layers in the lower swash and shore break, LS or even HS layers high on the beach or offshore on top of a preexisting CG layer. These combinations are seen in the preserved lenses and strata in the cores and the cross-shore trench (Figures 3, 5 and 7).

5. Conclusions

It is important to know the range of grain sizes on a beach for sediment transport and morphological modeling, in part so that each fraction can be moved around accurately (e.g., [11,20]). As the redistribution of different grain sizes takes place, the local morphology changes owing to subtle changes in erodibility and transportability. In addition to this direct relationship, as local morphology, grain size and slope change, the hydrodynamics change in response. Thus, grain size, hydrodynamics (turbulence, roughness, infiltration, *etc.*) and morphology are all interdependent through feedbacks. Further, it is hypothesized that variations in size and sorting of sediment along the foreshore affect larger-scale profile morphology through feedback with forcing wave conditions [24]. The present conceptual model attempts to elucidate the three processes that work simultaneously (Figure 10) to sort and redistribute sediment on the foreshore and beach and explain typical beach stratigraphy. These processes create sedimentary layers, which are observed in the beach and, when exposed, become part of the hydro-, morpho-, and sediment dynamic feedback working on the beach. Further work will explore the relationships between stratigraphy and the feedbacks between related hydrodynamics and morphodynamics.

Acknowledgments: The authors would like to acknowledge support from the National Science Foundation, Geomorphology and Land Use Dynamics (EAR 0952225, EAR 0952163), the Office of Naval Research (N00014-13-1-0621), and the U.S. Army Corps of Engineers (Coastal Field Data Collection program; Morphology work unit). The authors would also like to thank the staff of the Field Research Facility, especially J. Pipes, B. Scarborough, and M. Preisser, for their dedicated efforts and expertise. Thanks to C. Hagen for field support and photography. Special thanks to K. Brodie and N. Spore, USACE-CHL, for collection and processing of lidar data.

Author Contributions: E.G. and A.R. conceived the research. E.G., A.R., H.W. and J.M. designed and performed the experiments. All authors worked on data analysis (E.G. on DIS data, H.W. and M.K. on core analyses, J.M. on lidar data, A.R. on bathymetric and wave data). E.G. wrote the first draft of the manuscript and all authors contributed to the final text and figures.

Conflicts of Interest: The authors declare no conflict of interest.

References

1. Komar, P.D. *Beach Processes and Sedimentation*; Prentice-Hall: Upper Saddle River, NJ, USA, 1998.
2. Nielsen, P. *Coastal and Estuarine Processes*; World Scientific Press: Singapore, 2009.
3. Holland, K.T.; Elmore, P.A. A review of heterogeneous sediments in coastal environments. *Earth-Sci. Rev.* **2008**, *89*, 116–134. [CrossRef]
4. Buscombe, D.; Masselink, G. Concepts in gravel beach dynamics. *Earth-Sci. Rev.* **2006**, *79*, 33–52. [CrossRef]

5. Hay, A.E.; Zedel, L.; Stark, N. Sediment dynamics on a steep, megatidal, mixed sand-gravel-cobble beach. *Earth Surf. Dyn.* **2014**, *2*, 443–453. [CrossRef]
6. Ivamy, M.C.; Kench, P.S. Hydrodynamics and morphological adjustment of a mixed sand and gravel beach, Torere, Bay of Plenty, New Zealand. *Mar. Geol.* **2006**, *228*, 137–152. [CrossRef]
7. Mason, T.; Coates, T. Sediment transport processes on mixed beaches: A review for shoreline management. *J. Coast. Res.* **2001**, *17*, 645–657.
8. Otvos, E.G. Sedimentation-erosion cycles of single tidal periods on Long Island Sound beaches. *J. Sed. Pet.* **1965**, *35*, 604–609.
9. Duncan, J.R. The effects of water table and tide cycle on swash-backwash sediment distribution and beach profile development. *Mar. Geol.* **1964**, *2*, 186–197. [CrossRef]
10. Gallagher, E.L.; MacMahan, J.H.; Reniers, A.J.H.M.; Brown, J.; Thornton, E.B. Grain size variability on a rip-channeled beach. *Mar. Geol.* **2011**, *287*, 43–53. [CrossRef]
11. Reniers, A.J.H.M.; Gallagher, E.L.; MacMahan, J.H.; Brown, J.A.; van Rooijen, A.A.; van Thiel de Vries, J.S.M.; van Prooijen, B.C. Observations and modeling of steep-beach grain-size variability. *J. Geophys. Res. Oceans* **2013**, *118*, 577–591. [CrossRef]
12. Neal, A.; Pontee, N.I.; Pye, K.; Richards, J. Internal structure of mixed-sand-and-gravel beach deposits revealed using ground-penetrating radar. *Sedimentology* **2002**, *49*, 789–804.
13. Plant, N.; Holland, K.T.; Puleo, J.; Gallagher, E.L. Predictions skill of nearshore profile evolution models. *J. Geophys. Res.* **2004**, *109*, C01006. [CrossRef]
14. Castelle, B.; Ruessink, B.G.; Bonneton, P.; Marieu, V.; Bruneau, N.; Price, T. Coupling mechanisms in double sandbar systems, Part 1: Patterns and physical explanation. *Earth Surf. Process. Landf.* **2010**, *35*, 476–486. [CrossRef]
15. Roelvink, J.A.; Reniers, A.J.H.M. *Guide to Modeling Coastal Morphology*; World Scientific Press: Singapore, 2011.
16. Johnson, B.D.; Kobayashi, N.; Gravens, M.B. Cross-shore numerical model CSHORE for waves, currents, sediment transport and beach profile evolution. In Proceedings of the Great Lakes Coastal Flood Study, 2012 Federal Inter-Agency Intiative, Washington, DC, USA, September 2012.
17. Nielsen, P.; Callaghan, D.P. Shear stress and sediment transport calculations for sheet flow under waves. *Coast. Eng.* **2003**, *47*, 347–354. [CrossRef]
18. Roelvink, J.A.; Reniers, A.J.H.M.; van Dongeren, A.R.; van Thiel de Vries, J.S.M.; McCall, R.T.; Lescinski, J. Modeling storm impacts on beaches, dunes and barrier islands. *Coast. Eng.* **2009**, *56*, 1133–1152. [CrossRef]
19. Hoefel, F.; Elgar, S. Wave induced sediment transport and sand bar migration. *Science* **2003**, *299*, 1885. [CrossRef] [PubMed]
20. Blondeaux, P. Sediment mixtures, coastal bedforms and grain sorting phenomena: An overview of theoretical analyses. *Adv. Water. Res.* **2012**, *48*, 113–124. [CrossRef]
21. Soulsby, R. *Dynamics of Marine Sands*; Thomas Telford Limited: London, UK, 1997.
22. Gallagher, E.L.; Elgar, S.; Guza, R.T. Observations of sand bar evolution on a natural beach. *J. Geophys. Res.* **1998**, *103*, 3203–3215. [CrossRef]
23. Ruessink, B.; Kuriyama, Y.; Reniers, A.J.H.M.; Roelvink, J.A.; Walstra, D. Modeling cross-shore sandbar behavior on the timescale of weeks. *J. Geophys. Res.* **2007**, *112*, F03010. [CrossRef]
24. Gallagher, E.L.; Reniers, A.J.H.M.; Wadman, H.; McNinch, J.; MacMahan, J. Onservations and modeling of grain size variability on and in a steep beach. In Proceedings of the AGU Ocean Sciences Meeting 2014, American Geophysical Union, Honolulu, HI, USA, February 2014.
25. Rubin, D.M. A simple autocorrelation algorithm for determining grain size from digital images of sediment. *J. Sediment. Res.* **2004**, *74*, 160–165. [CrossRef]
26. Brodie, K.L.; Raubenheimer, B.; Elgar, S.; Slocum, R.K.; McNinch, J.E. Lidar and Pressure Measurements of Inner-Surfzone Waves and Setup. *J. Atm. Ocean. Tech.* **2015**, *32*, 1945–1959. [CrossRef]
27. Wadman, H.M. Controls on continental shelf stratigraphy, Waiapu River, New Zealand. Ph.D. Thesis, Virginia Institute of Marine Science, College of William and Mary, Williamsburg, VA, USA, May 2008.
28. Foxgrover, A.C. Quantifying the overwash component of barrier island morphodynamics, Onslow Beach, NC. Master's Thesis, Virginia Institute of Marine Science, College of William and Mary, Williamsburg, VA, USA, May 2008.

29. Wadman, H.M.; McNinch, J.E. Integrating subaerial and nearshore geologic metrics for predicting shoreline change: Onslow Beach. In Proceedings of the 7th International Symposium on Coastal Engineering and Science of Coastal Sediment Processes, Miami, FL, USA, 2–6 May 2011.

30. Stauble, D.K.; Cialone, M.A. Sediment dynamics and profile interactions: Duck94. *Proc. Inter. Conf. Coast. Eng.* **1996**, *4*, 3921–3934.

31. Puleo, J.A.; Lanckriet, T.; Blenkinsopp, C. Bed level fluctuatios in the inner surf and swash zone of a dissipative beach. *Mar. Geol.* **2014**, *349*, 99–112. [CrossRef]

32. Turner, I.L.; Russell, P.E.; Butt, T. Measurement of wave-by-wave bed-levels in the swash zone. *Coast. Eng.* **2008**, *55*, 1237–1242. [CrossRef]

33. Blenkinsopp, C.E.; Turner, I.L.; Masselink, G.; Russell, P.E. Swash zone sediment fluxes: Field observations. *Coast. Eng.* **2011**, *58*, 28–44. [CrossRef]

34. Brodie, K.L.; McNinch, J.E. Measuring Bathymetry, Runup, and Beach Volume Change during Storms: New Methodology Quantifies Substantial Changes in Cross-Shore Sediment Flux. In Proceedings of the American Geophysical Union, Fall Meeting 2009, San Francisco, CA, USA, December 2009.

35. Brodie, K.L. Processes driving storm-scale coastal change along the Outer Banks, North Carolina: Insights from during-storm observations using CLARIS. In Proceedings of the American Geophysical Union, Fall Meeting 2010, San Francisco, CA, USA, December 2010.

36. Roberts, T.M.; Wang, P.; Puleo, J.A. Storm-driven cyclic beach morphodynamics of a mixed sand and gravel beach along the Mid-Atlantic coast, USA. *Mar. Geol.* **2013**, *346*, 403–421. [CrossRef]

Journal of
*Marine Science
and Engineering*

MDPI

Review

Quantitative Estimates of Bio-Remodeling on Coastal Rock Surfaces

Marta Pappalardo [1,*], Markus Buehler [2], Alessandro Chelli [3], Luca Cironi [1], Federica Pannacciulli [4] and Zhao Qin [2]

[1]   Department of Earth Sciences, Pisa University, Pisa 56126, Italy; l.cironi@studenti.unipi.it
[2]   Department of Civil and Environmental Engineering, Massachusetts Institute of Technology, Cambridge, MA 02139-4307, USA; mbuehler@mit.edu (B.M.); qinzhao@mit.edu (Q.Z.)
[3]   Department of Physics and Earth Sciences "M. Melloni", Parma University, Parma 43100, Italy; alessandro.chelli@unipr.it
[4]   Marine Environment Research Centre (ENEA)-Santa Teresa P.O., Pozzuolo di Lerici 19100, Italy; federica.pannacciulli@enea.it
*   Correspondence: marta.pappalardo@unipi.it; Tel.: +39-050-2215748

Academic Editor: Gerben Ruessink
Received: 5 March 2016; Accepted: 2 May 2016; Published: 26 May 2016

Abstract: Remodeling of rocky coasts and erosion rates have been widely studied in past years, but not all the involved processes acting over rocks surface have been quantitatively evaluated yet. The first goal of this paper is to revise the different methodologies employed in the quantification of the effect of biotic agents on rocks exposed to coastal morphologic agents, comparing their efficiency. Secondly, we focus on geological methods to assess and quantify bio-remodeling, presenting some case studies in an area of the Mediterranean Sea in which different geological methods, inspired from the revised literature, have been tested in order to provide a quantitative assessment of the effects some biological covers exert over rocky platforms in tidal and supra-tidal environments. In particular, different experimental designs based on Schmidt hammer test results have been applied in order to estimate rock hardness related to different orders of littoral platforms and the bio-erosive/bio-protective role of *Chthamalus* ssp. and *Verrucariaadriatica*. All data collected have been analyzed using statistical tests to evaluate the significance of the measures and methodologies. The effectiveness of this approach is analyzed, and its limits are highlighted. In order to overcome the latter, a strategy combining geological and experimental–computational approaches is proposed, potentially capable of revealing novel clues on bio-erosion dynamics. An experimental-computational proposal, to assess the indirect effects of the biofilm coverage of rocky shores, is presented in this paper, focusing on the shear forces exerted during hydration-dehydration cycles. The results of computational modeling can be compared to experimental evidence, from nanoscopic to macroscopic scales.

Keywords: bioerosion; bioprotection; rocky coasts; rock hardness; materials science; computational modeling; geomorphology

1. Introduction

The influence of biological agents in rocky coastal landforms shaping has been recognized for a long time [1–8]. Virtually, biota may exert a bioerosive, bioprotective, or bioconstructional role on rocky coasts. In this paper, we focus on bioerosional and bioprotective effects of biota on coastal rocks, considering both of them as part of the "bio-remodeling" effect, and discuss the importance of a quantitative understanding of their role. Bioconstructors as well as borers operating on reefs are neglected, as processes and rates are quite different [9]. A valuable quantitative approach to erosion on biogenic rocks is provided by Moses [10].

Due to the high population density in coastal areas and the cost of coastal planning, bio-remodeling represents a relevant problem which has been neglected in the recent literature on coastal management, focused mainly on beach erosion [11]. Providing an updated conceptual model of biotic agency on coastal rocks, Naylor *et al.* [12] suggest that future research in this field should be aimed at providing quantitative estimates of this process in different contexts, in order to assess its contribution to the global sedimentary budget. In this sense, it is crucial to identify the best methodologies suitable for tackling this issue with a strictly quantitative approach.

The aim of this paper is thus to revise the different methodologies employed in the quantification of the effect of biotic agents on rocks exposed to coastal morphologic agents, comparing their efficiency and inferring specific indications useful for the wide community of scientists working in the coastal environment (engineers and, in a broad sense, all environmental scientists). Literature on biotic agents on coastal rocks has been revised, focusing on quantitative assessments and on methodological approaches. In some cases, quantification should be considered in a relative sense as, e.g., the estimate of a percentage due to bioerosion of overall weathering. As a complement for this review, we provide new and partly unpublished data, suggesting possible additional methodologies, based both on experimental and computational activities.

Rocky shorelines are shaped by a suite of weathering processes (physical, chemical, and biological) that operate on them, reducing the resisting force of rock (FR). Weathering processes affect the coastal profile being scaled in importance according to their elevation with respect to mean sea level (see [13], Figure 7.3). Among them, bioerosion and its counterpart bioprotection (on the whole "bio-remodeling") are recognized as playing a relevant role in the intertidal and lower midlittoral zones [12], but their efficiency long- and cross-shore still needs to be thoroughly defined. In the midlittoral zone, biodiversity is maximized due to the density of biomass colonizing rock surfaces [14]. Here, a variety of biota displays different activities on rocks, the rate and magnitude of each depending on both ecological factors and environmental constraints.

Since the term "bioerosion" was introduced by Neumann [1] for the coastal environment, the contribution of biota to the weathering processes affecting coastal rocks has been evaluated by many authors. The papers of Schneider [2], Trudgill [4], and Torunski [3] still represent the classical reference studies and constitute the fundamentals for any bio-process approach to geomorphological studies of rocky coasts, although limited to limestone shores. These works assess the concept of biological zoning of rocky shorelines, *i.e.*, the distribution of groups of biota in horizontal bands, scaled with respect to elevation from sea level (Figure 1). Sea level being, as a matter of fact, a virtual concept, as its elevation with respect to an extraterrestrial reference system is permanently fluctuating in space and time [15], we refer in this work to "mean sea level", *i.e.*, the reference ordnance datum of the official Italian Elevation Network. Biological zoning implies a cross-shore variation in the rate of erosion due to biota, which is determined by strictly ecological factors [16–18].

Figure 1. Biological community bands on a typical Mediterranean rocky shore (with a tidal range of 0.4–1 m).

2. Methods

The literature existing on bio-remodeling (see [12,19,20] for comprehensive reviewing) is huge, even if limited to the coastal environment. In this work, we focus on those papers that provide "quantitative" estimates of the effectiveness of the geomorphologic action of this process on rocky shorelines. The authors of this work having quite different cultural backgrounds, it was important to rigorously fix the criteria for the selection of those data that is considered in this review. In particular, it was necessary to assess what we accept as a "quantitative" datum due to the object of our investigation, which is complex, as all natural phenomena are. Those works investigating how different types of biological cover influence weathering rates and types and according to which patterns this interaction changes in time and space, albeit relevant, were properly considered qualitative and thus not included in this review. Many papers provide estimates of the effectiveness of bio-remodeling inferred from proxy data that may be represented by physical or chemical parameters not related to the rock itself, *i.e.*, not geotechnical in a broad sense. Although extremely relevant from a general point of view, these papers tackle issues that are beyond the focus of this work. Our selection includes only those papers that isolate the sole biological contribution from overall weathering in order to measure it, at least as a percentage of the overall weathering rate. In doing this, we follow the indications of [12] about aiming to unravel the contribution of bioerosion to coastal sediment budget.

In addition to this analysis of published works, an essay of some novel approaches is proposed to the reader based on original case studies. The specific working methods of each case study are illustrated in the dedicated paragraphs.

3. Review of Published Works

In this section, we summarize the popular research methods for estimating bio-remodeling in the coastal environment and highlight their efficiency and limitations in certain aspects. Bioerosion and bioprotection is examined separately; in fact, a number of works aimed at quantifying the former have been carried out through times, whereas bioprotection has seldom been highlighted and almost never quantified. The main problem in estimating bioerosion is unraveling its effect from other weathering processes acting on rocky coasts. The common approach to tackle this issue is focusing research in those parts of the coastal rock where bioerosion is thought to be the dominant process so that the other processes can be neglected. The most relevant quantitative work carried out in the past century is summarized by Schneider and Le Campion-Alsumard [21].

Different methods have been employed to quantitatively assess bioerosion. Apparently, grazers are biota with an overall easier quantifiable erosive role on rocky coasts. The first works on this topic date back to the 1960s [22]. The problem of estimating the bioerosive effect of grazers can be approached using a direct or an indirect method. The latter relates the amount of grazer fecal pellets production to rock erosion rate. Working on the Adriatic coast, [3] and [16] sampled fecal pellets from the different biota, such as the gastropods *Monodonta* and *Patella* and the sea urchin *Paracentrotus*. They produce abraded particles whose grain size is determined by their boring pattern. Erosion rates were estimated on average 1 mm/year with a variability pattern depending on coastal zonation. More case studies based on an indirect approach to the quantification of the effect of grazers are provided among others by Abensperg-Traun *et al.* [23] and Andrews and Williams [24]. Fornos *et al.* [25] worked on three different grazers living along the calcareous coasts of Mallorca (Spain), *i.e.*, the gastropods *Melarapheneritoides*, *Monodonta turbinate*, and *Patella rustica*. Through field observations and biota manipulation in the lab, they infer the amount of rock eroded by each of the grazers by measuring the amount of fecal pellets produced, in order to determine the amount of rock ingested. The daily quantity of rock eroded by an organism was converted into an erosion rate (mm/year) for a single species of grazer, multiplying the yearly amount of fecal pellets produced daily by an organism per the average density of that biota. *Patella* proved to be the most efficient eroder, determining erosion rates exceeding 0.5 mm/year. The validity of this approach was recently assessed by Vidal *et al.* [26]. Trudgill [4] and Trudgill *et al.* [6] worked out an indirect method to measure the contribution of grazers to overall erosion, combining an estimate of the grazed area with the average depth of incision. This method is efficient as it provides a quantitative estimate of the input into the sediment budget of mineral particles deriving from grazer bioerosion. Its limitations are that it can only take into account a specific portion of the global process of bio-remodeling.

A second methodology applied to quantitatively assess bioerosion can be used to estimate the overall contribution of biota, and not only of grazers. This approach employs the micro erosion meter (MEM, or TMEM in its instrumental evolution). This instrument provides high-precision measurements of erosion rates on exposed rocks [27]. Based on repeated measurement of oblique coordinates, this instrument proved suitable for getting records of coastal retreat, especially on limestones. For a review of destruction rates obtained in different environments on carbonate rocks, see Furlani *et al.* [28]. Apart from the technical constraints of the TMEM, the major limit of this method is that it is difficult to disentangle the contribution of bioerosion from that of all other weathering processes. Only a few authors, thus, used a micro erosion meter to measure merely bioerosion. Among them, Torunski [3] measured the destruction rate on limestone bedrock along the Adriatic coast (Central Mediterranean) obtaining values around 0.1–1 mm/year, depending on variable exposure to wetting of the rock. This approach is recommended only in those cases in which the effect of bioerosion is with no doubt overwhelming. Nevertheless, Stephenson and Finlayson [29] are rather positive in this sense, suggesting that experiments including specific manipulations (such as those with mesh bags) may be effective in polygenetic environments. More recently [18] used, in Southern Portugal, TMEM repeated measurements combined with other methods (measurement of the volume of macroborer-produced cavities and of the bedrock strength) to test the effect of biota on rocks. This approach compares rates of downwearing obtained from rock patches in the same environmental conditions but with different biological covers. Results highlight a negative correlation between TMEM downwearing rates and the amount of algae covering the rock. A recent work along the French Mediterranean coast [30], although not providing a quantitative assessment of erosion rates, quantitatively approaches the role of biofilm in mediating the degrading effects of microclimatic changes on coastal rocks. The quantification is provided by micro-topographic change of the rock surface measured through the TMEM.

A further method for quantifying bioerosion relates to the loss in weight of experimental trials, *i.e.*, rock chips that can be manipulated both in the natural environment [31] and in the laboratory [32]. The latter measured an increase in weight due to biologically produced calcium carbonate in freshly cut limestone and eolianite exposed in the laboratory to sea water for a few months, whereas previously exposed rock chips underwent, during the same time span of exposure in the laboratory, a reduction in weight, which was related to bioerosion. Similarly, Trudgill and Crabtree [5] performed conventional calculations of rates of downwearing by microborers of known age through exposure of test materials. Weathering assessment through weight changes of exposed rock blocks has been rather popular in the past (see [31]). However, in this case, the main concern with this method is disentangling the effect of different processes.

Another type of investigation involves microscopic scale observations and measurements. Radtke *et al.* [8] provide a critical review of the work that has been done to highlight bioerosion at the microscopic scale up to the end of the twentieth century. In this century, this type of approach has been employed by some authors, providing robust quantification of the action of borers. Naylor and Viles [19] used three different microscopy techniques (*i.e.*, optical light, scanning electron, and laser scanning microscopes) to highlight the nanoscale features created on the surface and in the immediate subsurface of trial limestone blocks exposed for seven months at sea level in the natural coastal environment along the coast of Falasarna (Isle of Crete, East Mediterranean). Microscopic observation highlighted three types of features that could be related to biological weathering processes: (1) microborings; (2) biological etching; and (3) chemical etching, which are all mostly related to the biofilm development on the rock surface and, in particular, to blue-green algae (cyanobacteria). Estimates of the efficiency of bioerosion relied on measurement of the average borehole diameter (ranging from 6.1 to 6.4 μm) and average number (12–14 every 25 cm^2), and on the percentage cover of biological etching (60%–86%). Sticking to those features revealed by SEM observations, Coombes *et al.* [33] quantified the amount of weathering due to biota-rock interaction through the thickness of the penetration of euendolithic microborers and through the surface width and density of borehole entrances. These authors also highlighted the occurrence of biochemical crusts and of biological cryptoendolithic growths, which represent agents of facilitative bioerosion (*sensu* Naylor *et al.* [12]) difficult to quantify. Finally, they relate the amount and type of biota-rock interaction to the bedrock type, applying their experimental design in study sites in Cornwall (UK), to natural bedrocks of different lithology as well as to concrete. Naylor *et al.* [12] extends to macro-scale borers the method of estimating the bioerosive effect of microborers, measuring the size and frequency of boreholes and their depth of penetration to a genus of polychaete, *Boccardia*.

Interestingly, microclimate elements monitoring [33–35] has recently been used to provide a quantitative insight on the relationship between physical and biological agents responsible for coastal rocks shaping. This approach, although unable to provide direct erosion rates, deserves to be considered very promising for improving our comprehension of geomorphic processes driven by biota and addressing future research aimed at quantifying such processes.

4. Bio-Remodeling: A Geological Approach

Employing geological methods to assess and quantify bioerosion has been suggested by Naylor *et al.* [12]. Rock hardness/strength changes assessment may provide a quantification of bioerosion [36], although mostly providing a relative estimate of the contribution of biota to surface weathering. In this section, we illustrate some experiments carried out by some of the authors of this paper in order to highlight the differences in rock hardness between coastal rock surfaces differently affected by the presence of biota. Our preliminary experimental evidence come from Schmidt hammer testing on a number of sites along the coast of NW Italy stretching for *ca.* 300 km; their main features are reported in Figure 2. Data collection followed a preordered spatial arrangement (Figure 3), worked out in order to facilitate the employment of some of the data in different experimental designs and maximize the efficiency of their subsequent statistical treatment. The whole coastal tract was split into two study areas, represented by rocky littorals differently shaped, separated by a tract of sandy coast, which was neglected. In each study area, a number of localities were selected, each of which displays homogeneous geological features (rock type, fracturing degree) and almost constant exposure to incoming waves. Inside each locality two to five sites were identified; each site represents a basic unit of landform, *i.e.*, a "shore platform". In geomorphology, this term is broadly used to indicate a polygenetic, sub-horizontal, or moderately seaward dipping rocky surface located in the midlittoral or lower supralittoral [37]. For rock hardness assessment, the methodology proposed by Aydin and Basu [38] was employed, based on an instrumental tool, the Schmidt hammer, which records the rebound distance of a piston connected to a plunger that is pressed against the rock surface by the operator, keeping a few cm away from rock discontinuities. The rebound distance is transformed into a dimensionless index known as the rebound value (R) that provides a measure of the rock surface hardness. In the detail, within each site, a number of testing points are selected, the position and density of which depends on the experimental design. Inside each 10×10 cm sampling area (quadrat), 35 readings were taken with the Schmidt hammer. These were processed removing the lowest 10 and averaging the remaining 25. The final value corrected, for the effect of the dip of the device, represents the surface hardness of the rock in the testing point.

It had been demonstrated [39] that, in one of the study areas (Eastern Liguria), small shore platforms can be subdivided in two portions, regardless of their bedrock type: a more seaward portion from mean sea level to an elevation of 0.5–4 m (depending on wave exposure), characterized by the extensive, often continuous patching of the surface by biofilm, and an upper portion in which biofilm is very scattered or absent. Visually, the two portions can be distinguished based on a chromatic difference; independently from the bedrock type, the lower portion is darker than the upper. Chelli *et al.* [39] tested the two portions of six shore platforms in Eastern Liguria with the Schmidt hammer and highlighted that the lower portion of each platform displays lower rock hardness than the upper one up to 17%, differences in most cases being statistically significant. This evidence suggest that weathering processes are more effective in the portion of the platform displaying a biofilm cover; it is thus sensible to hypothesize that biofilm may play a role in bedrock hardness lowering.

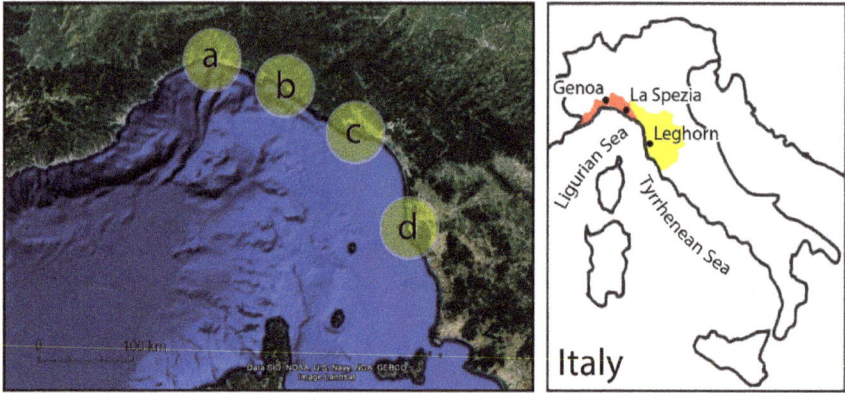

Figure 2. Location and main features of the study sites: (**a**) Pontetto (silicic flysch); (**b**) Lavagna (calcareous flysch); (**c**) Tellaro (carbonates); (**d**) Calafuria (sandstone).

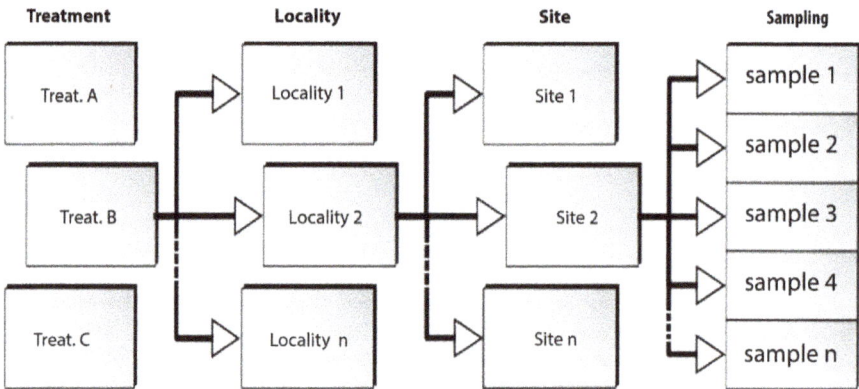

Figure 3. Spatial arrangement of experimental data (general design). Locations were set to minimize substrate differences (homogeneous features, rock type, exposure, *etc.*); sites (corresponding to shore platforms) were randomly selected within the localities.

Further testing carried out along cross-shore transects highlighted a seaward-landward positive gradient of the rebound value (R) within the portion of the platforms below the cyanobacteria patching upper boundary. Chelli *et al.* [40] tested this positive trend on five shore platforms in Eastern Liguria (Locality: Palmaria, Figure 2), two of which were reshaped by quarrying activity in the 19th century and thus had been exposed to coastal weathering for less than 200 years. The reduction in mean R value of the weathered rock surface was assessed separately for each upper and lower portion of each platform and compared to the values tested on artificially exposed, unweathered rock on the same landform. The reduction proved to be lower in the platform upper portion, ranging from 13% to 32%, whereas in the lower portion, extensively patched by biofilm, it ranged from 35% to 48%.

Similar results were obtained for the natural and for the quarried rock surfaces; this reveals that the weathering processes in the study area have a fast evolution rate.

The increase in rock hardness with elevation from sea level should be related not only to the presence of the biofilm, but also to the simultaneous colonization of the littoral by different types of biota, acting at different elevations according to their ecological needs. The displacement of the different biota creates what is known as the "biological zoning" of rocky shores [41–43]. In Eastern Liguria [39], the biota affecting the rocky shores are represented by macroalgae dominating the sublittoral (Figure 1), sharing their habitat with scattered sea urchins. A belt densely populated by a specific benthic community then characterizes the midlittoral, including grazers in the lower part (limpets, e.g., *Patella* spp. and other Gastropods) and barnacles in the upper part, belonging to the genus *Chthamalus* spp. The supralittoral is dominated by a continuous patina of endolithic Cyanobacteria, and, above the sharp, upper boundary of this, Cyanobacteria spots are mixed with the lichen *Verrucaria adriatica* in patches.

Preliminarily, a cross-shore variation in the rate of bio-remodeling can be hypothesized due to this biological zoning, but its pattern is not completely disclosed. In fact, for some of the biota colonizing the rock, a bioerosive effect has been demonstrated (in particular for the grazers and, partly, also for the Cyanobacteria within the biofilm, see Section 3), but others, such as barnacles, play an enigmatic role ([44]; see also Section 4.3).

4.1. Comparison of Measured Rock Hardness between Platforms of Different Order (i.e., Different Elevation)

The purpose of this experiment was providing further evidence of the effectiveness of biota in decreasing rock hardness through a comparison of the mean rebound value tested on a group of shore platforms constrained in elevation within the low supra-littoral (first-order platforms) and another group of platforms, in the high supra-littoral, within the same coastal tract, ranging in elevation from 11 to 13 m above sea level (a.s.l.) (second-order platforms, Figure 4). These should be considered inherited landforms [45], *i.e.*, a counterpart of mid-supralittoral shore platforms that had been closer to sea level in the past and that are now above the ordinary reach of sea spray due to either tectonic uplift or to a relative sea level lowering [46].

Figure 4. Staircase formed by different shore platform orders (*i.e.*, different elevation), Eastern Liguria-Villa Baldini locality.

The activities were carried out in three localities of the Eastern Liguria study area, shaped in a flysch bedrock, namely, Castello Cirlo (CC), Villa Baldini (VB), and Villa Beatrice (VBE). Two to three platforms (each one considered one "site" in the experimental design) from each locality were tested, three to four for each order. On each of them, a number of testing points was randomly selected (approximately one point each 2 m^2 of surface), each of which yielded a rebound (R) value. On the whole, 75 points were tested. Averaging the R values of all the testing points within a platform/site the mean R value for the platform/site was plotted against the platform elevation a.s.l. (Figure 5).

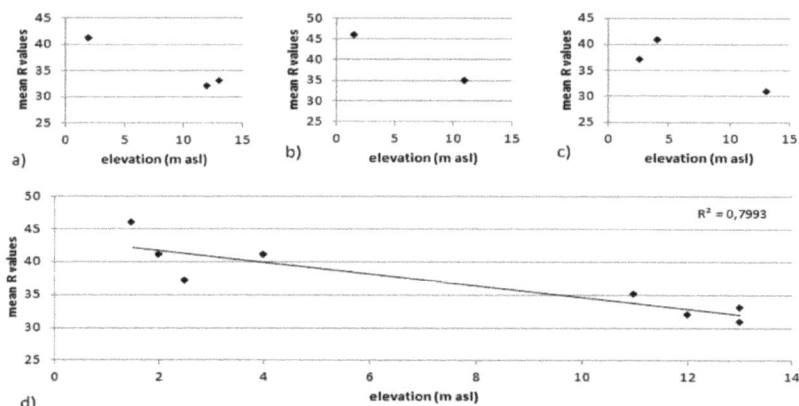

Figure 5. Mean Schmidt hammer rebound values (R) variation with elevation a.s.l. for localities considered in Section 4.1: (**a**) Castello Cirlo; (**b**) Villa Baldini; (**c**) Villa Beatrice; (**d**) a cumulative plot of the three localities together.

A negative correlation between elevation and rock hardness can be envisaged. R-values obtained from testing points of each platform/site were tested against those obtained from testing points of each other platform/site within the same locality using Student's *t*-test (95% confidence interval). Only mean R values obtained comparing platforms/sites of different order are significantly different (Table 1). This experiment demonstrates that platforms of a different order are significantly different from the point of view of weathering intensity and in particular that those of the first (lower) order, located in the low supra-littoral, are less weathered than those of the II (upper) order, constrained in the high supra-littoral. This difference can be quantified on average of the 20%. Being only the lower platforms affected by biota, we can confidently hypothesize that reduction in rock hardness in upper-order platforms is not due to the presence of biota (bioerosion), but to the persistence of abiotic weathering agents for a much longer time than they have been acting in the lower-order platforms.

Table 1. *T*-test results for differences between different platform orders within the same localities. Site codes contains the indication of the locality (CC—Castell Cirlo, VB—Villa Baldini, VBE—Villa Beatrice) and the position of the site within the locality (E—eastern, C—central, W—western).

sites	T Values (Calulated)	T Values (Tabulated)	Degrees of Freedom	Significance	Platform Order
CCE-CCC	3.9	2.26	9	S	different
CCW-CCC	3.4	2.20	11	S	different
CCW-CCE	0.6	2.14	14	NS	same
VBAW-VBAE	3.4	2.08	20	S	different
VBAW-VBAC	1.5	2.13	15	NS	same
VBAC-VBAE	2.16	2.09	19	S	different
VBCL-VBW	4.41	2.16	13	S	different

4.2. Preliminary Assessment of the Contribution of Barnacles and Lichens to the Overall Platform Remodeling

Through this experiment, a preliminary attempt was made to explore the contribution to the platform bio-remodeling of *Chthamalus* spp. (barnacle) and *Verrucaria* spp. (lichen), the most common sessile organisms of the upper-midlittoral and lower-supralittoral zones in the study area of Eastern Liguria.

A simple experimental design, focused in a single locality (Tellaro), where the bedrock is represented by Upper Triassic-Lower Jurassic dolomites and limestones, was worked out, affected by a complex template of different fault systems [47]. Testing was carried out on the two sites SS1 and SS5, each corresponding to a shore platform (Figure 6). Exposure to incoming waves was the same, and the biota displayed a similar zonation pattern with the same organisms present at the two sites. A well-defined barnacle belt, constituted by a mixture of *Chthamalus montagui* and *Chthamalus stellatus*, is confined to a relatively small area of the shore. In fact, barnacles are concentrated within a belt with a width of approximately 30 cm, as expected for sheltered sites in microtidal environments [48], which is located in the midlittoral slightly above it. Immediately above the barnacle belt, another area covered by cyanobacteria and *Verrucaria adriatica* is present, stretched for approximately 0.5 m up to the upper boundary of the low supra-littoral. The uppermost part of the platforms (high supra-littoral) is covered only by patches of *Verrucaria adriatica*. Rock hardness was recorded in 153 testing points with a Schmidt hammer according to the methodology illustrated above; 81 testing points were located across the barnacle belt upper boundary and 72 in the high supralittoral. Measures of the biota cover percentage were combined to them, *i.e.*, in 10 × 10 cm quadrats, centered on each testing point. Biota measures were performed using the point intercept method [49] that consists of counting the number of intercepts (out of 25 in total) that cover the target organism compared to bare rock. A stratified sampling strategy was adopted: The 10 × 10 cm quadrats were randomly selected in areas exhibiting an abundant (>50%) or scarce (<50%) cover of target organisms. Quadrats were distributed on open rock at two different heights on the shore: in the barnacle belt and in the upper part of the supralittoral zone where the lichen *Verrucaria adriatica* is the main representative. Measures were recorded by hammering directly on the substratum except for those areas exhibiting an abundant *Chthamalus* spp. cover. Here, prior to the employment of the Schmidt hammer, we carefully scraped off barnacles from the quadrat while making sure not to damage rock surface. *Verrucaria* was not scraped off as the very first piston impacts apparently destroyed it.

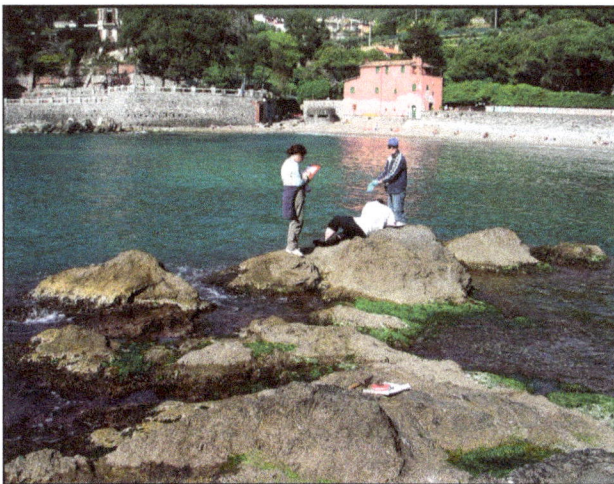

Figure 6. Site SS1 first-order platform.

Results show (Table 2, Figure 7) that at both sites mean *R* values are slightly lower where the barnacle cover is higher (>50%), meaning that the presence of *Chthamalus* spp. could negatively influence rock hardness. When compared with *t*-tests (Table 2), only data relative to SS1 provide significant results. At the SS5 site, differences in rock hardness between covered and uncovered quadrats do not show statistically significant results. This could be attributed to the lower number of replicate quadrats screened at SS5 than at SS1. The pattern observed at SS1 for *Verrucaria adriatica* resembles the one recorded for *Chthamalus* spp.: *R* values are lower where the lichen percentage cover is higher, meaning that its presence could weaken rock hardness. However, at SS5 no pattern is evident. In addition, the *t*-tests do not highlight significant results at any site, meaning that differences observed at SS1 could be due to mere chance.

As for the role of *Verrucaria adriatica* in influencing rock hardness, results suggest that there is no relationship between lichen cover and rock hardness (Table 2, Figure 7). At the SS1 site, *R* values are moderately lower, but not statistically significant, where cover is >50%. At site SS5, instead, no difference in hardness was revealed between covered and uncovered rock. This may be because our experimental design was not suitable for the purpose, while it was for barnacles, because *Verrucaria* has neither a bioprotective nor a bioerosive role, or because this lichen plays both roles at the same time [50–52].

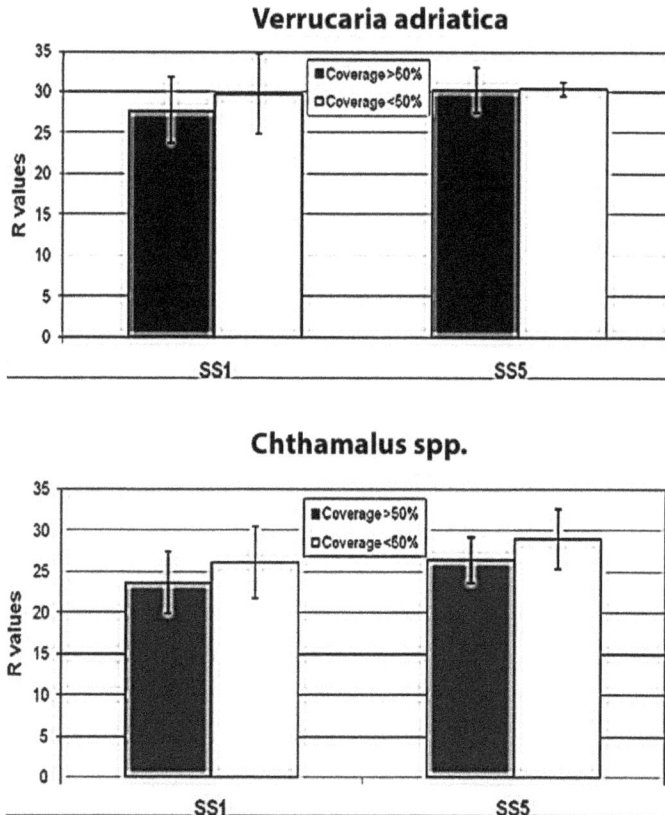

Figure 7. Mean *R* values at both sites, SS1 and SS5.

Table 2. *T*-test results for SS1 and SS5 for both biota.

Verrucaria adriatica

Site	Total n° Replicate Quadrats	Total n° R Values	Mean R and st. dev. When Cover >50%	Mean R and st. dev. When Cover <50%	T	df	P
SS1	36	1260	27.7 ± 4 (19–31)	29.7 ± 4.9 (19–34)	1.4	34	0.1806
SS5	36	1260	30.2 ± 2.7 (27–33)	30.3 ± 0.8 (29–31)	0.1	8	0.9203

Chthamalus spp.

Site	Total n° Replicate Quadrats	Total n° R Values	Mean R and st. dev. When Cover >50%	Mean R and st. dev. When Cover <50%	T	df	P
SS1	53	1855	23.6 ± 3.8 (17–31)	26.1 ± 4.3 (19–34)	2.3	51	0.0256
SS5	28	980	26.4 ± 2.8 (24–31)	29.0 ± 3.6 (24–34)	1.4	10	0.1875

4.3. Dependence of Rock Strength from the Abundance of Barnacles Cover

A more articulated test was designed in order to highlight the differences in rock hardness between coastal rocks differently affected by the presence of barnacles in comparable conditions of exposure to marine morphological processes.

Within the study area of Eastern Liguria, three localities were selected (Figure 2), each of which displayed homogeneous features (rock type, fracturing degree, exposure). One of the localities coincides with the one already tested in the experiment described in Section 4.2 (Tellaro, dolomite, and limestone bedrock); the new localities are both shaped in a terrigenous bedrock due to the late phases of sedimentation within the Tethys sedimentary basin. In the Lavagna locality, the dominant rock type is slate, whereas the Pontetto bedrock (Figure 8) is a marly limestone flysch. The platforms in Lavagna locality have been recently (*ca.*10 years) artificially reshaped to improve the coastal slope stability. For each locality, a number of testing points—26 in Tellaro, 30 in Lavagna, and 60 in Pontetto—were randomly selected within a portion of the rock surface, *ca.* 50 cm wide, located across the barnacle band upper boundary, slightly above the midlittoral-supralittoral boundary. Half of the replicate quadrats displayed an abundant (>50%) barnacle (*Chthamalus* spp.) cover, whereas in the other half the cover was scarce (<50%). All the points were tested according to the methodology applied in Section 4.2, measuring biota percentage cover within a 10 × 10 cm quadrate (replicate quadrate) centered in each of them. In the Tellaro locality, 26 more points were also tested in the supralittoral, half of which displayed a lichen cover >50% (*Verrucaria adriatica*), whereas in the other half the lichen cover was <50%.

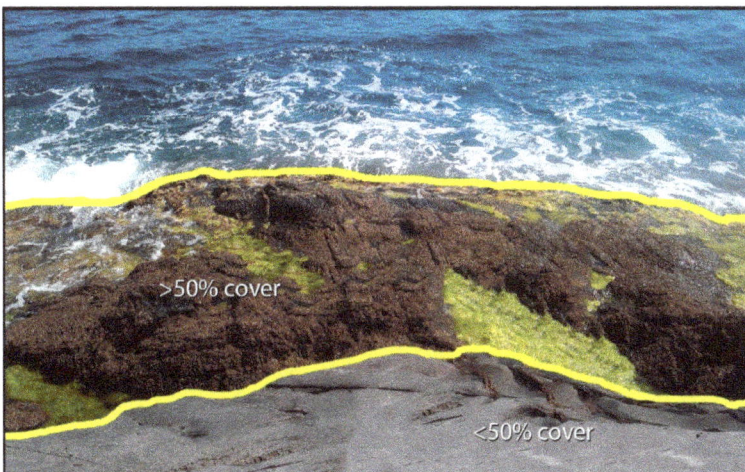

Figure 8. First order platform *Chthamalus* spp. coverage, Pontetto locality.

Data were analyzed separately for each locality and for the two biota. In Pontetto and Tellaro localities, where the average R value is very similar, the quadrats with barnacle cover >50% (covered) displayed (Figure 9) a lower R value than those with barnacle cover <50% (uncovered). The reduction in hardness is 6%. These results are consistent with those from the preliminary experiment described in Section 4.2. In the Lavagna locality, the overall R value is higher than in the previous localities due to recent anthropic rock reshaping; here, testing on covered quadrats show a harder bedrock than on uncovered ones, suggesting a bioprotective role of biota.

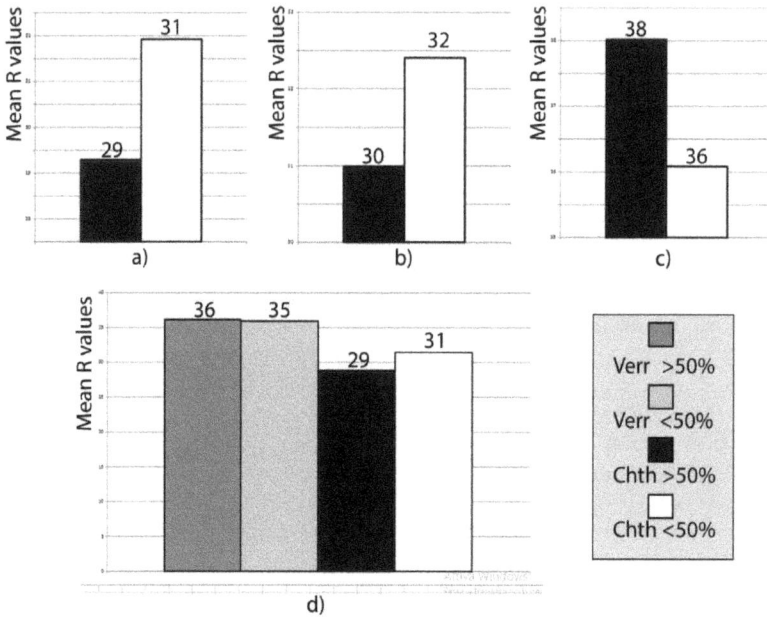

Figure 9. Rebound (R) values related to *Chthamalus* spp. covers on different localities. (**a**) Tellaro locality; (**b**) Pontetto locality; (**c**) Lavagna locality; and (**d**) comparison between *Verrucaria* and *Chthamalus* spp. in Tellaro locality.

In all the three localities, the two datasets cannot be differentiated on a statistical basis, as the results of the *t*-test indicate (Table 3). This could be due to the moderate difference in hardness of covered rocks with respect to uncovered ones; the resolution of the measuring tool could be too low to highlight such small differences [53]. Nevertheless, our evidence is indicative of a possible bioerosive role of barnacles in natural contexts. It is instead unlikely that the reduction in hardness between uncovered and covered quadrats is due to their differing elevation from sea level, as it is not relevant to control the efficiency of sea closeness-dependent weathering processes, the quadrats being constrained within a narrow belt and the upper barnacle boundary being undulated.

Table 3. *T*-test results for Pontetto, Tellaro, and Lavagna localities and sites.

Locality	Site	Significance (P)
	1	NS
Pontetto	2	0.008358
	3	NS
Tellaro	1	NS
Lavagna	1	NS

For the lichen cover, instead, no bio-remodeling effect was highlighted. Our data suggest (Figure 9) that the number of replicate quadrats and data scatter (indicated by standard deviation and by the level of significance from the *t*-test) are negatively correlated. It is possible that a consistent increase of test points (*i.e.*, of tested replicate quadrats) in each locality would enable the differentiation of the two datasets.

In order to investigate the source of variability of the *R* value, the ANOVA statistic test was applied for the three localities. Within each locality the replicates were subdivided into three sites, based on spatial contiguity. The experimental design (Figure 3) adopted in this work employs three factors: the fixed factor "treatment" represented by two levels (>50% and<50%) and the two random factors "locality" and "site." The response of this text shows a probability value which is highly significant for the "site" factor. This demonstrates that the variability between sites of the same locality is great; thus, a remarkable variability exists within each site. These results prevent the possibility of drawing conclusions about the differences between treatments (>50% and <50%) as well as between localities.

4.4. Minimizing the Environmental Differences between Sites in Order to Disentangle the Effect of Barnacles Cover from Other Weathering Processes

The results of the experiment illustrated in Section 4.3 suggest that moderate reduction in rock hardness is likely to be due to the effect of the barnacle cover along the rocky shores of Eastern Liguria. To get reliable evidence of the action of barnacles on littoral rocks in our study area, it should be necessary to increase the resolution of the rock hardness measuring tool and to minimize disturbance from other factors affecting this parameter. In particular, data processing with ANOVA demonstrates that a great spatial variability of rock hardness exists between sites at the small scale.

In order to minimize environmental differences between sites, a new study area, Calafuria, was selected, 120 km apart from Eastern Liguria (Figure 2). Here, the rocky shore is extremely uniform, being shaped in a sandstone bedrock belonging to a single member of the Macigno Formation (Oligocene), a silicic turbidite sandstone, medium-coarse grained, moderately sorted, interbedded with very tiny fine conglomerate beds [54]. In this area, an experimental design was worked out (Figure 3) based on the random selection of nine sites (*i.e.*, platforms) subdivided into three localities. For each site, ten sampling points and corresponding replicate quadrats were selected with the same principle applied to the experiment in Section 4.3: biota cover percentage and rock hardness were measured in all quadrats, located across the barnacle (*Chthamalus* spp.) band (midlittoral to low supralittoral) upper boundary, displaying one half of an abundant (>50%) barnacle cover, and one half of the scarce (<50%) cover. Five additional quadrats were tested for each site, in which barnacles were scraped off from the rock one month before testing, and the rock surface was then left exposed to air and sea water with no protection. On the whole, 135 quadrats were tested.

Rebound data are remarkably more scattered than in the previous experiment (Figure 10), although the bedrock was lithologically and structurally homogeneous and the replicate quadrats were more numerous in the Calafuria area. Averaging all the *R* values obtained from quadrats equally treated, we obtain a negligible difference in the *R* value, as the histograms in Figure 10 show. Data processing with *t*-test confirms the randomness of such differences, and ANOVA processing highlights that uncovered and covered quadrats (treatments) do not display significant *R* differences, regardless of the way they are grouped (by area, locality or site).

It should be concluded, thus, that the methodological approach adopted in this Section highlighted a possible effect of biota in coastal rocks shaping, but it was not appropriate to quantify the effect of bio-remodeling. In fact, we found overwhelming differences of *R* values between sites of the same locality and between replicas within the same site, for each treatment, due to the variability within each site. For this reason, simply using a higher resolution instrument for rock hardness assessment, such as the Equotip, is not likely to improve our results, and a different approach, such as the one described in the next paragraph, seems more promising.

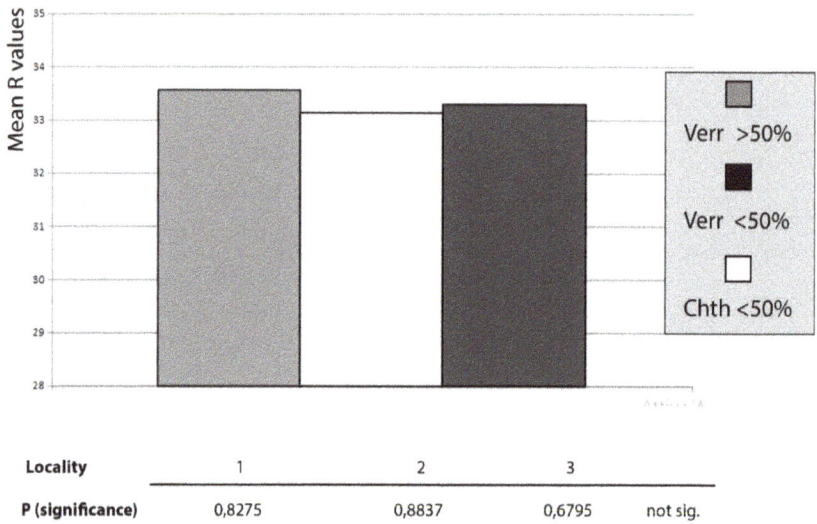

Locality	1	2	3	
P (significance)	0,8275	0,8837	0,6795	not sig.

Figure 10. Mean rebound (*R*) values related to *Chthamalus* spp. and *Verrucaria* covers on Calafuria locality and significance (*P*) values related to different treatments and initial biota coverage.

5. Biofilm EPS and Bio-Remodeling: A Computational Approach Proposal

A new approach to provide wide insights in the bio-remodeling process of coastal landforms is based on computational modeling. Biofilm is widely distributed on any type of substrate (even anthropogenic ones) as a fundamental component of the ecosystem, and is well studied and monitored for scientific purposes. Biofilm structure is usually based on different cyanobacteria and diatom communities attached to the bedrock and linked together in a complex gelatinous structure of biochemical compounds, water and ions (with a wide species variability in percentages and composition) known as extra-cellular polymeric substances (EPS) (Figure 11).

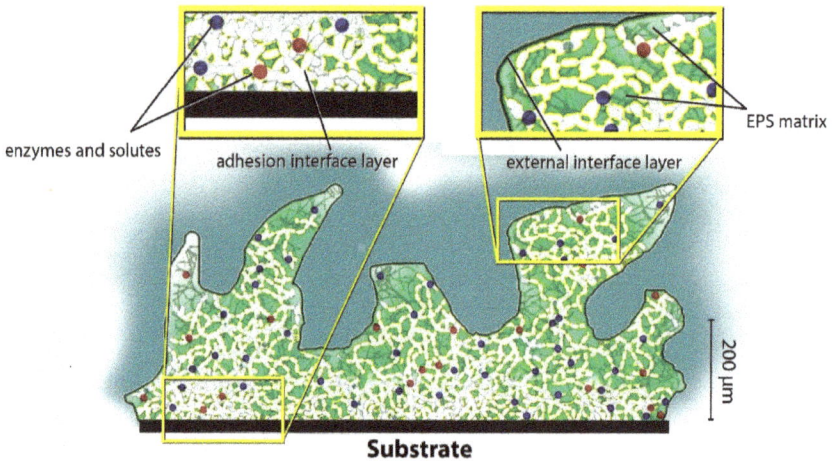

Figure 11. Main features and components of a biofilm layer.

Many scientific publications have involved experimental approaches to quantify the impact of biofilms over substrates in underwater and subaerial environments, mainly quantifying the direct action exerted by endolithic/epilithic cyanobacteria in controlled conditions [55] or field testing/monitoring of areas of interest with geotechnical procedures and instruments [35]. In particular, researchers have focused on the capability of cyanobacteria to operate like borers at the micro-scale and thus weaken the rock surface. Instead, however, in order to investigate the role of biofilm in mediating the interaction between coastal rocks and weathering agents (*i.e.*, indirect action), a computational modeling approach can be used.

Biofilm, likeother life forms, adapts to stress conditions lowering metabolism and dehydrating EPS (normally rich of water up to 97%); the high variability and hierarchical nature of organic EPS structures and the inherent difficulties disentangling the indirect effects of cyanobacteria presence from other shaping forces acting over the coastal landforms permits a multidisciplinary approach involving biology and computational modeling. This field of study is known as materiomics [56] and is based on the assumption that a material system (like biological materials) is hierarchically organized in sub-components leading to emerging non-linear behavior that cannot be explained analyzing the single components of the system. The characterization of the composition and structure is necessary to build an accurate continuum computational model; analyzing samples of bedrock with scanning and optical microscopes leads to the acquisition of key information on the texture and structural features of the rock (especially mineral grain size and sorting, porosity, joint width, and spacing). Moreover, the determination of the biofilm community is necessary to gain a specific knowledge of the mean EPS compound composition. The following step is taken to determine the abundance of the biomass present on the bedrock area. As known, the cyanobacteria are photosynthetic organisms, rich in a-chlorophyll, which has strong absorption in red–near-infrared (NIR) bands; a strong correlation is found by NIR remote sensing photographic analysis and analytically detected biomass [57]. The key approach for a multiscale computational model for the EPS is to build or retrieve the models of substrate and EPS main constituents from online open source databases (such as RCSB.org) and run a molecular dynamics (MD) simulation to study mechanical properties of the compounds involved using VMD and LAMMPS open source software (http://www.ks.uiuc.edu/) therefor. The simulation needs to be progressively scaled to wider assemblies from nanoscale to a continuum model for stress material testing.

Scaling up with thousands of compounds acting in a simulation simultaneously and retaining the full atomistic (fine-graining) information currently is not possible, as it would require huge amounts of time and enormous computer calculation power. The solution to such a problem is a simplification of the behavior of the single compound to a more manageable model with the same chemical and mechanical characteristics but with fewer degrees of freedom. The process is known as coarse-graining and is normally employed in materials science to investigate the interactivity of discrete particle systems (Figure 12). Moreover, the simulation gives clues about molecular forces involved, adhesion dynamics, and interaction behaviors in solutions or other compounds.

To quantify the forces exerted over the surface rock layer by EPS hydration/dehydration cycles and other weathering effects on the rock, a further approximation of the behavior of the material as a continuum homogeneous mass rather than a discrete particles system is necessary; the simulation is based on a coupled continuum material with shear forces acting on the micro-fractures and interstitial spaces replicating the mean rock texture previously seen.

Shear stress is induced by the volume expansion and contraction respectively of mineral grains and EPS, and simulation can thus be adapted to the magnitude and frequency of weathering processes actually taking place in the field.

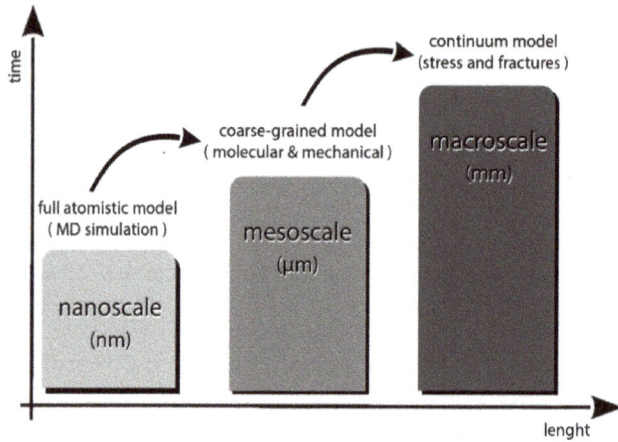

Figure 12. Multiscaling process from molecular dynamics to a continuum theory passing by coarse-grained models for mechanical and dynamical simulations.

The computational model has to be validated with direct measures that are able to quantify, in controlled experimental cases, the real shear forces applied and the substrate behavior and mechanical properties. This information is necessary to confirm simulation efficacy and the related hypothesis; as seen in other works [58–60], tensile machines and sensors are used to quantify adhesion forces or failure loads of byssus filaments and spider webs. For this case in particular, we suggested a procedure using a tensile machine in order to study the behavior of the unaltered rock sample under traction load conditions. This procedure may be used to calibrate extensometer sensors too in order to avoid data fluctuation due to external factors (temperature, moisture, *etc.*) at low intensity loads, using an extensometric bridge in the experimental setup. Such variations can be quantified with a digital signal output (Figure 13).

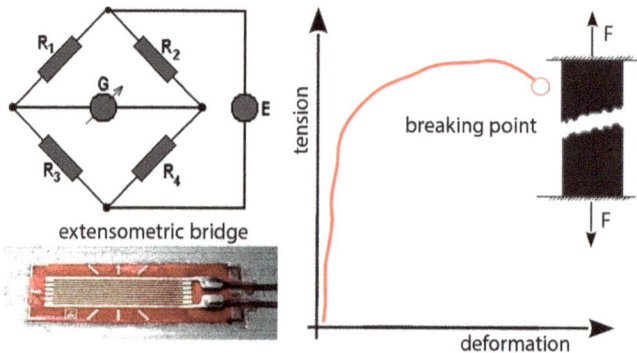

Figure 13. Low-intensity tensil test measures required for the calibration of the extensometric sensor. All measurements must be taken with extensometric bridge to avoid data fluctuation.

The extensometer is pasted on the surface of the body, generally using instant adhesives such as cyanoacrylate, following the deformations of the surface to which it is bonded, elongating and shortening together, and causing a variation in the electrical resistance of the wire. Moreover, it is possible to test an unaltered rock sample with a compressive machine in order to calibrate Equotip *L* measurements previously collected in the field (Figure 14).

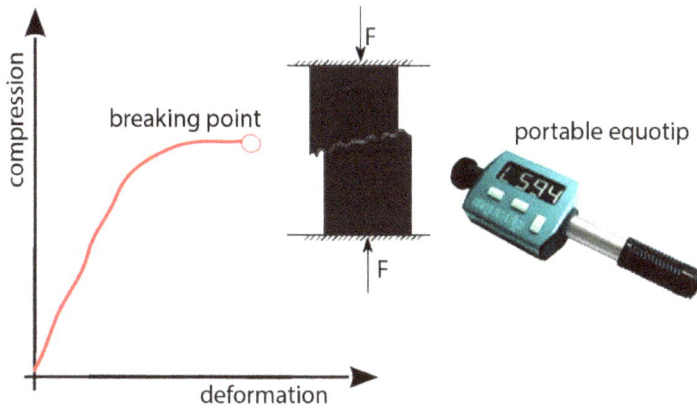

Figure 14. Low-intensity compressive test measures required for the calibration of Equotip *L* values.

After sensor calibration, it is possible to proceed with the validation test under stress conditions (changing moisture levels) for the colonized sample; slices of progressive thicknesses are tested in order to evaluate the magnitude order and the maximum depth penetration of the forces acting on the colonized surface, which are necessary to interpolate a mean surface deformation value (Figure 15).

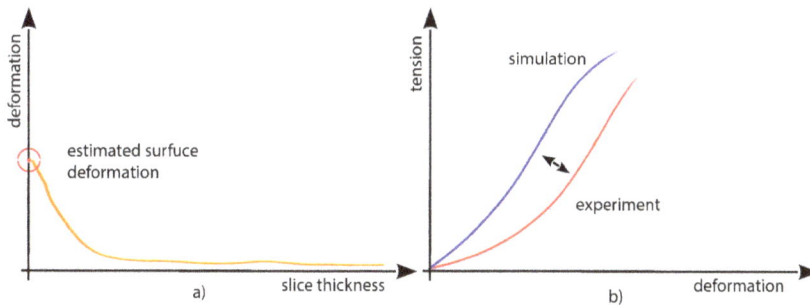

Figure 15. (**a**) Data interpolation of deformation over slice thickness to estimated surface force penetration and (**b**) data comparison of measured shear forces with computational model behavior to validate the the procedure. For any slice thickness, 10 obtained values are required to obtain reliable measure of deformation using semi-ex-post analysis and with maximal semi-dispersion for the field of uncertainty.

A further issue is to test the behavior of EPS in the presence of a single pollutant (for example PAHs or heavy metals), which may have a relevant impact on data collected in the field, gaining clues over the impact human pollution may lead to the trophic chain and indirectly to erosion rates of coastal landforms.

The computational approach displays a number of advantages compared to the experimental approach. Among them are the possibility of setting arbitrary boundary conditions to the system (which are really difficult to control in experimental procedures), the possibility of performing accelerated computational simulations of dynamics that may occur in geological time spans in nature (such as erosion dynamics), and the possibility of understanding how dynamics propagate differently, scaling the hierarchy of the system (cross-scaling).

6. Conclusions

Bioerosion and bioprotection (here indicated as bio-remodeling) have been extensively investigated in marine and subaerial contexts. Many efforts have also been made to provide quantitative estimates of the efficiency of bio-remodeling compared to that of other shaping agents, and some results have been achieved in assessing the contribution of bioerosion to the littoral sediment budget.

Our experimental evidence, as presented in this paper, suggests that geological methods may provide some useful insight into the subject of bio-remodeling quantification. In particular, rock hardness/strength assessment tools provide quantitative estimates of the efficiency of weathering agents and thus, with an appropriate experimental design, also of that of bioerosive and bioprotective effects. Nevertheless, it is evident that there are some local factors that limit our capability of disentangling the effect of biota when we test rock hardness/strength on a rocky surface. Widening the spatial scale of analysis is not sufficient to remove this site effect. This happens because there is a lack of knowledge on how biota, especially sessile ones, interact with the surfaces of coastal rocks. In particular, it would be relevant to learn how the covered rock surface changes its chemical composition and mechanical property, as well as quantify this change for different types of biota and surface coverage. Both biota and rock are very complex, and more than a single method to explore the problem is required. We suggest that it would be worthwhile to combine experiments and simulations to overcome this problem. In fact, experimentally assessing the quantitative contribution of biota to the overall rock surface remodeling, differentiating it from the contribution of other weathering agents, may be prevented by the difficulty of working out a suitable experimental design, whereas a computational approach allows for the extrapolation of the role of biota from that of other weathering agents. Moreover, materiomics consider the hierarchical contributions of each of the sub-components constituting biological materials.

We propose combining the materials science and environmental science approaches at the intersection of mechanics and biology and hypothesize that the activity of sessile biota alters the surface property of coastal rock by changing both its mechanical strength and chemical composition. We hypothesize that bioerosion occurs primarily at the interface between biota and coastal rock and cause the rock surface to lose its strength, which couples with physical and chemical weathering and wave erosion to shape landforms effectively.

The parallel experimental and computational modeling approach may offer a synergic solution able to accelerate the processes of identification of single components acting in complex dynamic systems; moreover, it is a viable way to validate experimental results or hypotheses in simplified environments, allowing the calibration of new methodologies to save time and funds, which is a fundamental issue in any research field.

The effects of bioerosion and bioprotection have remarkable practical value in bridging the knowledge gap between the evolution of coastal landforms and the environmental changes due to human activities. Lack of such knowledge prevents us from understanding the significance of the impact of biota on the long-term evolution of coastal landforms and human activities connected to them.

Acknowledgments: The authors kindly acknowledge E. Rosa, M. Cattafesta, L. De Fabritiis, and M.C. Nanni for experimental data collection. This work was performed in the framework of the MIT-UniPi Joint Project "Assessing the Effect of Biota on Coastal Rock Surfaces: a Quantitative Approach."

Author Contributions: M. Pappalardo and Z. Qin conceived the paper; A, Chelli, F. Pannacciulli, and M. Pappalardo designed and performed the experiments and analyzed the data; M. Buehler and Z. Qin proposed the computational tools; M. Pappalardo and L. Cironi wrote the paper.

Conflicts of Interest: The authors declare no conflicts of interest.

References

1. Neumann, A.C. Observations on coastalerosion in Bermuda and measurements on the boring rate of the sponge, ClionaLampa. *Limnol. Oceanogr.* **1966**, *11*, 92–108. [CrossRef]
2. Schneider, J. Biological and inorganic factors in the destruction of limestone coasts. *Contrib. Sedimentol.* **1976**, *6*, 1–112.
3. Torunski, H. Biological erosion and its significance for the morphogenesis of limestone coasts and for nearshore sedimentation (Northern Adriatic). *Senckenberg. Marit.* **1979**, *11*, 193–265.
4. Trudgill, S.T. Bioerosion of intertidal limestone, Co. Clare, Eire—3. Zonation, process and form. *Mar. Geol.* **1987**, *74*, 111–121. [CrossRef]
5. Trudgill, S.T.; Crabtree, R.W. Bioerosion of intertidal limestone, Co. Clare, Eire—2. Hiatellaarctica. *Mar. Geol.* **1987**, *74*, 99–109. [CrossRef]
6. Trudgill, S.T.; Smart, P.L.; Friederich, H.; Crabtree, K.W. Bioerosion of intertidal limestone. *Mar. Geol.* **1987**, *74*, 85–98. [CrossRef]
7. Spencer, T. Coastal Biogeomorphology. In *Biogeomorphology*; Blackwell: Oxford, UK, 1988; pp. 255–318.
8. Radtke, G.; Hofmann, K.; Golubic, S. A bibliographic overview of micro- and macroscopic bioerosion. *Cour. Forsch. Inst. Senckenberg.* **1997**, *201*, 307–340.
9. Hutchings, P. *Encyclopedia of Modern Coral Reefs*; Springer: Rotterdam, The Netherlands, 2011; pp. 139–156.
10. Moses, C.A. Tropical rock coasts: Cliff, notch and platform erosion dynamics. *Prog. Phys. Geogr.* **2013**, *37*, 206–226. [CrossRef]
11. French, P.W. Coastal zone management. In *The Encyclopedia of Coastal Science*; Springer: Rotterdam, The Netherlands, 2005; pp. 313–319.
12. Naylor, L.A.; Coombes, M.A.; Viles, H.A. Reconceptualising the role of organisms in the erosion of rock coasts: A new model. *Geomorphology* **2012**, *157*, 17–30. [CrossRef]
13. Furlani, S.; Pappalardo, M.; Gómez-Pujol, L.; Chelli, A. The rock coast of the Mediterranean and Black seas. In *Rock Coast Geomorphology: A Global Synthesis*; Kennedy, D.M., Stephenson, W.J., Naylor, L.A., Eds.; Geological Society: London, UK, 2014; Volume 40, pp. 89–122.
14. Maggi, E.; Milazzo, M.; Graziano, M.; Chemello, M.; Benedetti-Cecchi, L. Latitudinal and local scale variations in a rocky intertidal interaction web. *Mar. Ecol. Prog. Ser.* **2015**, *534*, 39–48. [CrossRef]
15. Stiros, S.C.; Pirazzoli, P.A. Direct determination of tidal levels for engineering applications based on biological observations. *Coast. Eng.* **2008**, *55*, 459–467. [CrossRef]
16. Schneider, J.; Torunski, H. Biokarst on limestone coasts, morphogenesis and sediment production. *Mar. Ecol.* **1983**, *4*, 45–63. [CrossRef]
17. Palmer, M.; Fornós, J.J.; Balaguer, P.; Gómez-Pujol, L.; Pons, G.X.; Villanueva, G. Spatial and seasonal variability of the macro-invertebrate community of a rocky coast in Mallorca (Balearic Islands): Implications for bioerosion. *Hydrobiologia* **2003**, *501*, 13–21. [CrossRef]
18. Moura, D.; Gabriel, S.; Gamito, S.; Santos, R.; Zugasti, E.; Naylor, L.A.; Gomes, A.; Tavares, A.M.; Martins, A.L. Integrated assessment of bioerosion, biocover and downwearing rates of carbonate rock shore platforms in Southern Portugal. *Cont. Shelf Res.* **2012**, *38*, 79–88. [CrossRef]
19. Naylor, L.A.; Viles, H.A. A new technique for evaluating short-term rates of coastal bioerosion and bioprotection. *Geomorphology* **2002**, *47*, 31–44. [CrossRef]
20. Carter, N.E.A.; Viles, H.A. Bioprotection explored: Thestory of a little known earth surface process. *Geomorphology* **2005**, *67*, 273–281. [CrossRef]
21. Schneider, J.; Le Campion-Alsumard, T. Construction and destruction of carbonates by marine and freshwater cyanobacteria. *Eur. J. Phycol.* **1999**, *34*, 417–426. [CrossRef]
22. McLean, R.F. Measurement of beach rock erosion by some tropical marine gastropods. *Bull. Mar. Sci.* **1967**, *17*, 551–561.
23. Abensperg-Traun, M.; Wheaton, G.A.; Eliot, I.G. Bioerosion, notch formation and micromorphology in the intertidal and supratidal zones of calcareous sandstone stack. *J. R. Soc. West. Aust.* **1990**, *73*, 47–56.
24. Andrews, C.; Williams, R.B.G. Limpet erosion of chalk shore platforms in southeast England. *Earth Surf. Process. Landf.* **2000**, *25*, 1371–1382. [CrossRef]

25. Fornós, J.J.; Pons, G.X.; Gómez-Pujol, L.; Balaquer, P. The role of biological processes and rates of downwearing due to grazing organisms on Mallorcan carbonate coasts (western Mediterranean). *Z. Geomorphol. Suppl.* **2006**, *144*, 161–181.

26. Vidal, M.; Fornós, J.J.; Gómez-Pujol, L.; Pons, G.X.; Balaguer, P. Exploring rock coast bioerosion: Rock fragment intestine transit time and erosion rates computation of the gastropod *Monodontaarticulata* (Lamarck, 1822). *J. Coast. Res.* **2013**, *65*, 1704–1709. [CrossRef]

27. Stephenson, W.J. *The Micro and Traversing Erosion Meter. Treatise on Geomorphology*; Shroder, J., Ed.; Elsevier: Amsterdam, The Netherlands, 2013; Volume 10, pp. 164–169.

28. Furlani, S.; Cucchi, F.; Forti, F.; Rossi, A. Comparison between coastal and inland karst limestone lowering rates in the northeast Adriatic Region (Italy and Croatia). *Geomorphology* **2009**, *104*, 73–81. [CrossRef]

29. Stephenson, W.J.; Finlayson, B.L. Measuring erosion with the micro-erosion meter. Contributions to understanding landform evolution. *Earth Sci. Rev.* **2009**, *95*, 53–62. [CrossRef]

30. Mayaud, J.R.; Viles, H.A.; Coombes, M.A. Exploring the influence of biofilm on short-term expansion and contraction of supratidal rock: An example from the Mediterranean. *Earth Surf. Process. Landf.* **2014**, *39*, 1404–1412. [CrossRef]

31. Moses, C.A. Field rock block exposure trials. *Z. Geomorphol.* **2000**, *120*, 33–50.

32. Dalongeville, R.; Le Campion, T.; Fontaine, M.F. Bilanbioconstruction-biodestruction dans les roches carbonatées en mer Méditerranée: Étude expérimentale et implications geomorphologiques. *Z. Geomorfol.* **1994**, *38*, 457–474.

33. Coombes, M.A.; Naylor, L.A.; Thompson, R.C.; Roast, S.D.; Gómez-Pujol, L.; Fairhurst, R.J. Colonization and weathering of engineering materials by marine microorganisms: An SEM study. *Earth Surf. Process. Landf.* **2011**, *36*, 582–593. [CrossRef]

34. Coombes, M.A.; Naylor, L.A. Rock warming and drying under simulated intertidal conditions, part II: Weathering and biological influences on evaporative cooling and near-surface micro-climatic conditions as an example of biogeomorphic ecosystem engineering. *Earth Surf. Process. Landf.* **2012**, *37*, 100–118. [CrossRef]

35. Coombes, M.A.; Naylor, L.A.; Viles, H.A.; Thompson, R.C. Bioprotection and disturbance: Seaweed, microclimatic stability and conditions for mechanical weathering in the intertidal zone. *Geomorphology* **2013**, *202*, 4–14. [CrossRef]

36. Coombes, M.A.; Feal-Pérez, A.; Naylor, L.A.; Wilhelm, K. A non-destructive tool for detecting changes in the hardness of engineering materials: Application of the Equotipdurometerin the coastal zone. *Eng. Geol.* **2013**, *167*, 14–19. [CrossRef]

37. Stephenson, W.J.; Dickson, M.E.; Trenhaile, A.S. Rock Coasts. In *Coastal Geomorphology*; Academic Press: San Diego, CA, USA, 2013; Volume 10, pp. 289–307.

38. Aydin, A.; Basu, A. The Schmidt hammer in rock material characterization. *Eng. Geol.* **2005**, *81*, 1–14. [CrossRef]

39. Chelli, A.; Pappalardo, M.; Arozarena, L.I.; Federici, P.R. The relative influence of lithology and weathering in shaping shore platforms along the coastline of the Gulf of La Spezia (NW Italy) as revealed by rock strength. *Geomorphology* **2010**, *118*, 93–104. [CrossRef]

40. Chelli, A.; Pappalardo, M.; Rosa, E. Le piattaforme litorali dell'Isola Palmaria(Golfo della Spezia): Un esempio di forme ereditate nella costa rocciosa della Liguria orientale. *Atti Soc. Tosc. Sci. Nat. Mem. Ser. A* **2010**, *115*, 39–54.

41. Peres, J.; Picard, J. Les corniches calcaires d'origine biologique en Mediterranee occidentale. *Rec. Trav. Stn. Mar.* **1964**, *4*, 2–33.

42. Laborel, J.; Laborel-Deguen, F. Biological indicators of relative sea-level variations and of co-seismic displacements in the Mediterranean Region. *J. Coast. Res.* **1994**, *10*, 395–415.

43. Kelletat, D. Zonality of modern coastal processes and sea-level indicators. *Palaeogeogr. Palaeoclimatol. Palaeoecol.* **1988**, *68*, 219–230. [CrossRef]

44. Coombes, M.A.; Naylor, L.A.; Roast, S.D.; Thompson, R.C. Coastal defences and biodiversity: The influence of material choice and small-scale surface texture on biological outcomes. In Proceedings of the ICE Conference on Coasts, Marine Structures & Breakwaters, EICC Scotland, Edinburgh, UK, 16 September 2009; Volume 2, pp. 474–485.

45. Trenhaile, A.S.; Alberti, A.P.; Cortizas, A.M.; Casais, M.C.; Chao, R.B. Rock coast inheritance: An example from Galicia, northwestern Spain. *Earth Surf. Process. Landf.* **1999**, *24*, 605–621. [CrossRef]

46. Biagioni, F.; Cipolla, F.; Pappalardo, M. The problem of using marine terraces with unclear inner edge for palaeo sea-level determination: A case study from Liguria (NW Italy). *Atti Soc. Tosc. Sci. Nat. Mem. Ser. A* **2011**, *116*, 23–32.

47. Arozarena, L.I. Factors in the development of rocky coasts between Lerici and Tellaro (Gulf of La Spezia, Liguria, Italy). *Geogr. Fis. Dinam. Quat.* **2006**, *29*, 71–81.

48. Pannacciulli, F.G.; Relini, G. The vertical distribution of *Chthamalusmontagui* and *Chthamalusstellatus* (Crustacea, Cirripedia) in two areas of the NW Mediterranean Sea. *Hydrobiologia* **2000**, *426*, 105–112. [CrossRef]

49. Meese, R.J.; Tomich, P.A. Dots on the rocks: An evaluation of percent cover estimation methods. *J. Exp. Mar. Biol. Ecol.* **1992**, *165*, 59–73. [CrossRef]

50. Nimis, P.L.; Pinna, D.; Salvadori, O. *Licheni e Conservazione dei Monumenti*; CLUEB: Bologna, Italy, 1992; pp. 1–165.

51. Cucchi, F.; Furlani, S.; Burelli, G.; Zini, L.; Tretiach, M. Variazioni microtopografiche di superfici carbonatiche colonizzate da licheni endolitici (Carso Giuliano, Maiella). *Atti e Mem. della Comm. Grotte. E. Boegan.* **2009**, *42*, 67–84.

52. Carter, N.E.A.; Viles, H.A. Experimental investigations into the interactions between moisture, rock surface temperatures and an epilithic lichen cover in the bioprotection of limestone. *Build. Env.* **2003**, *38*, 1225–1234. [CrossRef]

53. Viles, H.A.; Goudie, A.S.; Grab, S.; Lalley, J. The use of the Schmidt Hammer and Equotip for rock hardness assessment in geomorphology and heritage science: A comparative analysis. *Earth Surf. Process. Landf.* **2010**, *36*, 323–333. [CrossRef]

54. Sciarra, N.; Marchetti, D.; Avanzi, G.D.A.; Calista, M. Rock slope analysis on the complex Livorno coastal cliff (Tuscany, Italy). *Geogr. Fis. Dinam. Quat.* **2014**, *37*, 113–130.

55. Coombes, M.A. Biogeomorphology of Coastal Structures: Understanding Interactions between Hard Substrata and Colonising Organisms as a Tool for Ecological Enhancement. Ph.D. Thesis, University of Exeter, Exeter, UK, 2011.

56. Cranford, S.W.; Buehler, M.J. *Biomateriomics*; Springer Science &Business Media: Dordrecht, The Netherlands, 2012; pp. 30–54.

57. Murphy, R.J.; Underwood, A.J.; Jackson, A.C. Field-based remote sensing of intertidal epilithic chlorophyll: Techniques using specialized and conventional digital cameras. *J. Exp. Mar. Biol. Ecol.* **2009**, *380*, 68–76. [CrossRef]

58. Qin, Z.; Buehler, M.J. Spider silk: Webs measure up. *Nat. Mater.* **2013**, *12*, 1–3. [CrossRef] [PubMed]

59. Qin, Z.; Buehler, M.J. Impact tolerance in mussel thread networks by heterogeneous material distribution. *Nat. Commun.* **2013**, *4*, 1–8. [CrossRef] [PubMed]

60. Qin, Z.; Buehler, M.J. Molecular mechanics of mussel adhesion proteins. *J. Mech. Phys. Solids* **2014**, *62*, 19–30. [CrossRef]

Journal of
Marine Science and Engineering

MDPI

Article

Sandbar Migration and Shoreline Change on the Chirihama Coast, Japan

Masatoshi Yuhi [1],*, Masayuki Matsuyama [2] and Kazuhiro Hayakawa [3]

[1]   School of Environmental Design, Kanazawa University, Kakuma-Machi, Kanazawa 920-1192, Japan
[2]   Specified Nonprofit Corporation, The Society of Ocean Romantics, 1-18-11-1011 Kanda, Chiyoda-ku,
     Tokyo 101-0047, Japan; m-matsuyama@mtb.biglobe.ne.jp
[3]   Public Works Department, Ishikawa Prefectural Government, 1-1 Kuratsuki, Kanazawa 920-8580, Japan;
     haya1122@pref.ishikawa.lg.jp
*   Correspondence: yuhi@se.kanazawa-u.ac.jp; Tel.: +81-76-234-4609

Academic Editor: Gerben Ruessink
Received: 29 February 2016; Accepted: 13 May 2016; Published: 3 June 2016

Abstract: Sandy beaches play a key role in regional tourism. It is important to understand the principal morphological processes behind preserving attractive beaches. In this study, morphological variation on the Chirihama Coast, Japan, an important local tourism resource, was investigated using two sets of field surveys. The objective was to analyze and document the multi-scale behaviors of the beach. First, long-term shoreline changes were examined based on shoreline surveys over the last two decades. Then, the middle-term behavior of multiple bar systems was analyzed based on the cross-shore profile surveys from 1998 to 2010. An empirical orthogonal function (EOF) analysis was conducted to capture the principal modes of the systematic bar migration. The shoreline analysis indicated a long-term eroding trend and showed that the seasonal variation has recently tended to increase. The profile analysis demonstrated that net offshore migrations of bars have been repeated with a return period of approximately four years. This general behavior of the bar system is similar to the net offshore migration phenomena observed at other sites in the world. EOF analysis revealed a relationship between bar configuration and middle-term variations in shoreline location; when a new bar is generated near the shoreline and a triple bar configuration is established, the shoreline tends to temporarily retreat, whereas the shoreline experiences an advance when the outer bar has most evolved.

Keywords: multiple sandbars; periodic migration; shoreline change; beach profile

1. Introduction

Many of the sandy beaches in the world have played an important role in regional tourism. The Chirihama Coast in Japan, which is located on the northwestern coast of Honshu (Figure 1), contains such popular beaches [1]. The long sandy beaches consist of very fine sand and are accessed by the 7 km-long 'Nagisa Driveway', a marine drive for automobiles on the backshore (Figure 2). More than 800,000 tourists visit the coast every year, mainly from spring to autumn, and enjoy driving automobiles on the beach. To maintain this attractive tourism resource, it is necessary to preserve the wide sandy beaches. However, the beach has recently been suffering from accelerated erosion problems under both direct and indirect influences of human interference in the coastal-river watershed. Historically, the long sandy beach developed in this area as a segment within a large littoral cell stretching approximately 75 km along the Japan Sea coast. Our study area is located at the downdrift (northeast) end of this cell. The principal source of sediment here is the Tedori River, the mouth of which is located approximately 45 km southwest of the study area. In recent decades, however, the sediment discharge from the Tedori River has decreased because of various anthropogenic modifications of the river basin, such as extensive sand mining and dam construction commencing in the 1960s [2,3]. In addition, this area was largely separated from the updrift (southwest) portion of the littoral cell following the construction and extension of the main breakwater at Kanazawa Port (which stretches out to water depths of approximately 13 m with a total length of 3.2 km) after 1970 [4]. The combined effects of a decrease in sediment discharge and the disruption of longshore sediment transport have resulted in an accelerated retreat of the shoreline. The shoreline retreated approximately 50 m during the last two decades when the erosion was most significant.

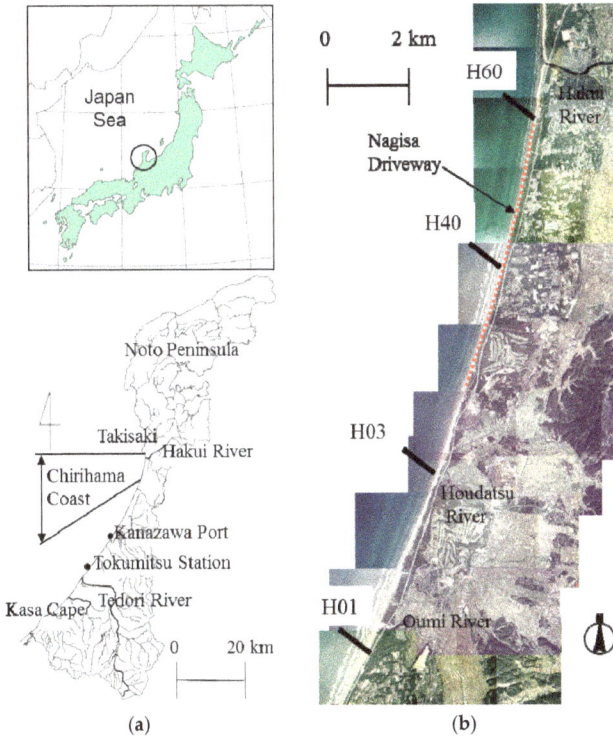

Figure 1. Field site: (**a**) location of the study site; (**b**) location of the cross-shore survey lines (H stands for the Hakui district where the Chirihama Coast is located.).

Figure 2. Nagisa Driveway on the Chirihama Coast.

Morphologically, the beach profile on the Chirihama Coast is characterized by the development of large-scale multiple sandbar systems [1,5–8]. The evolution of multiple sandbars can be inferred from an aerial photograph taken in 1975 (Figure 1b). Because the configuration of the bar system may indirectly influence morphological variations on the foreshore and backshore, it is essential, for the future preservation and development of the marine drive, to develop a physical understanding of ongoing morphological processes, including shoreline variations and changes in underwater bathymetries.

Longshore bars are known to play an important morphological role in the development of sandy beaches. For example, they dissipate extremely high wave energy during storms and significantly influence dynamic and ecological processes in the surf zone. In the past, a number of scientific and engineering investigations have been performed on the evolution of bars over a wide range of time scales [9–12]. Among them, the interannual behaviors of bars have recently been analyzed on the basis of long-term survey records at several locations around the world [13–18]. It has been reported that multiple bar systems exhibit systematic, cyclic net offshore migration (NOM) on a time scale of one to 20 years.

Although several studies have mentioned the existence and features of large-scale multiple sandbars on the Chirihama Coast [1,5–7], existing studies of the characteristics of temporal variation in these bars are very restricted. Recently, the authors have investigated the basic characteristics of bar evolution on the Chirihama Coast and demonstrated that the interannual movements of these bars are significant [8]. The bar system exhibits features that are consistent with previously reported NOM behavior at other sites [13–18]. However, the present understanding of the characteristics of temporal and spatial variations on the beach is still limited. Continued and more detailed reexamination is needed to clarify the physical processes and to obtain appropriate insight for planning future beach preservation. In this study, accordingly, the long-term variations of the shoreline and middle-term variation of seabed profiles on the Chirihama Coast are investigated using two sets of field surveys. First, the characteristics of the spatiotemporal variation of shoreline position are examined for the last two decades. The variations of sediment volume and depth contour lines are also discussed at a time scale of years. Then, we further examine the characteristics of systematic bar migration based on a profile record collected over a period of 13 years. Alongshore, as well as cross-shore variations are examined. Finally, the principal modes of bar migration are clarified using empirical orthogonal function (EOF) analysis. The relationship between the transitions in bar configurations and middle-term shoreline variations over several years is also examined.

The principal objective of this study is to analyze and document the morphological behavior of the multiple bar system and related variations on the Chirihama Coast. The results obtained in this study will contribute to an improved scientific understanding of morphologic features on this beach, such as shoreline movements, bar configurations and the relationships between them. In addition, the results will be useful for understanding and comparing the behavior of similar sandy beaches around the world.

2. Field Site

2.1. Location

The Chirihama Coast is located midway along the northwest coast of Honshu, Japan facing the Japan Sea (Figure 1). The coastline is nearly straight and has a general NNE-SSW orientation. The study area includes an approximately 13 km-long stretch of coastline (including Nagisa Driveway in the north part) that is part of a large littoral cell stretching approximately 75 km along the coast. The beach slope is typically in the range of 1/80 to 1/200. The seabed slope becomes more gradual as the alongshore locations move northward. Figure 1b is an aerial photograph of the Chirihama Coast taken in 1975. The evolution of three sandbars is observed along the shoreline. Referring to the morphodynamic classifications of Wright and Short [19], the beach is mainly in the intermediate stage with a longshore bar trough.

From 1984, beach nourishment has been conducted on an irregular basis (*i.e.*, not every year) [1]. The volume, length and location varied with each nourishment exercise. Typically, 3000 to 5000 m^3 of sand were placed per year onto the shoreface. The corresponding longshore length of nourishment was several hundred meters. Nourishment locations were determined based on alongshore variations in beach width each year; *i.e.*, the nourishment was conducted where the beach width was expected to be insufficient to maintain the viability of the marine drive. Construction of a submerged breakwater started in 2010 in the middle of the study area. During the period of data analysis (1998 to 2010), however, the coastal stretch under examination was almost free from the direct influences of engineering structures.

2.2. Wave Climate and Coastal Currents

Because the study area is microtidal with a maximum tidal range of 0.4 m, the littoral sediment transport near the shoreline is dominated by seasonal wave actions. The wave climate has been measured at the Tokumitsu observation station (Figure 1a) of the Hokuriku Regional Development Bureau of Japan's Ministry of Land, Infrastructure, Transport and Tourism (HRDB), which is located approximately 40 km southwest of the study area. The wave height and direction have been measured approximately 1500 m offshore where the water depth is approximately 15 m. According to the observation record from 1995 to 2005, the characteristics of the wave climate are summarized as follows (Figure 3). In the summer, waves are calm, and the significant wave height is usually less than 1 m. The dominant incoming wave direction is from the NNW, although the wave direction is relatively widespread. In the winter season, the dominant wave direction is from the NW to the WNW. High waves with significant heights in excess of 1 m are often observed. Annual maximum significant wave height off the beach is in the range of 5 to 8 m. The annual net longshore sediment transport is considered to be from the NNE to the SSW near the shore. Further offshore, where the water depth is more than approximately 10 m, sand transport is predominantly affected by the Tsushima ocean current, as well as wind-driven currents. The dominant directions of these currents are from the SSW to the NNE, which is opposite that of the wave-induced current near the shoreline. Although sediment transport in this area is relatively weak compared to that of the surf zone, the occurrence of strong currents near the bottom may contribute to the long-term evolution of coastal topography [20,21]. According to previous field observations, the northward current prevails during the passage of low-pressure systems across the sea, and the current velocity at depths of 10 and 15 m could reach 90 and 40 cm/s, respectively [21,22]. Sediments are considered to migrate from the SSW to the NNE by these currents in the offshore area.

2.3. Sediment Characteristics

Sediment grain size in the study area is generally in the fine sand range. Sediment diameters become smaller as the location moves northward. Typical sediment diameters near the shoreline are 0.15 to 0.20 mm. According to a previous field study by the Ishikawa Prefectural Government concerning the sediment size distribution at the location of line H60 (H stands for the Hakui district where the Chirihama Coast is located.) in Figure 1b, sediment size is concentrated in a narrow range between 0.15 and 0.18 mm. Another investigation of the cross-shore variation of the sediment diameter [1] near the northern boundary of the study area (close to line H60 in Figure 1b) also demonstrated that the cross-shore variation in grain size (d_{60}) is small up to water depths of 50 m.

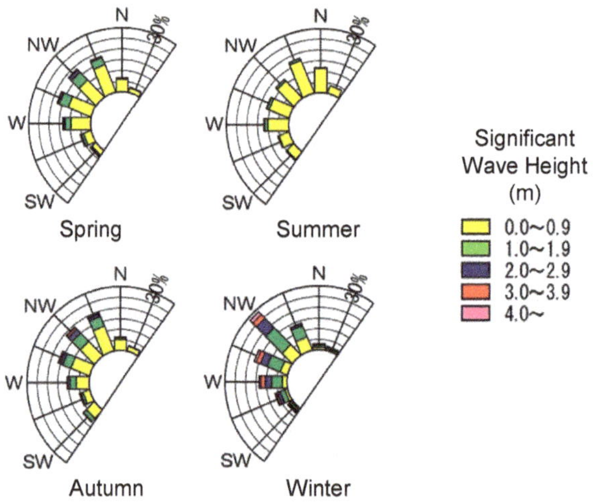

Figure 3. Wave climate observed at Tokumitsu Station (1995 to 2005).

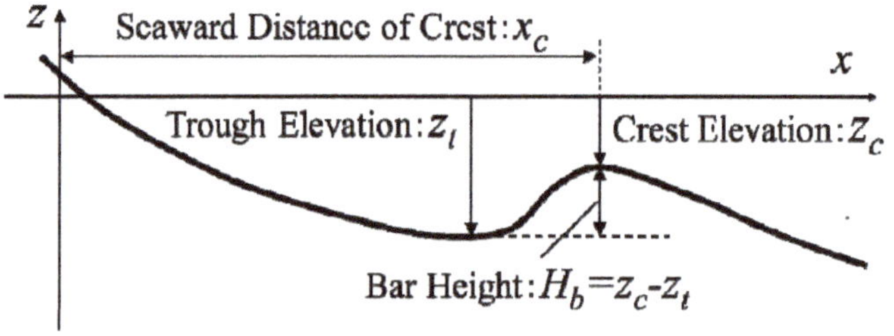

Figure 4. Definition of bar properties.

3. Datasets and Methods

3.1. Datasets

Surveys of the shoreline location have been carried out by the public works department of the Ishikawa Prefectural Government. The shoreline location has been measured twice a year, in March and September, over a 15.5-km alongshore stretch. In this study, the data from 1986 to 2006 are used. The long-term and large-scale trends of shoreline change were analyzed based on the shoreline survey.

Cross-shore profile surveys have been carried out since 1998 by HRDB along the four survey lines (H01, H03, H40 and H60) in Figure 1b. The longshore interval of these lines varies between 4000 and 4200 m. The cross-shore stretch of the survey varied each year in the range of 1 to 3.5 km. The strike of the survey lines is 307° (clockwise from north), which is slightly different from the shore normal direction. Subaqueous profiles have been measured using an echo sounder once a year, usually in autumn (from September to November). In this study, field data obtained from 1998 to 2010 were used. In the soundings collected during 2002, the bathymetry of the outer nearshore zone was not resolved on H01, H02 and H60. Bed elevations are described with respect to the Tokyo Peil (T.P.) datum, which is the standard ground elevation in Japan based on the Tokyo Bay mean sea level.

3.2. Quantification of Bar Properties

Variations in the sandbar properties have been examined as follows. First, the shoreline position was defined for each year as the location where the bed elevation is equal to the mean monthly lowest water level (T.P. + 0.01 m). The mean location of the shoreline during the study period was then computed for each line as a reference location. Hereafter, the cross-shore distance is taken as the distance from the mean location of the shoreline during the study period (denoted as $X = 0$ m), unless otherwise mentioned. The seaward distance of the crest (X_c) and vertical elevation of the crest (Z_c) and trough (Z_t) of bars were first examined to investigate the geometric features of the bar system (Figure 4). Bar height (H_b) was defined as the difference between the crest elevation and trough elevation. Bars with computed heights below a threshold value of 0.2 m (which corresponds to sounding accuracy) were considered to be unreliable and, hence, were excluded from the datasets. The alongshore variability has then been investigated in detail through a comparison between survey lines.

3.3. EOF Analysis

In the original survey record, the locations of survey points were not fixed during the study period. Accordingly, we first computed the bed elevation at fixed locations with a uniform cross-shore interval (20 m) using a linear interpolation for each transect. The cross-shore range of the analysis lies between a point slightly landward (40 to 100 m) of the mean location of the shoreline and a point located 2400 m offshore, which corresponds to a water depth of 12 to 13 m.

The characteristics of morphological variation on each survey line were then examined using EOF analysis. Formerly, EOF analysis has been used to describe changes within beach profiles (e.g., [23–25]). The EOFs correspond to a statistically optimal description of the data with respect to how the variance is concentrated in the eigenmodes, where the variance explained decreases with the mode number. The first eigenfunction (with the largest eigenvalue) explains most of the variation in bed-level changes. Each successively higher eigenfunction explains most of the variation left unexplained by the preceding eigenfunctions. In this study, the non-demeaned bed level $Z(x,t)$ was explained by the summation of various eigenmodes. Each mode was represented by the product of the time coefficient $C_n(t)$ and the spatial function $e_n(x)$, where the subscript n denotes the n-th mode:

$$Z(x,t) = \sum_n c_n(t)e_n(x) \tag{1}$$

The forms of eigenfunctions were determined empirically. In the present study, a real type of EOF analysis was carried out.

4. Results and Discussion

4.1. Characteristics of Shoreline Change

The long-term variation of the shoreline location was examined. Figure 5 shows the temporal variation of the alongshore-averaged shoreline location along a 7.2-km stretch of the marine drive region from 1986 to 2006. Over that period, the shoreline locations in March and September indicate a retreating trend of 1.1 m/year and 0.8 m/year, respectively. On average, the beach width has decreased approximately 20 m in the last two decades. The overall eroding trend is considered to be related to the construction of Kanazawa Port and the development of the Tedori River basin [2–4]. It is quite important to examine the possible causes of accelerated erosion, but such an examination is beyond the scope of the present study. The shoreline retreat was most accelerated during the periods 1988 to 1991 and 2001 to 2004.

The features of temporal variation are generally similar in spring and autumn surveys. Quantitatively, the year-to-year variation is more significant in September, and the anomaly from the long-term trend reached 10 m in some years. The seasonal variations from September to March and from April to September are plotted in Figure 6. Note that the shoreline retreat during autumn and winter (September to March) is slightly greater than the shoreline advance during spring and summer (April to September), and this difference has increased recently. This is related to the aforementioned difference in the rate of shoreline retreat in March and September. Moreover, the magnitude of shoreline change has been becoming larger recently. Figure 7 demonstrates the temporal variation of mean significant wave heights observed at Kanazawa Port in contrasting seasons (autumn and winter: from the beginning of October to the end of March; spring and summer: from the beginning of April to the end of September). The wave height in the autumn and winter indicates a weakly increasing trend. This is considered to be related to the recent increase in shoreline variation during winter.

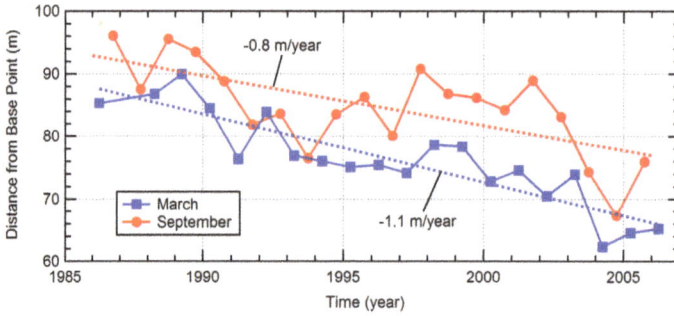

Figure 5. Long-term trend of shoreline change.

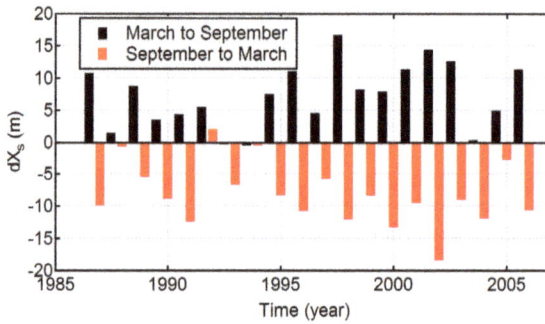

Figure 6. Seasonal variation of shoreline location.

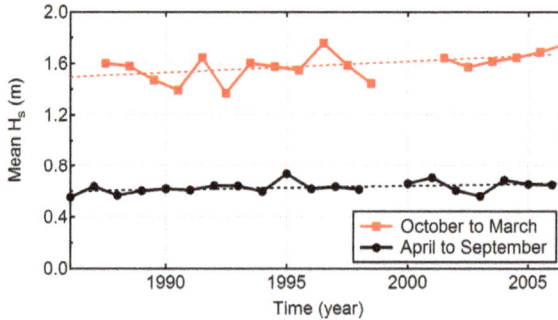

Figure 7. Variation of mean significant wave height during autumn and winter (from the beginning of October to the end of March) and spring and summer (from the beginning of April to the end of September).

The variation of shoreline location was not uniform over the stretch and period; several kinds of variation with different spatial and temporal scales coexist. Figure 8 represents the alongshore comparison of the rate of shoreline variation in the first (1986 to 1995) and second (1996 to 2005) halves of the study period. Note that a three-point moving-average was conducted in space to smooth out the local variation. The rate of variation was computed by linear regression. In general, a significant erosional trend is observed, except for the northern end of the study area where an accretionary trend is observed related to the influence of the construction of nearby Taki Port. In the first half of the study period, areas with shoreline advance and retreat appear alternatively. In the second half, the erosional trend became quite dominant. The locations where strong retreat of the shoreline was observed in the second half of the study period roughly correspond to the areas of shoreline advance in the first half. Shoreline retreat over the whole period was more significant in the northern part of the study area. The rate of shoreline retreat was not uniform during the study period. In some locations, it varied significantly between the first and second halves of the study period.

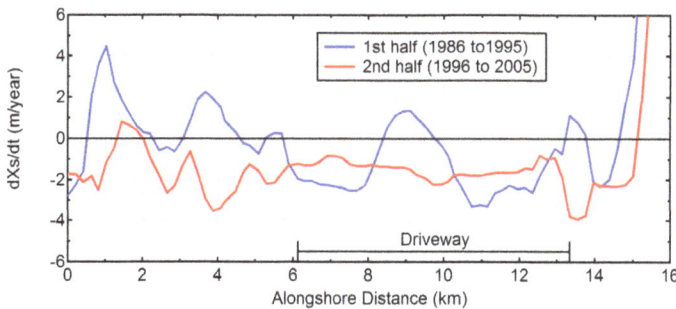

Figure 8. Alongshore comparison of shoreline change rates.

On the basis of the cross-shore survey record, the variation of shoreline location and sediment volume per unit width was computed for each line over the period 1998 to 2010. Although the shoreline analysis indicated a strong retreating trend in the range from 1.5 to 3.0 m/year, the results of volumetric change indicated quite different features; the change in sediment volume is generally small, and in fact, even a weakly increasing tendency was observed. This result implies that the trends in the variation of seabed-level change differ between nearshore and offshore regions. To clarify the difference related to the cross-shore locations, temporal variations of depth contour lines were examined for each line. As shown in Figure 9, the cross-shore seabed slope increases for lines that are located more to the south, which is consistent with the previous field observations [1,6]. This feature is more pronounced in areas where the water depth exceeds 7 m. On lines H40 and H60, located in the central and north parts of the study area, the depth contour lines offshore are slightly advanced with time, whereas the contour lines near the shoreline indicate a retreating trend. As a result, the slope of the seabed is tending to decrease. Because the data coverage of cross-shore survey data used in this study (approximately 10 years) is not sufficiently long and the magnitude of change is not very different from the order of sounding accuracy, further efforts are needed to continuously monitor bed-level changes and to clarify the possible cause of these peculiar changes.

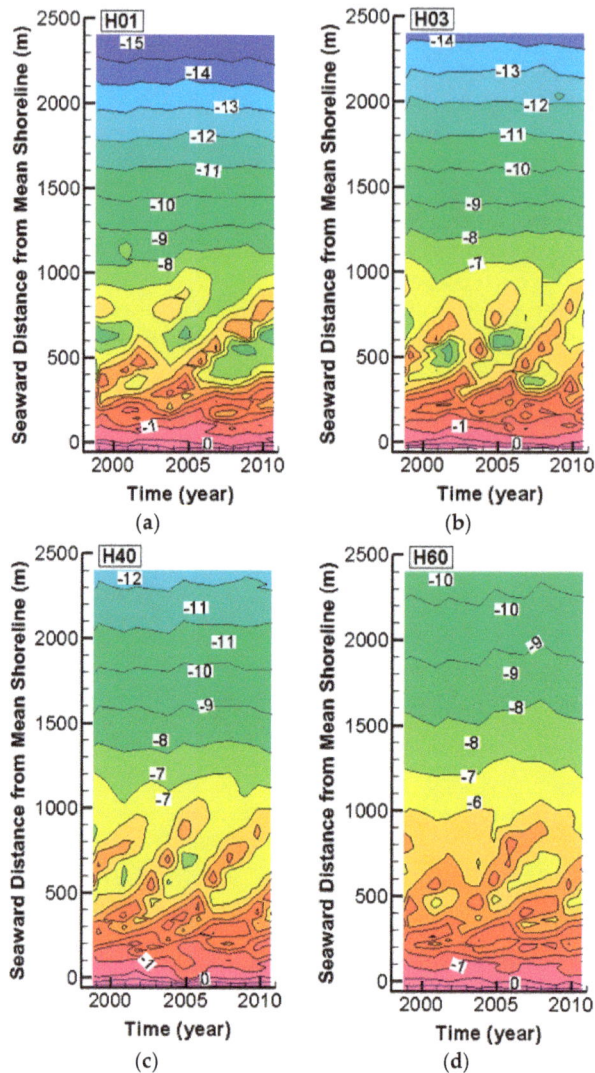

Figure 9. Variation of seabed level.

4.2. Characteristics of Systematic Sandbar Migration

Large-scale multiple sandbars have evolved in the study area. In Figure 9, it is clearly seen that interannual movements of these bars are significant, and the migration of the bars has been repeated several times. Figure 10 presents typical examples of observed interannual variations of cross-shore profiles on line H40 from November 2000 to November 2007. The bars migrate in a net offshore direction, and the transition of an individual bar consists of generation, seaward migration and decay. When the bar evolves most significantly, the bar height can reach 3 to 4 m. Similar types of temporal variation are observed on other lines. This kind of periodic movement of multiple sandbars is similar to the net offshore migration of bars reported at various locations around the world including the Netherlands [13,14,17], New Zealand [15,17] and Japan [16–18].

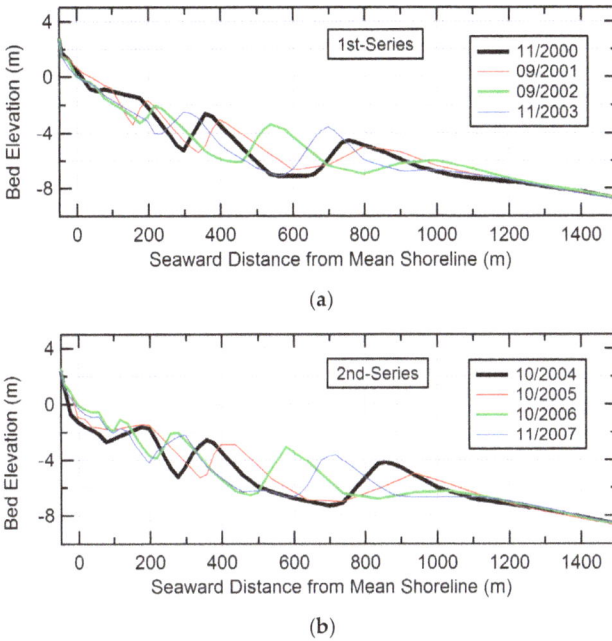

Figure 10. Example of the cyclic migration of sandbars (line H40).

The relationship between the crest and trough elevations is plotted in Figure 11. During the evolution of the bar (roughly corresponding to a crest in elevation of $Z_c > -4$ m), the water depth at the crest increases nearly proportionally to the depth at the trough. The slope of a linear regression line based on the least square method is in the range from 1.8 to 1.9. These values are slightly larger than those determined for the west coast of the United States (= 1.69) [26] and for various Japanese coasts (= 1.47) [5]. In the area where $Z_c < -4$ m, the trough elevation slightly increases with an increase of crest depth on the southern lines (H01 and H03), whereas on the northern lines (H40 and H60), a nearly constant value is taken around the closure depth. As a result, the bar height initially increases with depth at the crest (*i.e.*, seaward distance) as shown in Figure 12. It reaches its maximal value around $Z_c = -4.2$ m. The maximum bar height observed is as large as 4.8 m. Further offshore, the bar height decays progressively. This is consistent with a previous report for various Japanese coasts [5]. The temporal variation of the crest location and bar height is shown in Figure 13 for each line. In this figure, the centers of the circles indicate the locations of bar crests, and the areas of the circles are proportional to the bar heights. These figures clearly indicate that the bars have repeated systematic interannual migrations on all of the lines. The periodic cycles consist of generation, seaward migration and offshore decay. Bars were first generated at a distance of 100 to 200 m from the shore and then migrated seaward. The bar height increased with cross-shore distance until it reached 400 to 700 m. When the offshore bar evolved most significantly, the height became as large as 3 to 4 m. Then, the bar migrated further seaward and diminished at an offshore distance of around 1000 m. After the outermost bar diminished, the next bar replaced the former one. The velocity of offshore migrations was approximately 100 m/year within 400 m of the coast. In the area lying approximately 400 to 800 m from the shore, it was accelerated to 150 to 200 m/year. Further offshore, the migration velocity was not more than 100 m/year. The life span of an individual bar from generation to offshore decay is approximately eight years. The return period, which is defined as the interval between each cycle, is estimated to be three to four years. During the study period, three or four cycles of periodic movement were recognized.

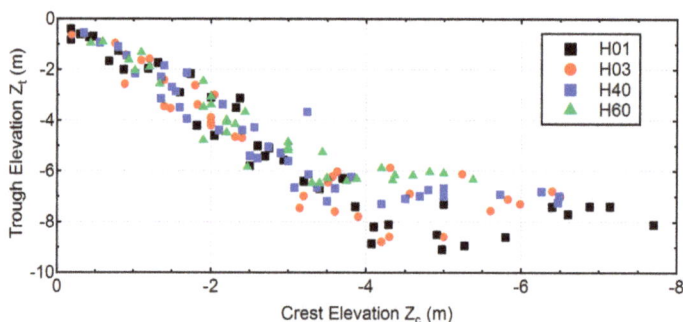

Figure 11. Relationship between crest and trough elevation.

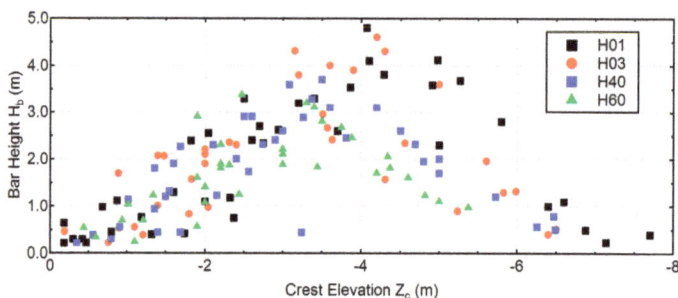

Figure 12. Relationship between bar height and crest elevation.

An alongshore comparison of bar migration was conducted next to examine the plan-view shape and migration characteristics of the bar system. Figure 14 displays examples of the comparisons of cross-shore profiles on the survey lines observed during 2003. In the figure, triple- or four-barred profiles are observed on each transect. From the observation of aerial photographs, the individual bars are generally considered to be continuous in the alongshore direction. To facilitate the example of an alongshore continuous bar configuration, estimated crest lines are plotted by a broken line and open circles in Figure 14. The seaward distance of the bar crests from the shoreline decreases as the location of the line moves southward (*i.e.*, the seabed slope becomes larger). As a result, the plan shape of an individual bar is slightly curved with respect to the shoreline. In Figure 14, the alongshore variability can be clearly seen. For example, the outermost bar on line H01 is substantially decayed on line H03. Further north on lines H40 and H60, the bar is considered to have disappeared already. On the contrary, the third bar (from the offshore) on line H60 has not evolved yet on H40, H03 and H01. The interannual movements of the longshore bar, which corresponds to the crest line in Figure 14, were tracked and compared on the four lines. The comparison shows that the seaward migration of the bar progressed from north (H60) to south (H01). In summary, the longshore bar is slightly closer to the shoreline as the alongshore location moves southward. The individual bar migrates net seaward, keeping the slightly curved plan shape with respect to the shoreline, and finally decays from the north to the south.

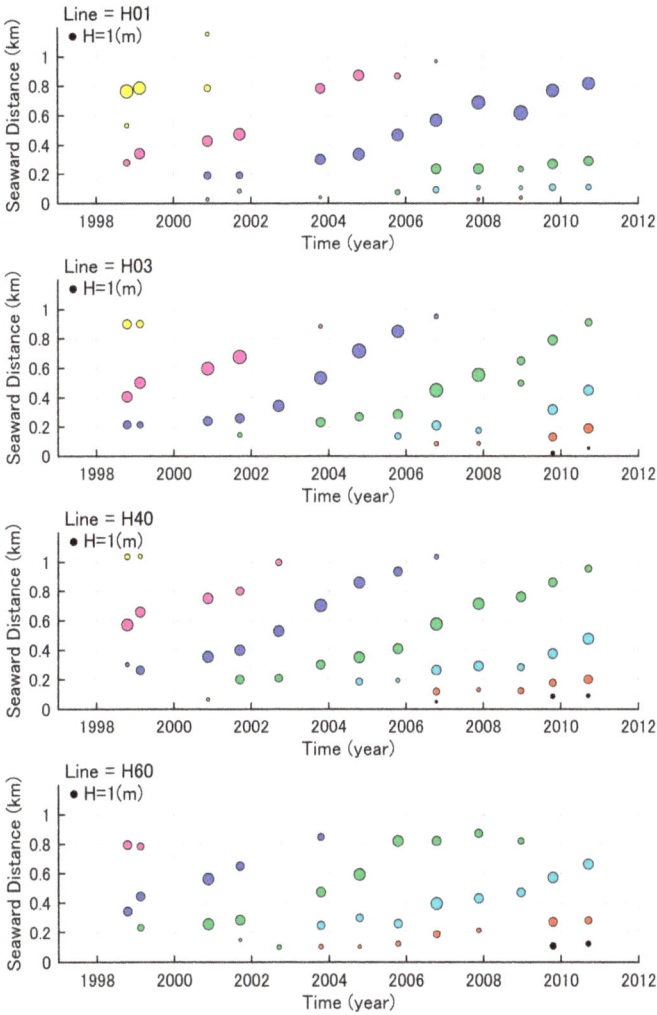

Figure 13. Cross-shore migration of bars on each transect.

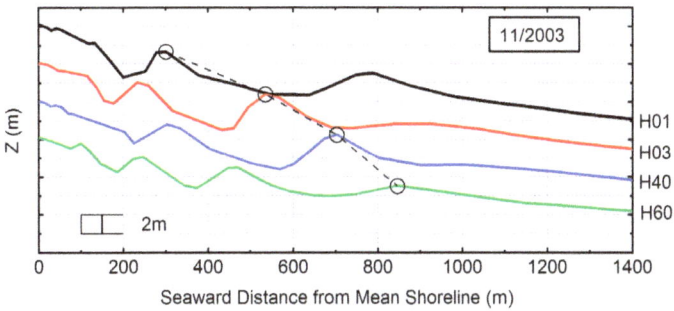

Figure 14. Alongshore comparison of the sandbar configuration.

The cyclic movements of the crest lines are plotted in Figure 15. The centers of the circles indicate the bar crest, and the areas of the circle are proportional to the bar height. To distinguish a series of periodic sequences of generation, migration and decay, crest markers that are considered to belong to the same generation of bar (*i.e.*, generated in nearly the same period) are indicated by the same color in the figure. For 1998, the bar indicated in yellow evolved as the outermost bar. Behind it, the next (purple) and the third (blue) bars also evolved. As time goes on, the outermost bar (yellow) migrates seaward and finally disappears from the north to the south. The cyclic movement indicates common features among the four lines. After one cycle, a configuration similar to that of 1998 reappeared in 2001. Another cycle was observed during 2001 and 2005, when the transition seems quite similar to the previous one. The bar behavior in the third cycle is different in some ways to the first and second cycles. Around 2007, the outermost bar (blue) disappeared in most of the area, and then, the next bar (red) started accelerating seaward migration. The migration is significant on the three lines H03, H40 and H60. In contrast, the old outermost bar (blue) remains on H01, and therefore the movement of the next generation of the bar (red) on H01 is much smaller. It is deduced that such an alongshore difference in bar migration resulted in the disconnection of the longshore bar between H01 and H03. From 2008 to 2009, the new bar (red) was considered to have realigned with the old bar (blue). It is assumed that the bars located more landward (light blue and red) have also realigned at the same time. The disconnection and realignment of a bar system has been reported at several NOM sites as 'bar switching' (e.g., [15]).

4.3. Principal Modes of Sandbar Migration

Next, EOF analysis was applied to cross-shore profiles on the four lines. The first empirical eigenfunctions explain over 99% of the total variation. The second and third eigenfunctions explain 70% to 80% of the remaining depth variation. The variance explained by higher modes is much smaller. Therefore, the focus is mainly placed on the top three modes of variation. Furthermore, variation of cross-shore profiles appears most clearly in the EOF results up to $X = 1300$ m, so even though the cross-shore range of EOF analysis extends to $X = 2400$ m, plots are limited to the shorter distance.

The temporal and spatial eigenfunctions for the first mode of long-term variation are shown in Figure 16. The spatial function of the first mode (e_1) expresses the mean profile of the seabed during the study period and is generally similar among the four lines. Detailed inspection indicates that, on line H01, the profile of e_1 has local maxima that correspond to the preferential locations of bar crests. The temporal coefficient of the first mode (C_1) is shown and compared to the cumulative change of sediment volume per unit width (ΔV) with respect to the value at 1998. It is noted that the time coefficients are plotted after being multiplied by -1 ($-C_1$) in the figure for comparison. Note that on all four lines, C_1 is very closely related to the volumetric change induced by the imbalance of alongshore sediment transport or cross-shore exchange of sediment with the backshore and dunes or offshore regions.

The combination of the second and third modes corresponds to the cyclic behavior of the sandbar system. Figure 17 describes the temporal and spatial eigenfunctions for the second and third modes. Both of the temporal functions, C_2 and C_3, show periodic variations. The phase difference between these two curves is approximately a quarter of the variation period. Similar features of phase shift are seen for the spatial eigenfunctions; the spatial functions e_2 and e_3 have features such that when one curve becomes a local maximum or minimum, the other crosses the horizontal axis. This implies that the relative difference in space between e_2 and e_3 is approximately a quarter wavelength. By synthesis of the characteristics of temporal and spatial functions, the combination of the second and third modes expresses the periodic seaward migration of bars.

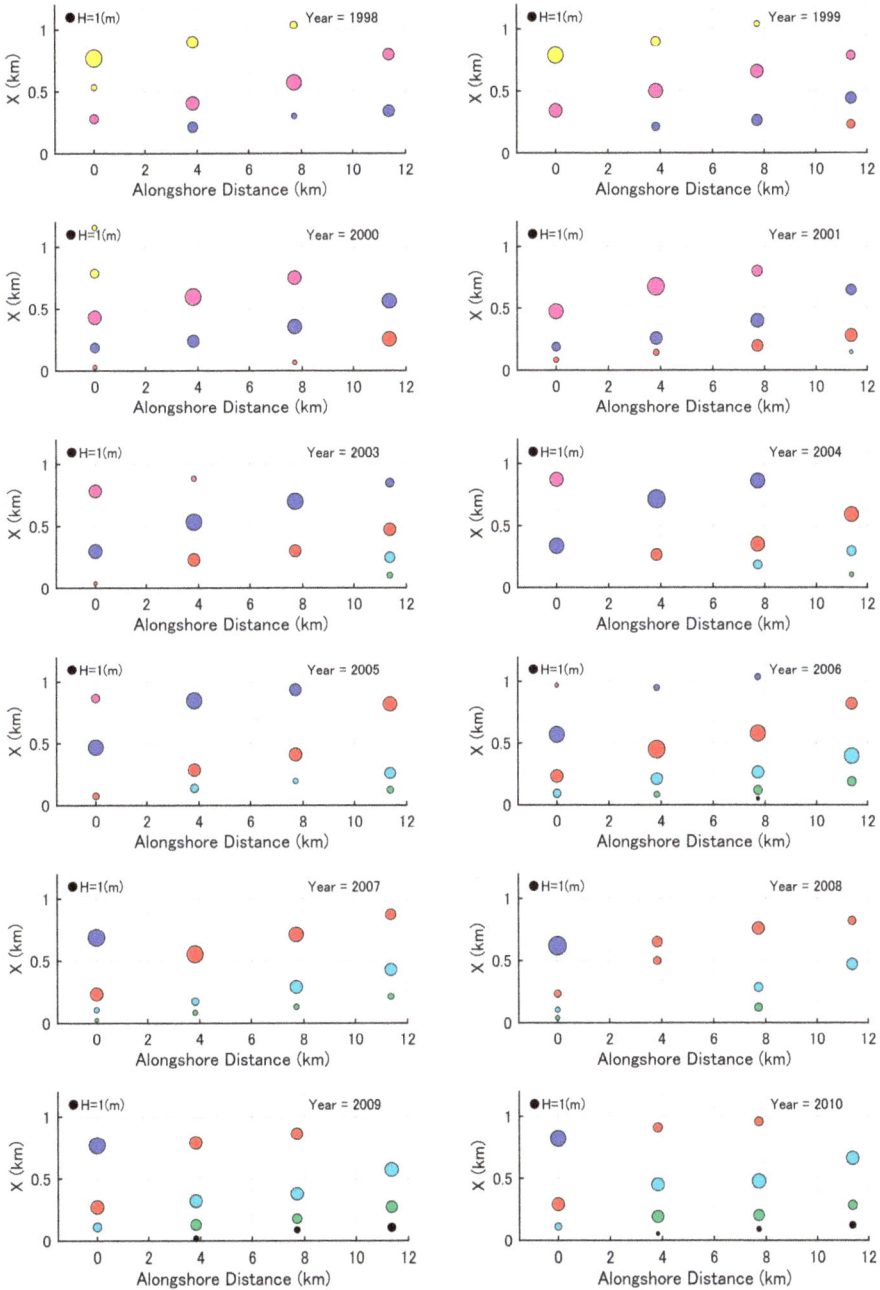

Figure 15. Comparison of the plan shape of the crest lines.

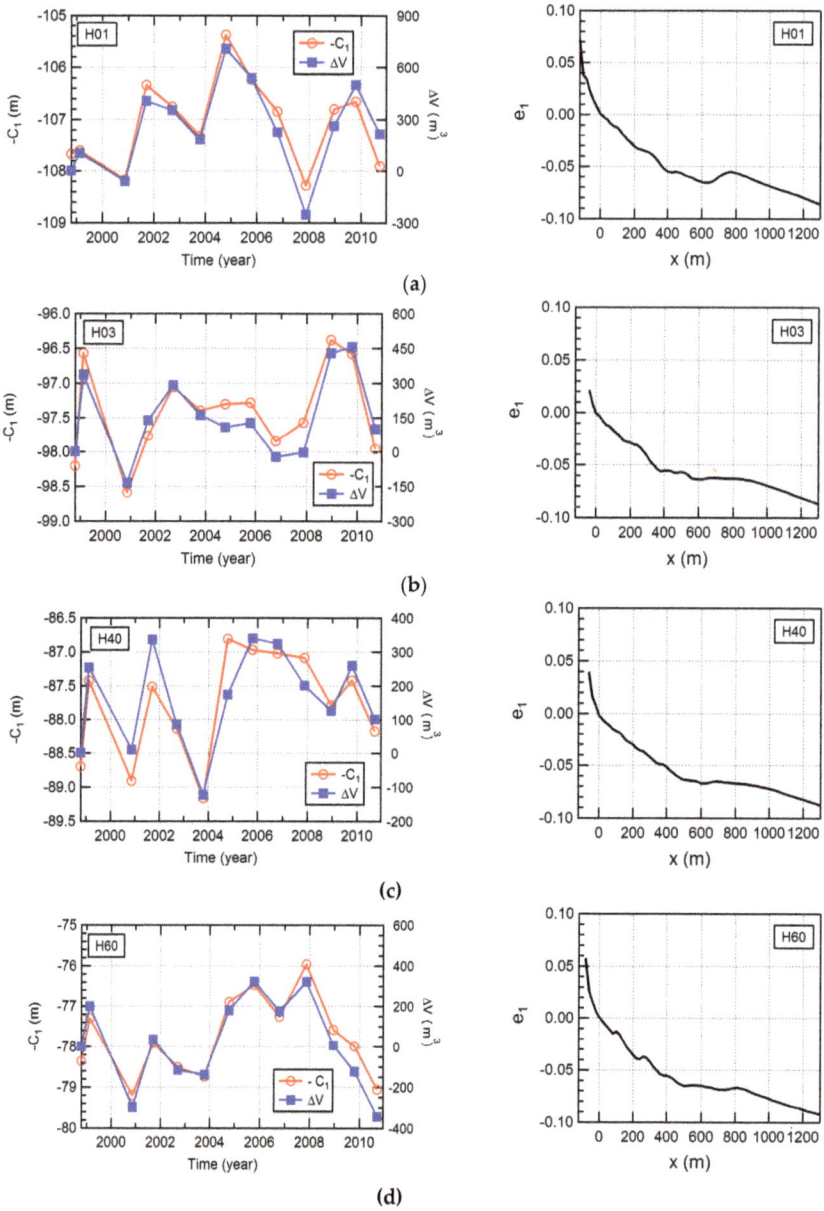

Figure 16. Temporal and spatial eigenfunctions of the first mode.

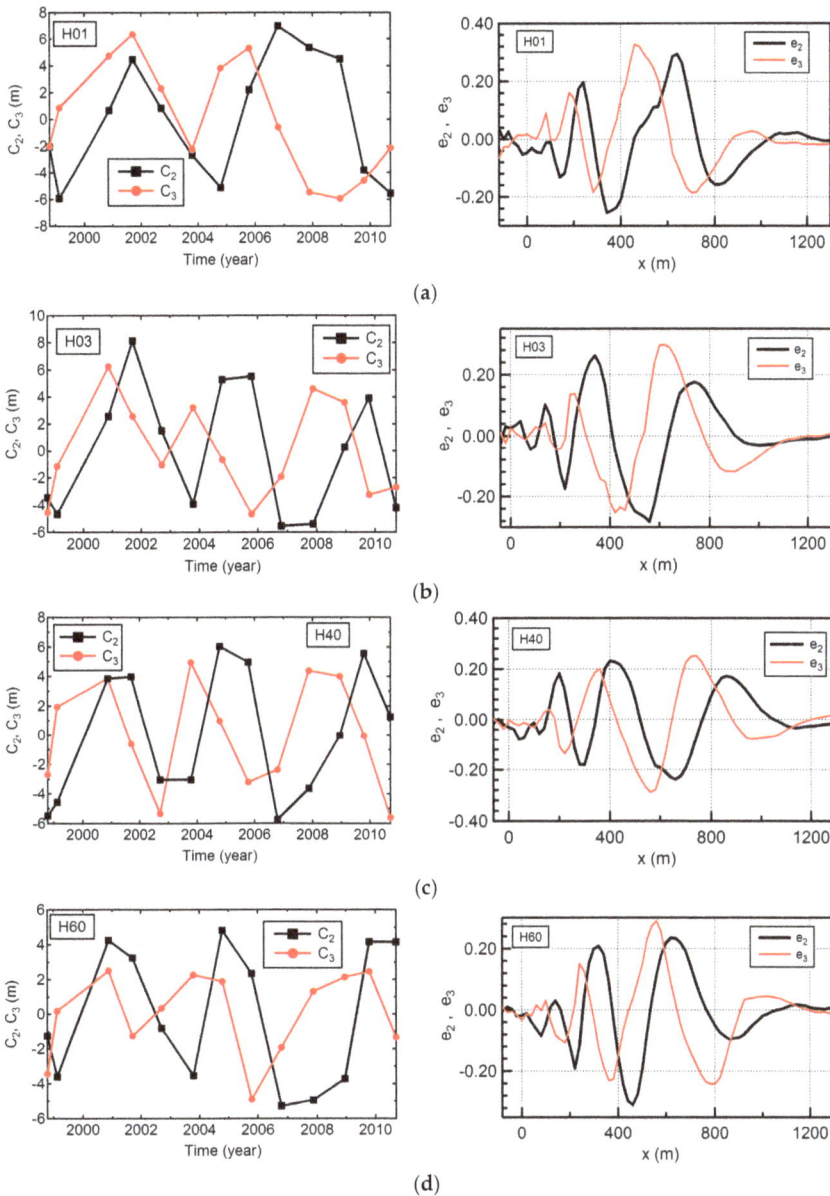

Figure 17. Temporal and spatial eigenfunctions of the second and third modes.

Figure 18 shows the summation of these two modes. The reconstructed data clearly reproduce the cross-shore periodic movement of the outer bar with an approximately four-year cycle. Note also that each bar first appears near the shoreline and then migrates seaward. The size of the bar increases with offshore migration. The net offshore movement continues throughout the bar cycle at this site. After the crest location exceeds 700 to 800 m, the bar begins to degenerate and finally disappears. The EOF results indicate that the life span of individual bars is around eight years on these lines. Although these two modes of bar migration are accompanied by short-term changes in sediment volume among the sub-areas in the profile, it can be deduced that these two modes do not contribute to the long-term variation of sediment volume over a whole transect because there are no long-term trends in the variations of C_2 and C_3.

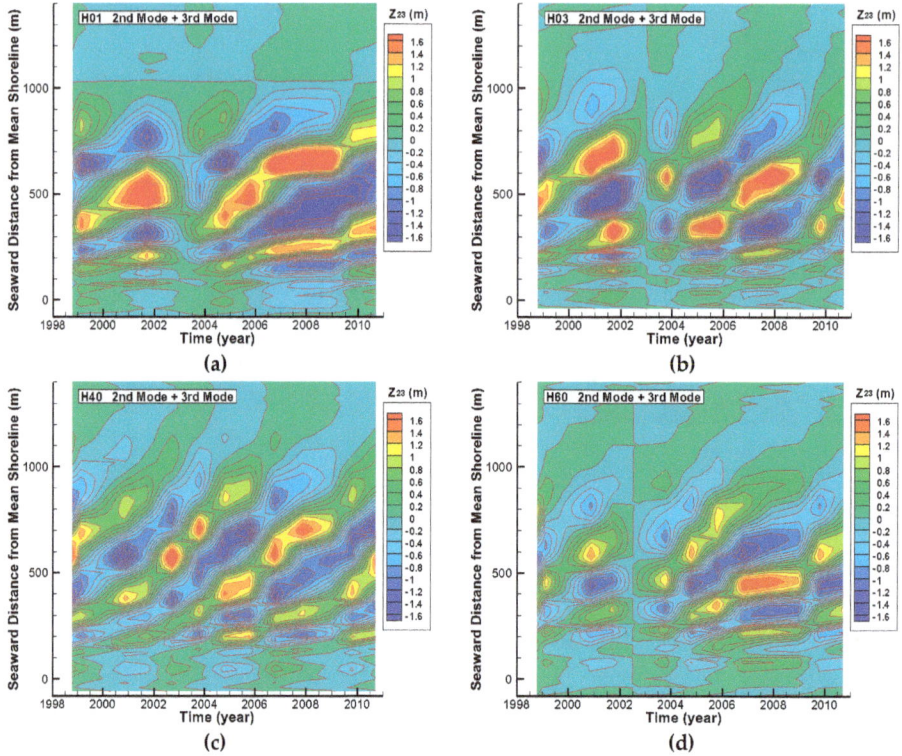

Figure 18. Reconstructed variation by the superposition of the second and third modes ($Z_{23} = C_2e_2 + C_3e_3$).

In the EOF analysis results, the magnitudes of the temporal functions of the second and third modes (C_2 and C_3) are generally larger than those for higher (fourth to sixth) modes (C_4 to C_6). Concerning the spatial functions, the magnitudes of the second to sixth modes (e_2 to e_6) are comparable in the bar migration zone where large-scale bed-level variation occurs. Because the variation in each mode is expressed by the product of the temporal and spatial functions, the second and third modes are dominant compared with higher modes in the bar migration zone. Near the shoreline, however, the magnitudes of e_4 to e_6 become larger than e_2 and e_3 in relatively many cases. Accordingly, the influences of the fourth to sixth modes are not negligible over the local area near the shoreline. Inspection of the spatial and temporal functions of higher (fourth to sixth) modes revealed that the long-term retreating trend of the shoreline is expressed by the fifth mode for lines H01, H40 and H60 and the sixth mode for line H03. It was also seen that the short-term (year-to-year) variation is substantial in the fourth to sixth modes. These higher modes should be taken into account when the long-term trend and year-to-year variation in the shoreline location need to be reproduced.

Finally, the relationship between bar migration and shoreline movement is examined. In general, the location of the shoreline is affected by various factors with different time scales. The annual survey record reflects the combination of the long-term (decadal) trend of erosion, with middle-term (several years) cyclic variations related to interannual bar migration and short-term variations (seasonal variation and short-term changes induced by storms). In the EOF analysis, these factors are decomposed into different eigenmodes. The middle-term shoreline movement related to NOM behavior appears in the second and third modes. Corresponding to the progress of NOM cycles, the cyclic variation of bed level is observed around $X = 0$ in Figure 18a to Figure 18d. Figure 19 compares shoreline location changes in the original survey with reconstructed EOF results for the first three ($Z_{1-3} = C_1e_1 + C_2e_2 + C_3e_3$) and five ($Z_{1-5} = C_1e_1 + \ldots + C_5e_5$) modes for line H40. The reconstructed variation expressed by Z_{1-5} adequately reproduced the long-term retreating trend of the shoreline. Among the overall variation, Z_{1-3} extracts oscillatory variations with amplitudes of approximately 8 m without containing the long-term trend. Hereafter, this kind of cyclic variation accompanied by the NOM phenomena is examined to clarify the relation between bar configurations and shoreline movement.

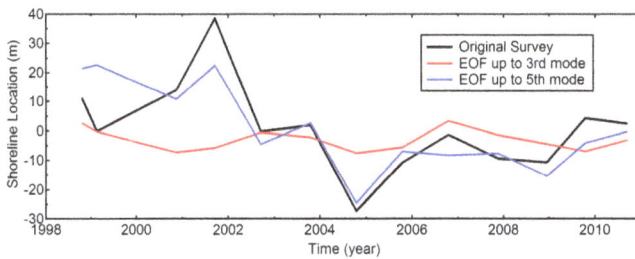

Figure 19. Comparison of shoreline variation within the original survey record, reconstructed variation by the superposition of empirical orthogonal function (EOF) results up to the third ($Z_{1-3} = C_1e_1 + C_2e_2 + C_3e_3$) and fifth ($Z_{1-5} = C_1e_1 + \ldots + C_5e_5$) modes.

Figure 20 and Figure 21 demonstrate typical examples of cross-shore profiles corresponding to times when the reconstructed EOF results up to the third mode (Z_{1-3}) indicate the retreat (Figure 20) and advance (Figure 21) of the shoreline. The shoreline tends to retreat when the outermost bar is decaying and a new bar is being generated at a seaward distance of some 200 m (*i.e.*, when a triple bar configuration is newly established). This feature is common in lines H03, H40 and H60. In contrast, the shoreline experiences an advance when the outer bar is most evolved. More specifically, on lines H40 and H60, shoreline advances are observed when the old outermost bar remained, but with substantial decay, and the next bar evolved dominantly as the middle bar in a triple bar configuration. On lines H01 and H03, the shoreline experiences an advance when the new outermost bar has evolved significantly in the double bar configuration (the old outermost bar has almost completely diminished). On lines H40 and H60, the shoreline location is neutral when the old outermost bar has completely decayed, and the profile has a double bar configuration.

As a typical example, consider the variation on line H40. The spatial functions in Figure 17c show that the second mode on line H40 corresponds to a triple bar configuration with a new bar generated at around $X = 200$ m, whereas the third mode indicates the double bar configuration (the third bar has not evolved yet). As was mentioned previously, the temporal coefficients of these modes have a phase shift of approximately a quarter of a period. Accordingly, when the second mode is dominant (*i.e.*, a large C_2 magnitude), the influence of the third mode decreases (*i.e.*, corresponding to a small C_3 magnitude). Therefore, the bar system is close to a double bar configuration when the third mode is dominant (e.g., 11/2003 in Figure 10), while the transition to a well-developed triple bar system is observed when the second mode is dominant instead (e.g., 10/2004). The cross-shore distribution of e_2 indicated that the third bar evolves at around $X = 200$ m when C_2 is maximum. At the same time, over the landward area corresponding to $X < 200$ m, where e_2 is negative, the bed level decreased, and correspondingly, the shoreline location retreated (e.g., 10/2004 in Figure 10). Therefore, the generation of a new bar and resulting transition into a triple bar configuration in a bar system may induce a temporal retreat of the shoreline, whereas the evolution of the outer bar accompanies an advance in the shoreline (e.g., at 10/2006 when C_2 had a negative peak). Note here that this kind of middle-term shoreline change over a time scale of years is cyclic and does not contribute to long-term changes in shoreline location within a decadal time scale.

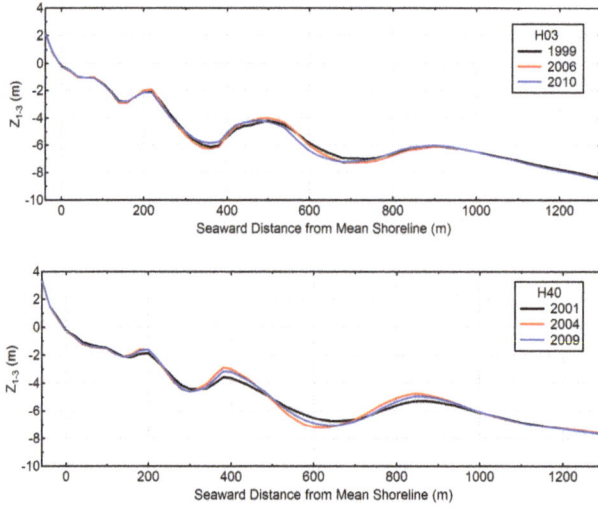

Figure 20. Examples of cross-shore profiles when net offshore migration (NOM)-related modes indicate shoreline retreat.

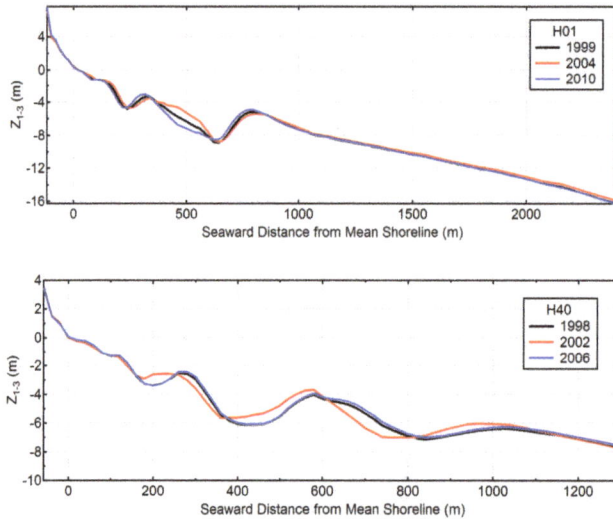

Figure 21. Examples of cross-shore profiles when NOM-related modes indicate shoreline advance.

These results imply that on sandy beaches similar to those on the Chirihama Coast where large-scale bar systems have evolved, the temporal variation of shoreline locations with a time scale of several years may appear with the systematic transition to a multiple bar system configuration.

Accordingly, it is very difficult on such sandy beaches to estimate the variation of sediment volume over the transect from only the change in the cross-shore shoreline location with an assumption of parallel translation in the cross-shore profile. Care should be taken on this point from an engineering point of view.

5. Summary Remarks

Representative morphological changes on the Chirihama Coast of Japan have been investigated based on two sets of field survey records. The analysis of shoreline survey records spanning two decades indicated a long-term eroding trend. Recently, an increasing seasonal variation has been revealed. Profile surveys undertaken over 13 years demonstrate that the depth contour lines near the shoreline have been progressively retreating, while the locations of contour lines in the offshore area are stable or have been slowly advancing recently. The implication for the seabed slope is that it is becoming gentler. Typical profile configurations are characterized by the presence of multiple bars. The height of the outer bar exceeds 4 m when the bar evolves most significantly. The cross-shore movements of the bars are significant. For the temporal variations, the net offshore migrations of bars have been repeated systematically. During the study period, three or four cycles consisting of the generation, seaward migration and offshore decay of bars have been recognized. The total lifespan of an individual bar is approximately eight years, and the interval between each cycle is approximately four years. It is also recognized that the seaward migration of individual bars propagates from north to south. The principal modes of sandbar migration obtained by EOF analysis clearly captured the main features of the systematic bar behavior. The EOF results suggest that the generation of a new bar at a distance of approximately 200 m from the shoreline, and a corresponding change in bar configuration into a new triple-bar system, results in the temporal retreat of the shoreline. In contrast, the shoreline experiences an advance when the outer bar has most evolved. This type of middle-term shoreline change does not contribute to the long-term changes at a decadal time scale.

Acknowledgments: Field survey records were provided by HRDB. The shoreline survey records were provided by Ishikawa Prefectural Government. This study was partially supported by Grants-in-Aid for Scientific Research by the Japan Society for the Promotion of Science (Nos. 25420517 and 16K06505). The authors wish to acknowledge the contributions of Messrs. Kawashima, Ura and Uehata (former students of Kanazawa University) in the initial stage of this study.

Author Contributions: M.Y. conceived of the general concept of the paper and wrote the manuscript. M.M. and K.H. analyzed the data and contributed analysis tools.

Conflicts of Interest: The authors declare no conflict of interest.

Abbreviations

The following abbreviations are used in this manuscript:

EOF	Empirical orthogonal eigenfunction
HRDB	Hokuriku Regional Development Bureau, Ministry of Land, Infrastructure, Transport and Tourism, Japan
T.P.	Tokyo Peil
JSCE	Japan Society of Civil Engineers

References

1. Ishida, H.; Takase, N.; Nagahara, H.; Ura, R. Current status of erosion problems on Chirihama Beach and Nagisa Driveway. *Proc. Coast. Eng. JSCE* **1984**, *31*, 355–359. (In Japanese)
2. Yuhi, M. Impacts of anthropogenic modifications of river basin on surrounding coasts: A Case Study. *J. Waterw. Port Coast. Ocean Eng. ASCE* **2008**, *134*, 336–344. [CrossRef]
3. Yuhi, M.; Dang, M.H.; Umeda, S. Comparison of accelerated erosion in riverbed and downstream coast by EOF analysis over a decadal scale. *J. Coast. Res.* **2013**, *SI65(1)*, 618–623. [CrossRef]

4. Mizumura, K.; Yamamoto, T.; Fujikawa, T. Prediction of sand movement near port of Kanazawa. *J. Waterw. Port Coast. Ocean Eng. ASCE* **1997**, *123*, 215–222. [CrossRef]

5. Ohmori, M.; Mogi, A.; Hoshino, T. *Geology in Shallow Sea*; Tokai University Press: Tokyo, Japan, 1971; pp. 209–217. (In Japanese)

6. Kato, K. On the relation between infragravity waves and multiple sandbars. *Proc. Coast. Eng. JSCE* **1984**, *31*, 441–445. (In Japanese)

7. Sunamura, T.; Takeda, I. Regional difference in the number of submarine longshore bars in Japan: An analysis based on breaking wave hypothesis. *Trans. Jpn. Geomorphol. Union* **2007**, *28*, 381–398.

8. Hayakawa, K.; Yuhi, M.; Ishida, H. Migration of multiple sandbars on the Chirihama Beach, JAPAN. In Proceedings of the Coastal Dynamics 2009: In Impacts of Human Activities on Dynamic Coastal Processes, Tokyo, Japan, 7–11 September 2009; Mizuguchi, M., Sato, S., Eds.; World Scientific: Singapore, 2009; p. 140.

9. Komar, P.A. *Beach Processes and Sedimentation*, 2nd ed.; Prentice-Hall Inc.: Upper Saddle River, NJ, USA, 1998; pp. 292–302.

10. Wijnberg, K.M.; Kroon, A. Barred beaches. *Geomorphology* **2002**, *48*, 103–120. [CrossRef]

11. Masselink, G.; Hughes, M.G. *Introduction to Coastal Processes and Geomorphology*; Hodder Arnold: London, UK, 2003; pp. 214–217.

12. Greenwood, B. Bars. In *Encyclopedia of Coastal Science*; Schwartz, M., Ed.; Springer: Dordrecht, The Netherlands, 2005; pp. 120–129.

13. Ruessink, B.G.; Kroon, A. The behavior of a multiple bar system in the nearshore zone of Terschelling, The Neterlands: 1965–1993. *Mar. Geol.* **1994**, *121*, 187–197. [CrossRef]

14. Wijnberg, K.M.; Terwindt, J.H.J. Extracting decadal morphological behavior from high-resolution, long-term bathymetric surveys along the Holland coast using eigenfunction analysis. *Mar. Geol.* **1995**, *126*, 301–350. [CrossRef]

15. Shand, R.D.; Bailey, D.G.; Shepherd, M.J. An Inter-site comparison of net offshore bar migration characteristics and environmental conditions. *J. Coast. Res.* **1999**, *15*, 750–765.

16. Kuriyama, Y. Medium-term bar behavior and associated sediment transport at Hasaki, Japan. *J. Geophys. Res.* **2002**, *107*, 3132. [CrossRef]

17. Ruessink, B.G.; Wijnberg, K.M.; Holman, R.A.; Kuriyama, Y.; Enckevort, I.M.J. Intersite comparison of interannual nearshore bar behavior. *J. Geophys. Res.* **2003**, *108*, 3249. [CrossRef]

18. Yuhi, M.; Okada, M. Long-term field observations of multiple bar properties on an eroding coast. *J. Coast. Res.* **2011**, *SI64*, 860–864.

19. Wright, L.D.; Short, A.D. Morphodynamic variability of surf zones and beaches: A synthesis. *Mar. Geol.* **1984**, *56*, 93–118. [CrossRef]

20. Niedroda, A.W.; Swift, D.J. Shoreface processes. In *Handbook of Coastal and Ocean Engineering*; Gulf Publishing Company: Houston, TX, USA, 1991; pp. 735–770.

21. Sato, S. Effects of winds and breaking waves on large-scale coastal currents developed by winter storms in Japan Sea. *Coast. Eng. Jpn.* **1996**, *39*, 129–144.

22. Tanaka, S.; Satoh, S.; Kawagishi, S.; Ishikawa, T.; Yamamoto, Y.; Asano, G. Sand transport mechanism in Ishikawa Coast. *Proc. Coast. Eng. JSCE* **1997**, *44*, 661–665. (In Japanese)

23. Winant, C.D.; Inman, D.L.; Nordstrom, C.E. Description of seasonal beach changes using empirical eigenfunctions. *J. Geophys. Res.* **1975**, *80*, 1979–1986. [CrossRef]

24. Dean, R.G.; Dalrymple, R.A. *Coastal Processes with Engineering Applications*; Cambridge University Press: Cambridge, UK, 2002; pp. 139–145.

25. Larson, M.; Capobianco, M.; Jansen, H.; Rozynski, G.; Southgate, H.N.; Stive, M.; Wijnberg, K.M.; Hulscher, S. Analysis and Modeling of Field Data on Coastal Morphological Evolution over Yearly and Decadal Time Scales. Part 1: Background and Linear Techniques. *J. Coast. Res.* **2003**, *19*, 760–775.

26. Keulegan, G.H. *Depths of Offshore Bars*; Engineering Notes No. 8; Beach Erosion Board, U.S. Army Engineer Waterways Experiment Station: Vicksburg, MS, USA, 1945.

MDPI AG

St. Alban-Anlage 66

4052 Basel, Switzerland

Tel. +41 61 683 77 34

Fax +41 61 302 89 18

http://www.mdpi.com

Journal of Marine Science and Engineering Editorial Office

E-mail: jmse@mdpi.com

http://www.mdpi.com/journal/jmse

www.ingramcontent.com/pod-product-compliance
Lightning Source LLC
Chambersburg PA
CBHW051842210326

41597CB00033B/5747